TALES FROM AN ENGINEERING CAREER

Kenneth M. Elovitz

Copyright © 2019 by Kenneth M. Elovitz

All Rights Reserved

Preface

This book reprints 447 newsletters that my firm, Energy Economics, Inc., sent to family, friends, colleagues, and clients on the back of monthly pocket calendars from January 1980 through March 2017.

Sometimes the message was a case history; other times it was a bit of news or philosophy. Either way, we aimed to bring a smile, inspire some thought, and let people know what we had been doing. Of course, we also hoped the calendars reminded people about the engineering services we provide.

I dedicate this book to my business partner and father, David Elovitz, who taught me much, both technically and working with people. But more than that, he gave me a chance to prove myself - something that others earlier in my career had been unwilling or unable to do.

There were times when I sat down to talk to my business partner and got an answer from my father. That's no doubt a hazard of any family business. But on the whole, the pluses far outweighed the minuses. Looking back and looking at today's market, I probably made less money than I would have working for a traditional engineering or manufacturing firm, but I had a whole lot more freedom, a far greater variety of experience, and a whole lot more fun.

My brother and I still practice engineering as Energy Economics, Inc., where we encounter and try to solve the types of problems that these newsletters describe.

It's been a great run. Enjoy the memories!

TABLE OF CONTENTS

January 1980 26
   CRACKED ICE 26
February 1980 27
   AN ILL WIND 27
March 1980 28
   LEMON-AID 28
April 1980 29
   EBTR 29
   Incorporation and Philosophy 29
May 1980 30
   EXORICISM 30
June 1980 31
   ORDINARY CLASSIC 31
July 1980 32
   THE END OF YOUR NOSE 32
August 1980 33
   TEMPEST IN A TEACUP 33
September 1980 34
   K I S S 34
October 1980 35
   NOW WE ARE TWO 35
   JUNK Mall 35
November 1980 36
   BUY NOW .... SAVE LATER 36
December 1980 37
   THE PUZZLE OF THE 6000 GALLON BALLOON 37
January 1981 38
   APOLOGIES TO CLEMENT MOORE 38
February 1981 39
   EXPENSIVE SAVINGS 39
March 1981 40
   HEADQUARTERS GETS A HOSING 40
April 1981 41
   FIGURES DON'T LIE, BUT 41
May 1981 42
   LITTLE RED HEN REVISITED 42
June 1981 43
   ALHPABET SOUP 43
July 1981 44
   VERSES! FOILED AGAIN! 44

August 1981	45
ON FELINE EXCORIATION	45
September 1981	46
THANK YOU, MR. OHM	46
October 1981	47
TWO RIGHTS CAN MAKE A WRONG	47
November 1981	48
FIRE!	48
December 1981	49
THE ALCHEMIST	49
January 1982	50
WORSE VERSE	50
February 1982	51
METER MYSTERY	51
March 1982	52
POLTERGEIST	52
April 1982	53
JABBERWOCKY	53
May 1982	54
BOILER EGGS	54
June 1982	55
MURPHY'S LAWS	55
(and other wisdom)	55
July 1982	56
PIRATED PINS	56
August 1982	57
655-2556	57
LABOR RELATIONS	57
September 1982	58
COMMUNICATION	58
October 1982	59
"IMPOSSIBLE"	59
November 1982	60
HOT TODAY, CHILLER TOMORROW	60
December 1982	61
DOUBLE TROUBLE	61
January 1983	62
JINGLE BELLS	62
February 1983	63
AIN'T IT SO!	63
March 1983	64
OUT OF SIGHT	64

April 1983	65
WHAT, A BARGAIN?	65
May 1983	66
WHAT, ME WORRY?	66
June 1983	67
CASE DISMISSED	67
July 1983	68
MORE FROM MURPHY	68
August 1983	69
BUDGET MENDER	69
September 1983	70
SHOEMAKERS' CHILDREN	70
October 1983	71
WHEN I WAS A CLIENT	71
November 1983	72
CRACKING UP	72
CRACKED ICE	72
December 1983	73
BUSY AS A BEE	73
January 1984	74
DECK THE HALLS	74
February 1984	75
MOTOR MATTER	75
March 1984	76
DONKEY PLANNING	76
April 1984	77
SENTENCE SENSE	77
May 1984	78
A ROSE BY ANY OTHER NAME	78
June 1984	79
WHOOMP!	79
July 1984	80
CHASING YOUR TAIL	80
August 1984	81
SERVICE	81
N.A.F.E.	81
September 1984	82
COMMISSIONING	82
October 1984	83
TRUE STORY	83
November 1984	84
STILL MORE MURPHY	84

December 1984	85
HOW MUCH?	85
January 1985	86
FROSTY THE SNOWMAN	86
February 1985	87
TROLL CONTROL	87
March 1985	88
APOLOGIES TO DR. SEUSS	88
April 1985	89
97.8% SOLUTION	89
May 1985	90
THREE GREAT TRUTHS	90
June 1985	91
BACK TO BASICS	91
July 1985	92
THE NOISE THAT ANNOYS	92
August 1985	93
CONVERSION FACTORS	93
September 1985	94
SAVVY SOLUTION	94
October 1985	95
ALTERNATIVE ENERGY	95
November 1985	96
DAMPER SCAMPER	96
December 1985	97
THERMO-TRIVIA	97
January 1986	98
SANTA CLAUS IS COMIN' TO TOWN	98
February 1986	99
WHO ARE THE ELOVITZS?	99
WHAT DO THEY DO?	99
March 1986	100
TO SHADE OR NOT TO SHADE	100
April 1986	101
I REMEMBER MURPHY	101
May 1986	102
THE ENGINEER AND THE DEVIL	102
June 1986	103
ASHRAE	103
July 1986	104
C'EST LA VIE	104
August 1986	105
7 WAYS TO WASTE MONEY	105

- September 1986 106
  - FAST TRACK 106
  - ON A PERSONAL NOTE 106
- October 1986 107
  - SHAKE 'N' BAKE 107
- November 1986 108
  - EXPERIENCE 108
- December 1986 109
  - DEBT COLLECTION 109
- January 1987 110
- February 1987 111
  - LESS CAN BE MORE 111
- March 1987 112
  - STANDARD PROGRESS REPORT 112
- April 1987 113
  - THE DAMPERS WERE HAMPERED 113
- May 1987 114
  - PUNABRIDGED DICTIONARY 114
- June 1987 115
  - 'COS WAS THE CAUSE 115
  - A Personal Note 115
- July 1987 116
  - TOO EFFICIENT? 116
- August 1987 117
  - A PROMIS IS A PROMISE 117
- September 1987 118
  - THE CONSULTANT ON A DESIGN-BUILD PROJECT 118
- October 1987 119
  - CAN'T WIN 'EM ALL 119
- November 1987 120
  - PROCRASTINATION - THE THIEF OF TIME 120
- December 1987 121
  - CIRCUMSTANTIAL EVIDENCE 121
- January 1988 122
  - TO THE TUNE OF 122
- February 1988 123
  - Rx 123
- March 1988 124
  - GETTING OUT OF HOT WATER 124
- April 1988 125
  - PLASTERED 125
- May 1988 126
  - OFCI 126

June 1988	127
THE IDEAL CONSULTANT	127
July 1988	128
BEST OF MURPHY	128
August 1988	129
THE ENGINEER	129
September 1988	130
ANGUISHED ENGLISH	130
October 1988	131
BUT, MR. LINCOLN	131
November 1988	132
LONG ARM	132
December 1988	133
A MODEL SOLUTION	133
January 1989	134
THE TWELVE DAYS OF CHRISTMAS	134
February 1989	135
HOT AIR STORY	135
March 1989	136
INDOOR AIR QUALITY	136
April 1989	137
FURTHER ADVENTURES	137
May 1989	138
HATFIELD COPY & McCOY CAFE	138
June 1989	139
GAS PAINS	139
July 1989	140
CFCS AND OZONE	140
August 1989	141
FIELD REPLACEABLE UNITS	141
September 1989	142
DARK SUCKERS	142
October 1989	143
RED HERRING	143
November 1989	144
I'VE BEEN WORKING ON THE RAILROAD	144
December 1989	145
A LETTER FROM MURRAY	145
January 1990	146
ANGELS WE HAVE HEARD ON HIGH	146
February 1990	147
TOO MUCH OF A GOOD THING	147

March 1990	148
DID YOU KNOW	148
April 1990	149
QUIZ	149
May 1990	150
LAST THING YOU NEED	150
June 1990	151
THE GREAT AIR HUNT	151
July 1990	152
EYE TO EYE	152
August 1990	153
HAVE AN ICE DAY	153
September 1990	154
POSITION IS EVERYTHING IN LIFE	154
October 1990	155
YOU CAN'T TAKE IT WITH YOU	155
NOW WE ARE THREE	155
November 1990	156
DON'T BELIEVE EVERYTHING YOU HEAR	156
December 1990	157
SONS OF MURPHY	157
January 1991	158
SEASON'S GREETINGS	158
February 1991	159
GOOD CONNECTIONS	159
March 1991	160
OVER THE RAINBOW	160
OUR NEW LOOK:	160
April 1991	161
IF I ONLY HAD A BRAIN	161
May 1991	162
PART TWO	162
June 1991	163
BACK IN CONTROL	163
July 1991	164
DAFFYNITIONS	164
August 1991	165
NEW ELEMENT	165
September 1991	166
HOW DO YOU SPELL RELIEF	166
October 1991	167
APOLOGIES TO GILBERT & SULLIVAN	167

November 1991	168
GREMLINOLOGY	168
December 1991	169
AND TO THINK THAT I SAW IT ON MULBERRY STREET	169
January 1992	170
FAREWELL TO '91	170
February 1992	171
WHEN I WAS A BOY ...	171
March 1992	172
KEEP ON PUMPIN'	172
April 1992	173
GUIDE TO SAFE FAX	173
ON A VERY DIFFERENT NOTE	173
May 1992	174
TO THE LIMIT	174
June 1992	175
THIN AIR	175
July 1992	176
SHAKE, RATTLE, & ROLL	176
August 1992	177
HIAWATHA	177
September 1992	178
SALESMAN	178
October 1992	179
FAX FACTS	179
A WORD OF CAUTION	179
November 1992	180
EVERYONE WINS	180
(except the lawyers)	180
December 1992	181
THE TURKEY ZONE	181
January 1993	182
AULD LANG SYNE	182
February 1993	183
ONCE UPON A TIME	183
March 1993	184
THE RETURN OF SALESMAN	184
April 1993	185
STILL MORE FROM MURPHY	185
May 1993	186
COOKING ODOR RAP	186
June 1993	187
GOVERNMENT	187

July 1993	188
PUFFED UP	188
August 1993	189
INADVERTENT SAUNA	189
September 1993	190
TECHNOLOGY TRANSFER	190
October 1993	191
CLEARING THE SMOKE	191
November 1993	192
SELECTION	192
December 1993	193
THE ONE THAT GOT AWAY	193
January 1994	194
THE BALLAD OF EEI	194
February 1994	195
THE "EXPERT" LEARNS A LESSON	195
March 1994	196
A REAL BULL STORY	196
OPINION	196
April 1994	197
A CURRENT TOPIC	197
May 1994	198
HOT WATER	198
June 1994	199
THE CARIBOU FACTOR	199
July 1994	200
MEDIALYSIS	200
August 1994	201
NEW INVENTION	201
September 1994	202
TEN TO ONE	202
October 1994	203
DILEMMA	203
November 1994	204
OSHA IAQ PROPOSAL	204
December 1994	205
GREEN LIGHTS	205
January 1995	206
RUDOLPH THE RED NOSED REINDEER	206
February 1995	207
CAN'T FIGHT CITY HALL	207
March 1995	208
NOT WITHOUT RISK	208

April 1995 209
   TROMPE L'OEIL 209
May 1995..210
   PLAIN TALK 210
June 1995..211
   LOOP THE LOOP 211
July 1995 212
   TWO OUT OF THREE 212
August 1995 213
   LANGUAGE BARRIER 213
   SURVIVING 213
September 1995 214
   BREAKER BREAKTHROUGH 214
October 1995 215
   SECOND CHANCE 215
November 1995 216
   DIGIT FIDGET 216
December 1995 217
   QUOTH THE MAVEN 217
January 1996 218
   CYBERTALE 218
February 1996..219
   PIPE DREAMS 219
March 1996..220
   CLUELESS 220
April 1996 221
   IT ALL DEPENDS HOW YOU LOOK AT IT 221
May 1996..222
   WHAT A GAS..222
June 1996..223
   ENGINEER STORIES 223
July 1996 224
   TOWER POWER..224
August 1996 225
   WHAT A RELIEF! 225
September 1996 226
   MORE ENGINEER STORIES 226
October 1996 227
   HOW TO BE A SMART CLIENT 227
November 1996 228
   RE-ENGINEERING 228
December 1996 229
   OUGHTA BE A LAW...229

January 1997	230
A LOOK BACK	230
February 1997	231
ANOTHER GREAT AIR HUNT	231
March 1997	232
ODE TO GIL CARLSON*	232
April 1997	233
AS THE HINGE CREAKS	233
May 1997	234
DR. SEUSS TECHNICAL WRITER	234
June 1997	235
HORSE TRADING	235
July 1997	236
QUOTABLE QUOTES	236
August 1997	237
THAT BIG SUCTION SOUND	237
September 1997	238
TRIVIA	238
October 1997	239
LINGUISTICS	239
November 1997	240
SURGERY	240
December 1997	241
THE YOUNG ENGINEER	241
January 1998	242
THE WALRUS and THE CARPENTER REVISITED	242
February 1998	243
RATS ON THE RUN	243
March 1998	244
NEW AIA DOCUMENTS	244
April 1998	245
TALK TO THE BOSS	245
May 1998	246
TRAPPED!	246
June 1998	247
YOGI BERRA	247
July 1998	248
PUMPED UP	248
August 1998	249
SALESMAN	249
September 1998	250
MAGIC WORDS	250

October 1998 251
   PUMP and PITCHER...251
November 1998 252
   ODD SPECIFICATIONS 252
December 1998 253
   AS THE PUMP TURNS 253
January 1999 254
   TO THE TUNE OF 254
February 1999...255
   A DATE WITH TIME..255
March 1999...256
   FLUFFERNUTTER 256
April 1999 257
   DOPEY KIDS 257
May 1999..258
   A BOOST FOR LESS BUCKS! 258
June 1999...259
   REDUCE, REUSE, RECYCLE..259
July 1999 260
   ANAGRAMS 260
August 1999...261
   NOAH IN THE NINETIES 261
September 1999 262
   PIPE DREAMS 262
October 1999 263
   WHOOSH 263
November 1999 264
   OREO COOKIES..264
December 1999 265
   PUMPING AWAY...265
January 2000 266
  to the tune of 266
   GOOD KING WENCESLAS...266
February 2000..267
   THERE'S NO AIR THERE 267
March 2000..268
   E-ENGINEERING...268
April 2000 269
   FLIGHT TEST 269
May 2000...270
   RISKY BUSINESS 270
June 2000..271
   CONTRACTORS and ENGINEERS...271

July 2000 272
   SOUNDS LIKE 272
August 2000 273
   URBAN LEGEND 273
September 2000 274
   DESIGN/BUILD 274
October 2000 275
   CHAPTER 13 275
November 2000 276
   THE DEVIL'S DICTIONARY 276
December 2000 277
   DILBERT'S CHRISTMAS 277
January 2001 278
   RHYME 278
February 2001 279
   DRYING OUT 279
March 2001 280
   DEREGULATION 280
April 2001 281
   THE CASE OF THE MISSING kW 281
May 2001 282
   DEAR MR. ENGINEER 282
June 2001 283
   SHAGGY PIRATE 283
July 2001 284
   WATER BEARER 284
August 2001 285
   A LITTLE HUMOR 285
   REST ASSURED 285
September 2001 286
   FACTS ABOUT WATER 286
October 2001 287
   HAS COMMISSIONING CHANGED? 287
November 2001 288
   DUAL ROLE 288
December 2001 289
   DON'T TREAT THE SYMPTOMS 289
January 2002 290
   2001 REPRISE 290
February 2002 291
   REWARD 291
   LYSISTRATA REVISITED 291

March 2002	292
DIVINE INSPIRATION	292
April 2002	293
HISTORY LESSON	293
May 2002	294
MONDEGREENS	294
June 2002	295
DRIP, DRIP, DRIP	295
July 2002	296
HOT STUFF	296
August 2002	297
MEMORY LANE	297
September 2002	298
BIG TROUBLE	298
October 2002	299
ACROSS THE POND	299
November 2002	300
THE VILLAGE HVAC-SMITH	300
December 2002	301
DID YOU KNOW	301
January 2003	302
PAUL REVERE and HENRY W. LONGFELLOW	302
February 2003	303
MICROSOFT HAIKU	303
March 2003	304
FOGGY, FOGGY, DEW	304
April 2003	305
NUMBERS	305
May 2003	306
SHARE AND SHARE ALIKE	306
June 2003	307
WOERTENDYKE vs. BUCK	307
July 2003	308
FIRE & ICE	308
August 2003	309
PIED PIPER	309
September 2003	310
TALE FROM A CHEM LAB	310
October 2003	311
IF THINGS GO WRONG	311
November 2003	312
ENGINEER JOKES	312

December 2003 313
   LEARN TO ADJUST 313
January 2004 314
   apologies to 314
   CLEMENT MOORE 314
February 2004 315
   NEW TECH 315
March 2004 316
   LOW FLOW WOE 316
April 2004 317
   DEMONOLOGY 317
May 2004 318
   PERSONNEL REFERENCES 318
June 2004 319
   BIGGER IS NOT BETTER 319
July 2004 320
   CREATIVE THINKING 320
August 2004 321
   DON'T JUMP TO CONCLUSIONS 321
September 2004 322
   BLOWING HOT AIR 322
October 2004 323
   THE TRUTH ABOUT CATS AND DOGS 323
   OUT OF THE MOUTHS 323
November 2004 324
   SPEED DEMONS 324
December 2004 325
   WHEN? 325
January 2005 326
   'TWAS THE NIGHT BEFORE 326
February 2005 327
   FREEZE DISEASE 327
March 2005 328
   ICE DAM DILEMMA 328
April 2005 329
   WRENCH RELIEF 329
May 2005 330
   I BLEW IT 330
June 2005 331
   QUOTES 331
July 2005 332
   OUT OF HOT WATER 332

August 2005	333
HIDDEN HEAT LOSS	333
September 2005	334
SIGNS OF OLD TIMES	334
October 2005	335
INSTRUCTIONS FOR LIFE	335
November 2005	336
DUCT TAPE	336
December 2005	337
THE ECSTASY AND THE AGONY	337
January 2006	338
TO THE TUNE OF	338
February 2006	339
RAIN FOREST	339
March 2006	340
WHAT GOES AROUND	340
April 2006	341
HISTORY	341
May 2006	342
WORDS FROM POPPY	342
June 2006	343
FLUSHED WITH PRIDE	343
July 2006	344
BORN TO GOOGLE	344
August 2006	345
DIFFERENTIAL DIAGNOSIS	345
September 2006	346
INVISIBLE HISTORY	346
October 2006	347
FLOOD!	347
November 2006	348
THE TELLTALE STAIN	348
December 2006	349
SWITCH HITTER	349
January 2007	350
APOLOGIES TO CASEY	350
February 2007	351
BOILER ODDITY	351
March 2007	352
HAPPINESS AND SUCCESS	352
April 2007	353
GLOWING SUCCESS	353

May 2007	354
FIVE STRANDS	354
June 2007	355
JUST A MOM	355
July 2007	356
SLEEPLESS IN SEATTLE	356
August 2007	357
TRUE OR FALSE?	357
September 2007	358
IN FROM THE COLD	358
October 2007	359
SHOEMAKERS' CHILDREN	359
November 2007	360
OUT WITH THE OLD ...?	360
December 2007	361
EATS, SHOOTS, AND LEAVES	361
January 2008	362
ENERGY GUYS ARE COMING TO TOWN	362
February 2008	363
THE BEAUTY OF NUMBERS	363
March 2008	364
THE NOSE KNOWS	364
April 2008	365
RANDOM KINDS OF FACTNESS*	365
May 2008	366
LIFE SUCKS	366
June 2008	367
A PERSONAL MESSAGE	367
July 2008	368
IDEAS TO PONDER	368
August 2008	369
STARVING UNITS	369
September 2008	370
FEEDING THE HUNGRY	370
October 2008	371
IF YOU WANT SOMETHING DONE RIGHT ...	371
November 2008	372
BLAME THE CROATIANS	372
RED TAPE AND PIPE	372
December 2008	373
COLD IN GEORGIA	373
January 2009	374
DIRGE ON 2008	374

February 2009	375
A FREEZIN' IN THE SUN	375
March 2009	376
DON'T BE SO SURE	376
April 2009	377
HOW THE STEAM GOT ITS HISS	377
May 2009	378
THE STRANDED ENGINEER	378
June 2009	379
INTEGRATED PROJECT DELIVERY	379
July 2009	380
SHEDDING LIGHT	380
August 2009	381
FROM THE LOVINS	381
September 2009	382
THE GHOST OF ENERGY USE	382
October 2009	383
BURMA SHAVE	383
November 2009	384
PREDICTION	384
December 2009	385
SHOULDA SAID	385
January 2010	386
JINGLE BELLS	386
February 2010	387
THE TRUTH, THE WHOLE TRUTH	387
March 2010	388
THE SCARECROW OF OZ	388
April 2010	389
FAVORITE THINGS	389
May 2010	390
GREAT MOLASSES FLOOD	390
June 2010	391
SHUT 'EM OFF!	391
July 2010	392
IT'S NOT EASY BEING GREEN	392
August 2010	393
TOLD YA SO	393
September 2010	394
TOO MUCH IS NOT ENOUGH	394
October 2010	395
CORRIDORS	395

November 2010 396
   PUNS FOR THE EDUCATED MIND 396
December 2010 397
   OL' MAN RIVER 397
January 2011 398
   WE THREE KINGS 398
February 2011 399
   THE COST TO BE GREEN 399
March 2011 400
   THE $124,000 RULE 400
April 2011 401
   PHILATELICA 401
May 2011 402
   REGULATIONS 402
June 2011 403
   GETTING OUT OF HOT WATER 403
July 2011 404
   WASTE OF MONEY 404
August 2011 405
   SPRING SHOWERS 405
September 2011 406
   BATTERY LIFE 406
October 2011 407
   HEATING TIPS 407
November 2011 408
   BOIL AWAY 408
December 2011 409
   TOO MUCH PRESSURE 409
January 2012 410
   to the tune of 410
   SANTA CLAUSE IS COMIN' TO TOWN 410
February 2012 411
   READ THIS BOOK 411
March 2012 412
   BIG QUESTION 412
April 2012 413
   REFLECTIONS 413
May 2012 414
   BLOWIN' IN THE WIND 414
June 2012 415
   EVER WONDER? 415
July 2012 416
   ANOTHER FREEZE 416

August 2012	417
PUMP IT UP	417
September 2012	418
COMPANY NEWS	418
October 2012	419
SHAKE, RATTLE, AND ROLL	419
November 2012	420
WINTER FUN	420
December 2012	421
CAN YOU HEAR ME NOW?	421
January 2013	422
ROCK OF AGES	422
(MA'OZ TZUR)	422
February 2013	423
RANDOM KINDS OF FACTNESS*	423
March 2013	424
YOU CAN'T FIGHT CITY HALL	424
April 2013	425
UP IN THE ATTIC	425
May 2013	426
PAPER OR AIR?	426
June 2013	427
THE NEXT GENERATION	427
July 2013	428
A QUICK TRIP	428
August 2013	429
UP OR DOWN?	429
September 2013	430
QUIET AND PEACE?	430
October 2013	431
THE ENERGY IN FOOD	431
November 2013	432
WHEN ALL ELSE FAILS	432
December 2013	433
A SWITCH IN TIME	433
January 2014	434
to the tune of	434
WHITE CHRISTMAS	434
February 2014	435
DIAL IT DOWN	435
March 2014	436
HUMIDOR	436

April 2014	437
BEING GREEN	437
May 2014	438
COLD IS RELATIVE	438
June 2014	439
DIVINE INTERVENTION	439
July 2014	440
STEP BY STEP	440
August 2014	441
NO FREE LUNCH	441
September 2014	442
WORDS OF WISDOM	442
WHAT'S UP	442
October 2014	443
WD-40	443
November 2014	444
PRESSURE!	444
December 2014	445
FREE LUNCH	445
January 2015	446
'TWAS THE NIGHT BEFORE	446
February 2015	447
LONG DISTANCE CALLING	447
March 2015	448
MORE RANDOM KINDS OF FACTNESS*	448
April 2015	449
INSPIRATION	449
May 2015	450
DID YOU KNOW	450
June 2015	451
STRANGEST DREAM	451
July 2015	452
COOL HISTORY	452
August 2015	453
SAVAGE WISDOM	453
September 2015	454
THE LOWLY TOILET	454
October 2015	455
FOLLOW THE BOUNCING BALL	455
November 2015	456
BE SURE BRAIN IS ENGAGED	456
December 2015	457
NO FLOW NO GO	457

January 2016 458
   JINGLE BELLS 458
February 2016 459
   GET STEAMED 459
March 2016 460
   PAPER OR PLASTIC 460
April 2016 461
   THE SAMURAI 461
   ELECTRICITY EXPLAINED 461
May 2016 462
   WASTE NOT, WANT NOT 462
June 2016 463
   BROKEN WINDOWS? 463
July 2016 464
   THE DEVIL IN THE DETAILS 464
August 2016 465
   THE HOLY MAN AND THE SNAKE 465
September 2016 466
   WHAT A STINK! 466
October 2016 467
   FIBONACCI 467
   GREAT BIKE 467
November 2016 468
   WHY SO HUMID? 468
December 2016 469
   BRAIN TEASERS 469
   THE SQUIRREL 469
January 2017 470
   CALENDAR GUYS 470
February 2017 471
   FIRST IMPRESSIONS 471
March 2017 472
   END OF AN ERA 472

January 1980

CRACKED ICE

When I first saw one of these shirt pocket monthly appointment calendars, I put it aside-----A busy guy like me simply couldn't get by without a separate line for every half-hour of the day, and I continued to carry my bulky "Week-at-a-glance". Until the day another consultant I was working with convinced me to at least give the shirt pocket version a try. You guessed it - I loved it. Much less bulky, always convenient in my shirt pocket (the appointment book usually stayed in my briefcase) and there really was enough room in the little blocks. I still kept a big calendar in the office, so Fran could keep track of me or set up meetings while I'm out, but I relied on the little card for carrying with me. It was the greatest thing since cracked ice! And with just a couple 'Of little changes, and a trip to my friendly neighborhood printer, I could make it even better.

Meanwhile, first one client, then another, saw my neat little shirt pocket calendar and asked where they could get one. As long as I was going to have them run off for me, a few more won't make much difference in the printer's bill, so I figured I might as well share the idea with all my friends and clients. So I'll be sending you one, like this, every month, when I get the monthly batch from the printer, and I'll try to find something you'll find of interest to put in this space each month, just to keep in touch.

I hope you'll enjoy these little calendars as much as I do, and find them just as handy; and that the coming year will be full of health, happiness and accomplishment for you and all your loved ones.

Dave Elovitz

February 1980

AN ILL WIND

The proverb says, "It's an ill wind that blows no good" and I am struck by how little good comes from the excessive airflow rates I find in many systems. Functionally, there are four factors that establish how much airflow is needed in any space:

(1) *Heat Transfer* - the amount of energy the airflow can carry varies with the quantity of air and the difference between the supply air temperature and the space temperature.

(2) Terminal velocity - there must be sufficient air motion so the occupied space doesn't feel "dead" or "stuffy;" this characteristic depends on outlet selection as well as air quantity.

(3) *Dilution* - primarily related to outside air quantity, to reduce the concentration of odors (or smoke, or moisture, or other contaminants) in the space.

(4) *Makeup* - to replace air removed from the space by exhaust systems.

Until recently, building codes and regulations sometimes required more airflow, especially more outside air, than was required by function, and systems were designed to satisfy those regulations, often with safety factors added at various stages of design, equipment selection and installation. Result: an ill wind indeed, that uselessly inflates energy bills. In the Boston area, for instance, each extra CFM of outside air brought in continuously and heated to 70F uses the energy in 2½ gallons of oil every year!

Many codes and regulations have been changed over the past several years to more realistic ventilation requirements as a result of the energy crunch. Smoking habits have changed, and fewer smokers may mean lower outside air requirements. When 2½ gallons of oil cost a quarter, you couldn't justify much investment in controls to vary outside air with occupancy, or to check air quantities periodically and adjust fan drives to keep them correct. It was easier just to crank the fan drives up for "plenty." A quarter won't buy 2½ gallons of oil any more ...... it takes more like two dollars now. And you may find it profitable to measure air quantities in your systems now, compare them to current regulations and design requirements, and cut your ill winds down to size.

Dave Elovitz

March 1980

LEMON-AID

After four years of struggling with it, the management of a Boston area health center decided the HVAC system in their 1974 building was a lemon. Now they have shown that, with the right recipe, it only takes a reasonable amount of sugar to make lemonade out of a lemon. The existing system was an unusual hybrid of multizone with variable air volume subzones. The temperature of the air supplied to each of four areas on each floor was varied by a thermostat in an exterior office. Branch duct systems to interior areas off each major zone were fitted with dampers controlled by thermostats in the interior areas. Detailed heating and cooling load requirements were calculated for each and every one of the 250 rooms in the two story health center and the existing duct system was compared to the requirements indicated by the load calculations.

As the saying goes, there was some good news and some bad news. The good news was that there was ample refrigeration and airflow capacity in the main units. The bad news was that it wasn't getting to where it was needed when it was needed. Much of the ductwork and most of the diffusers and registers were simply too small for the airflows required. But the BIG problem was how to change all that ductwork to a properly-sized, properly-routed, properly zoned system and still keep the health center in operation every day.

The answer lay in developing a scope of work made up of 67 individual modifications, subdividing most of the original ductwork for areas it could handle, in order to gradually convert the system to a new configuration with eight exterior heating/cooling zones (constant volume, variable supply air temperature) and 27 interior cooling zones (varying volume, constant supply-air temperature.) The entire conversion was completed without a single day's interruption of the health center's operating schedule, thanks to detailed preplanning and unusually close cooperation by the building management and the contractor's personnel. Temperatures are uniform and at the desired level throughout the building, ventilation continuous, energy consumption significantly reduced, and comfort vastly increased.

"Lemon-aid" turned a sour situation into a happy one for management, staff and patients alike.

Dave Elovitz

April 1980

EBTR

It is always risky to write in March about what the Government will do in April, but right now the Department of Energy expects the Emergency Building Temperature Restrictions to expire on April 16, without being renewed. Many of us will breathe a sigh of relief. The regulations were complicated to understand, hard to administer, and mandated working conditions outside the "comfort zone" developed through ASHRAE research. Happy as we might be to see EBTR gone, we should not forget them for several reasons. First, in the event of another severe supply disruption, they are likely to be imposed again, and we will be ready to implement them smoothly next time if we remember how we went about it the first time. Second, they have been an education to building owners and operators, and to tenants. Actual experience has shown them that winter temperatures below 76° and summer temperatures above 70° can be acceptable, and that changing temperatures can really save energy costs. Third, and perhaps most important, the basic idea behind the EBTR regulations does make sense: Arrange the system so no energy is used to heat above X° or cool below Y° EBTR said X was 65° and Y was 78°. After April 16 you can pick your own values for X and Y - perhaps 68° and 76° - that fall into what you consider acceptable comfort limits, and continue to use the EBTR strategies to save energy costs.

Incorporation and Philosophy

A few people have asked about the long term significance of my new corporate form of business, Energy Economics, Inc. Incorporating doesn't have to change my professional philosophy or my personal relations hip with all my clients. I just want to assure you that I continue to adhere to the belief that a client retains me instead of some big firm because he wants my own judgment -not the judgment of a faceless helper the client doesn't know and who doesn't know the client or understand his operations. I know I could do a lot more business and respond more quickly to your needs if I built up a large staff, but I am not going to pass your trust in me along to someone else unless I can have confidence that every piece of work that goes out with my name on it is as good, and as responsive to your individual needs, as I could make it myself. Keeping the books as Energy Economics, Inc. won't change that.

Dave Elovitz

May 1980

EXORICISM

It had to be evil spirits in the new ice rink. 100 feet of hard, shimmering ice, then 30 feet of mush, then good hard ice again. And it had been there for almost two years. The question of what caused it was now the subject of a lawsuit.

The rink operators had been able to keep the ice barely usable by running both brine pumps and bleeding air out of the rink piping in the soft area several times each day. No air appeared at other locations in the rink piping. Why did air only appear at one spot, and where was it coming from? Why did running both pumps help?

A few things about the piping differed from the usual ice rink design. The headers ran the length of the rink, and the "hairpins" under the slab ran the width. The connections to the hairpins came right off the top of the headers rather than at an angle from the side. 80th headers and hairpins were generously sized and the headers were full size the whole length of the rink. Some fancy arithmetic showed those differences were the source of the "evil spirits."

Calculating the velocity of the brine at various points in the system showed that air could collect at the top of a hairpin and cut off flow. That didn't happen in the first half of the rink because the initial velocity in the headers was high enough (with two pumps on) to carry all the air along with the brine. When enough brine went out through the hairpins to drop the velocity in the header, air separated out and collected at the top of the header, then vented from the top of the header up into the hairpins. That critical velocity occurred 100 feet from the beginning of the rink. In 30 feet more, all the air had escaped, no more "evil spirits." Science triumphs over black magic once again.

Dave Elovitz

June 1980

ORDINARY CLASSIC

It was an ordinary suburban industrial park building, typical of so many on the necklace of technology surrounding Boston. Designed for another tenant, the new one was staggered by the energy costs. They did all the usual things, took out lamps, shut things off nights and weekends, but it wasn't enough, so they asked for help. New design calculations based on the new tenant's use, and measurements of actual system performance showed the way to quite a few savings opportunities, but three stand out because they are so ordinary, they are classics:

(1) Discharge air temperature from the rooftop air conditioning units' was much lower than required, costing an extra $9500 a year for cooling and $8400 more than that for the extra reheat.

(2) Airflows from eight of the nine units were substantially higher than now required, which cost an extra $5400 a year for fan operation, $1500 for extra cooling, and $5000 for extra reheat.

(3) A large air conditioning system ran around the clock because a small computer room was in continuous use. The computer room had a standby air conditioning system in case the main unit failed. Simply running the standby unit nights and weekends, and relying on the main unit in case the standby failed, could save $1800 per year.

There were other things as well, but perhaps this little story of saving over $30,000 per year by getting the most out of what we have will remind all of us that sometimes we get so caught up in the exotica of the high technology sort of energy saving approaches that we lose sight of the basics .... the things that are so ordinary, they are classics.

Dave Elovitz

July 1980

THE END OF YOUR NOSE

"$16,000 a year for cooling in the winter! There must be a better way," said the college's utility manager. "It's as plain as the nose on your face." The science laboratories in the core of the building have a lot of heat producing equipment, and need steady temperatures year-round. The building was cooled with chi lied water fan-coils supplemented with conditioned ventilation air. Steam absorption chillers provided chilled water year-round for the fan coils, but the ventilation air temperature could be adjusted by varying the amount of outside air in the winter. The college had tried simply turning off the chiller and dropping the ventilation air temperature to provide winter cooling, but the results were uneven, unpredictable temperatures. The laboratories could not function.

Analysis of the design drawings showed that only the fan coils had temperature control. Four approaches for providing winter cooling were evaluated: Using outside air and providing control by using the fan coils as reheat; using cooling tower water through a heat exchanger; using cooling tower water directly in the chilled water circuit; and a small electric winter chiller.

The savings from a small winter chiller looked so good; would a larger chiller to carry the building spring and fall as well be even better?

It turned out that an even larger electric centrifugal chiller to replace the steam absorber entirely would almost pay for itself with the energy savings from the first year alone. The college would have been delighted with the savings from a smaller winter chiller, but they are even more pleased with 2½ times the savings for less than twice the investment. Sometimes it is worth looking a little beyond the end of your nose.

Dave Elovitz

August 1980

TEMPEST IN A TEACUP

"Maybe it isn't noise at all that is the problem - maybe it is vibration that is bothering him." The hospital's plant engineer had almost despaired of ever satisfying this particular office suite but decided to make one more try. You could almost feel, rather than hear, the air handler on the roof above the office, but the sound level meter showed within the acceptable range. At the high end, but within the range. Nevertheless, there clearly was a problem. The cup on the desktop showed tiny concentric circular tidal waves, a "tempest in a teacup."

Measurements with sophisticated vibration measuring equipment showed the office floor was, indeed, vibrating at the same frequency as the unit, but, oddly, the vibration at the floor was just as great as at the roof itself!

Something must be transmitting vibration directly from the roof to the floor slab, but what? The vibration readings showed where the "short circuit" occurred, and the only thing there was office partitions.

Inspecting the framing of the partitions showed many of the studs were jammed so tight between the roof deck and the floor slab that a 6/1000" feeler gauge would not slip in. At other locations in the suite where the studs were not jammed in so tightly, the vibration was not present, confirming the diagnosis.

Certainly the carpenter erecting partitions had not wedged the studs in that tightly; what had happened?

A little research into the history of the construction gave the answer. As in so many projects, the air handling unit was late, and all the interior partitions had been installed when the 6 ton machine was placed on the roof. The roof deflected normally under the weight of the unit, bearing directly on the longest studs, and our tempest in a teacup was born.

Dave Elovitz

September 1980

KISS

A fellow talked to me on the phone for a couple of hours the other day, trying to convince me a client should buy a computer to save energy. What he proposed was for the computer to turn off the air handlers for 15 minutes out of every hour. He said that would save 25% of the fan motor power, and 25% of the heat, but there were a few things that bothered me:

(1) Can we just arbitrarily shut off the outside air part of the time without knowing why the original quantity was selected? The only way the system could still do the job with 3/4 as much outside air is if the design called for too much to begin with.

(2) Bringing in less outside air saves heat, but interrupting heat to the space doesn't. Heat is lost from the space to outdoors whether the heating system is running or not, as long as indoors is warmer than outdoors. Certainly no one would notice the change in space temperature while the heat was off for 15 minutes, but there is no reduction in heat lost from the space, just a shift in when it is replaced.

(3) If you decide that you want to run the air handlers only 45 minutes out of every hour (or on any other regular cycle), you could do that with a $60 repeating timer instead of thousands of dollars worth of computer.

(4) Or, if you want 3/4 airflow, do it better by slowing the fans down. A steady 75% airflow would provide the same end result as cycling the fans, but provide constant ventilation without interruptions, save 58% of the fan horsepower rather than 25%, and require only a fraction of the cost of the computer system.

What's KISS got to do with all of that? It stands for the first rule of effective cost reduction: Keep It Simple, Stupid!

Dave Elovitz

October 1980

NOW WE ARE TWO

Two Registered Professional Engineers, that is. My son, Ken, has joined Energy Economics, Inc. and those clients that have worked with him already seem just as pleased about it as I am. Aside from being raised with engineering and mathematics the most common topics of dinner table conversation, Ken got a fine technical education at Lehigh University, and earned part of his way through school working in a contractor's engineering department.

After Lehigh Ken worked on production processes for Bethlehem Steel at their Bethlehem, PA headquarters, then at the big Burns Harbor plant in Indiana. His next post was Texas Instruments, where the processes got more exotic and the materials got more expensive, but things still weren't as fast paced and challenging as on the few assignments he had done for me. I was getting busier and busier - and behinder and behinder - so I was delighted to have him to come in with me. I hope he has as much fun solving your problems as I do.

Ken has his EIT registration in Pennsylvania and his Professional Engineering registration here in Massachusetts. While he has a broad background in all kinds of energy systems, because of his experience, he is especially knowledgeable on industrial processes and has done some fascinating failure analyses. Ask him to tell you about clinking failures in tool steel ingots if you get a chance. Or, better yet, ask him about the process energy question that's been bugging you.

JUNK Mail

We do hope that these calendars are useful, or at least interesting. If you'd rather we stop clogging up your in-basket with them, I'll be disappointed, but just let us know on the enclosed postal card. Or correct your address if we don't have it quite right, tell us what you think of the calendars, or just say "Keep 'em coming." But do let us know. Thanks.

Dave Elovitz

November 1980

BUY NOW .... SAVE LATER

If you were involved in a major construction project in 1973-4, you are not likely ever to forget it. Costs zoomed some 40% in 18 months. Although wages for mechanical and electrical trades skyrocketed, even mediocre mechanics were almost impossible to find. Equipment vendors quoted 24 to 36 months for air handlers, heat exchangers and switchgear. Most quotes included a cost escalation clause, or even "Price in effect at time of delivery."

We thought it unlikely it could ever happen again, but now there are signs that the same sort of pell-mell, no-holds barred grab for M/E equipment and skilled labor may be ahead for 1982-3. Southern California plumbers and pipefitters, who led the 1973-4 wage leap-frogging, just signed a new 3 year pact bringing average compensation above $50,000 a year by 1982. Lead times for items such as compressors, pumps, and electrical gear are already stretching out. Prices are stiffening. Engineering firms, particularly those specializing in the process industries, are finding it increasingly difficult to staff adequately.

Refinery upgrades, synfuel projects and similar super-projects are feeling the squeeze now, but, if history repeats itself, the squeeze will slowly move down through the construction industry until even the smallest mechanical and electrical jobs are hurt.

What can you do about it? You can take a look at what lies ahead for your company for the next five years and see what projects you should accelerate and get contracted with prices committed before things get really wild, and what could be put off to go to bid after things calm down again around 1985. In other words, position your company so it can stay out of the market during the period of instability when, and if, it comes.

Dave Elovitz

December 1980

THE PUZZLE OF THE 6000 GALLON BALLOON

The building was built in 1827. Most of the survey was typical for an older building. Loose wood sash. Single glazing. Incandescent lights. No heating zone controls. There was attic insulation, and that was ...

Puzzle #1: It was 40 degrees outside, yet we were comfortable in the attic despite the 6" blanket of insulation between us and the heated space below. Heat in an insulated attic is not unusual. Often there is a furred out space around a chimney, or a plumbing chase that lets warm air rise into the attic. But the usual paths seemed to be pretty well blocked off in this building.

Puzzle #2: The inside of the walls on the lower floors were cold. The insulating value of the heavy brick, stud space, and piaster calculated to a value that would give an inside surface temperature more like 65 degrees than 55. Cold spots on outside walls aren't all that rare, either. Outside air can leak in and cool the inside surface. But these cold spots didn't seem that localized.

A chance encounter with a magazine article on historical construction methods provided the clue. Years ago, outside walls used balloon framing. Studs ran nonstop from cellar to attic. No firestops at each floor as we have today.

Another trip to the attic, crawl out under the eaves, and there it was! Warm air pouring up into the attic from between studs that provided dozens of unobstructed chimneys from the basement up past four floors to the attic. Once those chimneys were sealed off, cold outside air drawn in on the lower floors no longer drew heat through the piaster and up into the attic - heat that used up about 6000 gallons of oil each year, which is why it was The Puzzle of the 6000 Gallon Balloon.

Dave Elovitz

January 1981

APOLOGIES TO CLEMENT MOORE

'Twas a cold winter night,
    and all 'round the block,
Not a burner was running,
    lest all be in hock.
The children were shivering,
    all huddled in bed,
And dreaming of 70,
    not 50 instead.
Mama in her blanket,
    and chilled to the liver,
Sat shiv'ring there,
    while the windows did quiver.
Then out on the lawn,
    there arouse such a clatter,
I sprang from my bed
    to see what was the matter.
There what to my wond'ring
    eyes did appear,
But the energy elf,
    and his two engineers!
More rapid than eagles,
    his remedies came.
He whistled and shouted,
    and called them by name.
"Weatherstrip! Insulate!
    Efficiency test!
Add timers, computers,
    Get maintenance best!
Turn off extra lights!
    Caulk up all the cracks!
Storm windows and doors
    can credit your tax."
And then from the boiler,
    I heard the elf roar:
"Your flue is 600!
    CO_2 should be more!"
A wink of his eye
    and a twist of his head,
"Don't shiver: Stop wasting!
    Conserve fuel," he said.
"Electricity, too,
    is too precious to fling.

If you use it all wisely,
    you'll live like a king."
He spoke not a word,
    but went straight to his work.
He audited everywhere,
    then turned with a jerk,
Made out a report,
    in clear, lucid prose,
Said, "Do all of these.
    End your energy woes!"
Then he called out to all,
    "Don't be OPEC-reliant."
"HAPPY NEW YEAR TO ALL,
    EVERY COLLEAGUE AND CLIENT"
From our house to yours
    for each family member,
"A wonderful year,
    from Jan. through December."

Dave and Ken Elovitz

February 1981

EXPENSIVE SAVINGS

"I have got to cut project costs. Interest rates are eating us alive on this job. The mechanical contractor suggests direct expansion cooling instead of central chilled water, but can I afford to pass up the energy savings?"

To answer that question, we compared the DX system to a 600 ton chilled water plant. The DX rooftop package system cost $85,000 less than the chilled water system, and saved another $40,000 for wiring and for building an equipment room for the chillers. The chilled water system used only 614 KW of electricity to produce 600 tons of cooling, 1.023 kW for each ton. But the DX units would use 1.34 kW for each ton. Even before paying the interest on the higher investment for the more efficient chilled water plant, it would take over seven years of energy savings (at current rates) to equal the extra cost of buying the chilled water plant compared to the rooftop DX units!

We asked the DX unit manufacturer to consider changes to make the units more efficient. By installing larger condenser coils we could reduce head pressure, improving efficiency, and reduce peak power use to 1.22 kW/ton.

These more efficient units would cost only $17,000 more than the standard ones, yet they would save about 130,000 kWh per year compared to the standard units. They would still use 210,000 kWh per year more than the more expensive (but more energy efficient) central chilled water plant, but it would take ten years of energy savings from the chilled water plant to equal the first cost savings from installing the "special" DX units.

About the developer's question .... I guess the answer is that sometimes you need to look carefully, and ask some questions, before you can figure out just how much you can afford to save!

Dave Elovitz

March 1981

HEADQUARTERS GETS A HOSING

"You can't type with mittens on! We want to help save gas, but if we keep the place at 55, no one can work! Isn't there something we can do?"

Gas-fired boilers for the handsome new corporate headquarters building had been great, compared to the cost of oil next door. But the gas shortage crisis had brought the Governor's order on Thursday: 55 degrees in all gas-heated industrial and commercial buildings.

Creative thinking, responsive vendors, and a terrific maintenance crew brought comfortable temperatures by Monday and saved even more gas than turning down to 55.

The oil-fired boilers in the building across the driveway had plenty of capacity, why not run hoses? Yes, the hose manufacturer could make 3" hoses for hot water at high pressures. Yes, he could make them by late Friday afternoon. The hose route was laid out carefully and measured. Across the boiler room, up and out through a roof vent, down across the lawn and driveway, then up the wall and across the roof to the penthouse. Vents at the two high points would air out as the hoses filled.

Insulation to wrap the hoses was located and picked up. A truckload of salt hay would provide protective cover from the weather. By Saturday morning all materials were on site.

Even though planning had barely kept ahead of doing, there was only one moment of panic. At two o'clock Saturday afternoon one set of fittings appeared to be missing, but an ingenious swapping of hoses made the connection possible. Filling the hoses started Sunday morning, slowly, venting all the air. Run hot water through a bypass to warm up the hoses, vent the last bits of air, then valve off the gas-fired boilers and open the valves to connect both systems. By three on Sunday, headquarters was being heated by oil, ending at least one company's gas crisis.

How often is a Plant Engineer a hero for giving headquarters a good hosing?

Dave Elovitz

April 1981

FIGURES DON'T LIE, BUT ...

"It seems too good to be true," the client said. "Will the savings really equal the investment in 16 months?"

The application seemed a natural for heat recovery: A drying oven that operates on three shifts, exhausting about 2500 CFM at 250°. That exhaust, plus other exhausts in the same area, used about 4000 CFM of makeup air. The proposal called for bringing 4000 CFM of outside air into the building heating system and using the heat recovered from the oven exhaust to heat the outside air up above room temperature, thus using it to heat the building as well.

The computer printout from the heat recovery unit could extract about 1600 million BTU per year, all right, and you would have to burn about $9400 worth of gas to make that much heat. What the manufacturer's computer overlooked was whether there was any use for all that recovered heat. We calculated what the supply air temperature from the heating system would be for each outdoor temperature. Then, using data from the heat recovery unit manufacturer, we calculated what the temperature of the outside air would be leaving the heat recovery unit. We found that above 27F outside, the air from the heat recovery would be so warm, it would overheat the building. That meant additional controls would be required to bypass some of the makeup air around the heat recovery unit, increasing the installation cost somewhat, but it also meant the building could use less heat than the unit would make available. So instead of saving $9400 worth of gas a year, the savings on heating the building and makeup air would only be about $4000.

Maybe $4000 savings per year from a $12-13,000 investment shouldn't be called "only" (especially the way the price of gas keeps going up), but you kind of wonder how often an impressive computer printout has sold an installation by telling the truth, but not the whole truth.

Dave Elovitz

LITTLE RED HEN REVISITED

Once upon a time there was a little red hen who discovered some grains of wheat in the barnyard. She called the animals together and said, "If we plant this wheat, we can make bread. Who will help me plant it?

"Not I," said the cow; "Nor I," said the pig; "Not I," said the duck; "Nor I," said the goose. "Then I will," said the little red hen, and she did.

The wheat grew tall and ripened into golden grain. "Who will help me reap the wheat?" asked the little red hen.

"Do you have an environmental permit?" said the duck? "I've too much seniority," said the cow. "I'd lose my unemployment compensation," said the goose. The pig said, "Not in my job classification."

"Then I will," said the little red hen, and she did.

The time came to bake the bread, and the little red hen asked "Who will help me bake the bread?"

"That would be overtime for me," said the pig. "Has OSHA approved your oven?" said the duck. "I'd lose my welfare benefits," said the cow. "If I'm the only helper," said the goose, "that's discrimination." "Then I will," said the little red hen, and she baked five loaves and held them up for her neighbors to see.

They all wanted some, even demanded a share, but the little red hen said, "No, I can eat them all by myself."

"Capitalist leech," hissed the goose. "Excess profits," rumbled the cow. "I demand equal rights," the duck squawked. The pig just grunted, while they painted "unfair" picket signs, and marched around and around the little red hen, shouting, until a government agent came.

"You must not be greedy," he said, and the hen replied, "But I earned the bread." "Exactly," said the agent, "in our wonderful free enterprise system, anyone can earn as much as he wants, but under our socially responsible government regulations, the productive workers divide their product with the idle."

And they all lived happily ever after, including the little red hen. But her neighbors wondered why she never baked any more bread.

Dave Elovitz

June 1981

ALHPABET SOUP

"We finally stopped the complaints about those hot outside offices," boasted the operating engineer. "Yeah, but now we're getting complaints that the interior spaces are too cold," the engineering manager answered, "and energy costs are going out of sight. Let's get someone to look at the whole system."

A check of room temperatures showed that the exterior offices were still warm, and the interior rooms were on the cool side. Supply air temperature even in the warm rooms was a cool 50F, which is lower than usually needed. The thing to do was calculate the cooling loads room by room to see if the available cooling equaled the load in individual areas, not just for the building as a whole.

Sure enough, the existing quantity of supply air to the outside offices just couldn't remove all the heat pouring through the single thickness glass walls.

When the operating engineer lowered the supply air temperature, he alleviated one problem (it kept the researchers quiet), but the penalties were increased operating costs and overcooled interior rooms: High airflows there delivered more than enough air to meet their cooling requirements. By increasing air quantities where more cooling was needed and decreasing where loads were lower, the supply air temperature could be increased to 55, and all the rooms kept comfortable.

One problem remained: a classroom filled to capacity all day was "too stuffy". Increasing the amount of supply air in that room to provide enough dehumidification would have overcooled it, unless expensive and wasteful reheat was added. An inexpensive but effective way to absorb more moisture was installing a transfer fan to draw air from adjacent areas. The supply air did the cooling and part of the dehumidification while the additional from next door picked up still more moisture, making the room comfortable.

Simple as ABC: once the CFMs match the BTUs, everything will be A-OK.

Ken Elovitz

July 1981

VERSES! FOILED AGAIN!

"This I will grant you,
The building's a jewel.
But as you *have* drawn it,
She'll burn too much fuel."

    The building inspector
    Began to explain
    The two fundamental
    Rules of the game.

"The most we allow
For U-O is point three.
Thirty-four is the max
For OTTV."

    Now this left the owner
    Deeply chagrinned.
    He simply didn't know
    Where to begin.

"I must get so me help.
Of that there's no doubt.
A good engineer
To figure this out!"

    The building was lovely -
    A new shopping center
    With great tall glass walls
    Where people would enter.

The traditional answer
Was use double glazing;
But the cost, my friend,
Was truly amazing.

    More insulation
    On roof *and/or* ceiling
    Was one simple way
    We thought of for dealing

With two giant skylights
And handsome glass walls
To be in compliance
For U Overall.

    Then came the problem
    Of OTTV:
    All of that glass!
    Alas! Woe is me.

Apply solar film;
Make the windows reflective!
Now that's an idea
That's real cost effective!

    So the heat loss in winter,
    As gaged by U-O,
    Finally came out
    Sufficiently low.

And as for that nemesis
OTTV,
(Summertime heat gain
Between you and me)

    The changes we worked out
    Reduced solar load
    And easily brought us
    Well within code.

And that got the monkey
Off of our backs
So me and my father
Could fin'ly relax.

Ken Elovitz

August 1981

ON FELINE EXCORIATION

An earlier study had recommended all the obvious things - relamping with energy saving fluorescent lamps, removing some lamps entirely, installing a separate water heater so the boiler could be shut down in summer, additional insulation on the outside walls.

The laboratories were crammed with equipment studded with knobs, blinking lights and flickering dials. There were big walk-in refrigerators and freezers and, down next to the boiler room, a vault jammed with sub-sub-zero freezers.

Could we find more energy savings that would not interfere with the stringent needs of the research activity? We expected to find that research equipment was the biggest electricity user, but we were still surprised to discover how big: almost 2/3 of the total electrical use! Lighting and cooling about 15% each, and air handling was only about 7%.

Analysis identified the usual ideas, of course, but one pair of unusual measures was kind of interesting: The freezer room must be continually cooled all year, to take away the heat of the freezer machinery. The first study had suggested bringing in outside air to cool the room in winter, but the cost of an opening in the thick concrete wall and an air shaft up to the ground level above had proven too high. We suggested adding another coil in the cooling unit piped to an air-cooled heat exchanger outdoors, and circulating anti-freeze to bring in outdoor cooling without any massive construction. And when it isn't cold enough outdoors to cool the freezer room for free, another energy saver can take over: A heat recovery unit on the freezer room cooling unit uses the heat rejected by that unit to preheat the large quantity of domestic hot water used for washing laboratory glassware.

Even more savings than the first report's recommendation, at half the cost. There really is more than one, way to skin a cat!

Dave Elovitz

September 1981

THANK YOU, MR. OHM

"6500 Kilowatts!! Just for heat? Cut it down!!

Even for a high-rise hotel, over 8.5 watts per square foot did seem like a lot of heating capacity. Where was it all going? Well, the biggest portion, about 3000 KW, heated outside air brought in for the many function rooms and kitchens, and close to another 1000 KW for outside air to replace exhaust from guest room toilets. Guest room fan coil heat accounted for almost 2000 KW more. What with some reheat coils, cabinet heaters and baseboard, it added up to 6459 KW.

Comparison with codes showed the exhaust systems removing more air than was necessary from many areas; new type energy efficient kitchen hoods would require less heat for kitchen makeup air; recognizing that the fans warmed the air a little could shave some more off heating coil capacities; and some zone reheat coils could be eliminated by allowing VAV boxes to go fully closed. Down to 5200 KW peak demand, with nothing left to reduce .... or was there?

Guest room fan coil units called for the "standard" size heater in each size unit, but, in many cases that was larger than necessary. A smaller stock size heater would do in several rooms on each floor, saving another 238 KW. Still more heating capacity in each room than needed, but custom sized heaters would be too expensive. Wait a minute! Ohm's law tells us KW equals volts squared over ohms. In industrial furnaces we lower the heat output by reducing voltage, why not here? Sure enough, an inexpensive buck/boost transformer would bring capacity in line with heat loss, and drop the peak demand another 190 KW! Down to 4761 KW! Not a bad job of demand reduction. And, as a bonus, about $150,000 a year off energy consumption as well. Thank you, George Simon Ohm.

Dave Elovitz

October 1981

TWO RIGHTS CAN MAKE A WRONG

It was kind of a hurry up job, so we were working on it together. Ken was calculating the heat gain and loss while I was setting up the energy use calculations.

"This architect really did things right," Ken said, admiring the generous insulation and judicious use of limited areas of strategically placed double glazing.

"So did the HVAC designer," I added. "This heat pump system won't need the electric boiler unless the temperature drops below zero!"

The building was heated and cooled with a couple of dozen small heat pumps. Units cooling the interior areas put heat into a common tower water loop, and units on the perimeter could take heat out for heating, or cool by putting heat in. If more heat went into the loop than was needed for heating, it would be rejected through the cooling tower. If the cooling units didn't put in enough heat, an electric boiler would add some. It really looked like a winner - energy efficient building and energy efficient heating.

But those two rights made a wrong. The building was so well insulated, even the perimeter offices needed heat only rarely; in fact they needed cooling until it got down freezing outside. Even then the large interior areas needed cooling. And, because they had expected the heat from the interior cooling units to be used to heat the perimeter, the designers had not provided any "free" outside air cooling. Our calculations of annual energy cost came to about $0.60 per square foot more than the budget, not because the heating system wasn't super efficient, but because the savings on heating didn't make up for the increased hours of mechanical cooling required even during the coldest weather.

The solution? Increase the size of the fresh air system and add controls to use outside air for cooling the interior in winter, except at very low temperatures when the system goes back to the heat pump mode. When two rights make a wrong, a third right can make it all right.

Dave Elovitz

November 1981

FIRE!

Three men - a plumber, university professor, and an engineer - live in three identical houses. At the same time during the night, the smoke alarm goes off in each house.

The plumber wakes up, leaps from his bed, gets all the water he can, and floods the entire house.

The professor wakes up, measures air temperature, wall thermal gradients, and CO and smoke concentration in the air. He enters all the data into his computer, uses Fourier's theorem to confirm that the house is on fire, and analyzes the reaction kinetics to calculate the exact amount of water required to extinguish a fire of that size. Meanwhile the house burns to the ground.

The engineer wakes up, smells for smoke, feels the door to see that it is safe to go into the hall, checks quickly and finds a small fire burning in a wastebasket. He puts it out with his home fire extinguisher, minimizing damage and disruption while accomplishing the job effectively. Before returning to bed he checks the carpet and drapes for any unextinguished sparks.

People frequently get warning signals or signs of trouble in all types of situations, both personal and professional. Some people, like the plumber, respond with overkill. They may achieve the desired end result (the fire is out), but the accompanying side effects of their efforts may be worse than the original problem.

Others, like the professor, bury themselves in theory and minutiae, losing sight of their objective. The result is an elegant analysis without noticeable progress toward a workable solution.

The astute troubleshooter is like the engineer in the story. He gathers enough data to be sure he understands the problem and verifies his conclusion with a quick double check. Then he uses the tools that fit the problem at hand and achieves effective and timely results.

Dave Elovitz

December 1981

THE ALCHEMIST

Once upon a time there was a Young Engineer - a metallurgist by training. After graduating from an engineering school in the hills of Pennsylvania, he went to work in the Steel Mill, the metallurgist's mecca. There his projects ranged from developing a computerized surface quality tracking system for a new continuous caster to devising techniques to extend the life of ingot molds and stools. He inspected cold rolled sheet by holding samples under special light and carbon and alloy steel plates by walking across them.

After several years, the little bit of alchemist inside every metallurgist overtook the Young Engineer, and he left the Steel Mill to work with gold, silver, and other precious metals. He used heat and pressure to inlay millionths of an inch of gold into copper and nickel and performed other feats of wizardry, but the Young Engineer remained restless.

Soon he joined forces with the Senior Engineer in a tiny consulting firm. There the challenge and variety of his assignments were matched only by the diversity among his clients. The two Engineers worked together zapping energy waste, hunting gremlins in existing systems and exorcising demons from new designs.

The Two Engineers wanted to keep in touch with their friends and clients, so they sent monthly messages on the backs of "handy pocket calendars". The Young Engineer knew he was a full partner when Another Consultant called to chortle over one of the Young Engineer's monthly messages. For in some circles the Two Engineers became almost as popular for their commenty calendars as for their engineering. In fact, the commenty calendars were so much fun that people started to clamor for reprints. You can clamor too. A note or a phone call will bring you a "handy pocket booklet" with some of our most popular messages.

Ken Elovitz

January 1982

WORSE VERSE

We've come to the end of the year,
And offer these fun verses here.
    We've tried not to cheat
    Or mess up the beat,
But some of the rhymes are quite queer!

An energy czar named McHolledge,
Slowed down all his fans with the knowledge,
    That moving less air
    Could get the heat there
And save lots of dough for his college.

There once was a high powered chiller
Much loved by the 'lectrical biller.
    A new cooling tower
    Reduced its horsepower.
The savings are really a thriller!

We've noticed that many a boiler.
Equipped with a hot water coil(er).
    Is kept much too hot
    When heat is used not,
Alas! an efficiency foiler.

A plant engineer name of Skinner.
Felt a draft after eating his dinner.
    He caulked all the sills,
    Cut energy bills,
Now Skinner is truly a winner.

There's so much that is new about lighting,
The savings are really exciting.
    Efficient new lamps
    Illumine like champs.
Less power for much better sighting.

Computer reports have got clout,
Accepted with nary a doubt.
    But we don't feel kindly
    Toward reading them blindly.
Beware garbage in, gospel out!

'Gainst energy bills so extensive,
He mounted a major offensive.
    Shut off extra light.
    Turn down heat at night!"
The savings he got were immensive.

To try to keep energy use,
From tying you up in a noose,
    Make energy audits.
    You'll get all our plaudits.
'Cause energy waste is uncouth!

A layer of roof insulation,
Storm windows and other creations,
    Add caulking and sealing,
    How warm you'll be feeling
From energy saving gyrations.

We want to make perfectly clear,
As we do at this time ev're year,
    This message now soars
    From our house to yours:
"A year of good health and great cheer!"

Dave and Ken Elovitz

February 1982

METER MYSTERY

Even with the latest in electronic transmitters, their computer said steam flow was 8000 lbs/hr when the utility measured 12,000. Steam flow for both metering systems was determined from the pressure differential across the same orifice plate. Both started with the same pressure differential and converted it with the identical transmitters. How could they get different answers?

The way to find out was to check the signals every step along the way. It only took a minute to find out that the milliamp outputs of the two sets of transmitters were not the same.

Could that be? Sure enough, a simple check by valving off the pressure inputs confirmed that something was awry. With zero differential pressure, the output of the new transmitters was not four milliamps! A caution in the transmitter manufacturer's literature gave a clue: Maximum voltage supply to the transmitter should not exceed 45 volts. Alas, the installer had missed that, and the applied voltage was 79. A quick adjustment got the voltage back in line, matching the milliamp signals from both sets of transmitters, but the flow rate readings still differed.

We had grappled with the gremlin, but only chased him deeper into the system. Where could he be? Hand calculations of flow from the transmitter output matched the computer's results. So the computer's algorithms must be OK. Or were they? Laborious calculations showed the orifice plate calibration constant had been calculated for steam at the standard 100 psig. But this steam was at 120 psig! Using a new orifice constant calculated for 120 psig matched the utility's readings almost exactly. And a phone call confirmed that the utility, expecting varying steam pressures, had incorporated automatic pressure compensation in their meter calculations. Just adding that feature to the user's new metering, and making up some graphs that made it easy to spot check the system would glue down that gremlin for good.

Ken Elovitz

March 1982

POLTERGEIST

"It's weird," the manager of the new building groaned. "Not a single problem with the HVAC since we opened in July, but ever since the weather turned cool, I can't keep the doors closed. They keep blowing open! Can you take a look?"

Sure enough, as I approached the building the doors were partly open, discharging a rush of warm air into the crisp October afternoon. I pushed them closed and they swung right out again.

Obviously, the building was being heavily pressurized by the outside air being brought in by the economizer cooling cycle for the interior zones. But I distinctly remembered two big relief hoods with gravity dampers on the HVAC drawings during pre-construction review. How could this building be being pressurized?

"Let's go up to the roof a look for those relief hoods. Could they have put in the wrong size or left them out?" No, they were there OK, and plenty big. But there was no air flowing out of them, even though it was clear from the position of the air handler dampers that large amounts of outside air were being brought in on the economizer cycle. And the building was under plenty of pressure. Even the spring balanced smoke relief skylights at the top of the stairwells were blowing open.

Perhaps something is blocking the flow of air from the plenum to the hoods. Let's take a look from underneath."

Twenty minutes of locating access tiles and peering above the ceiling finally found the dampers to the relief hoods, and at last we could open the ceiling just below them and get up on a ladder for a good look. The damper was tightly closed. But why? Wait a minute. As the flashlight swept across the underside of the damper its beam picked up a big red label. What did it say? "Install this damper THIS SIDE UP!"

Dave Elovitz

April 1982

JABBERWOCKY

I just started reading a magazine article that observes:

"In engineering as with any other exact field, mutual understanding is of utmost importance. To those in engineering, the margins of success allow only a very small tolerance for ambiguity. Certainly this is evidenced by the need for constant rhetoric over the problems of specifications and contract language, and even more dramatically, by the increasing deluge of court battles over intent and meaning. These activities not only point up the importance of a mutually comprehensible language and the use of it, but also the fact that significant deficiencies exist in that regard in the language used by engineers."

Now I must confess that I do like to throw in a few big words myself once in a while. But Ken has this game he plays, called "The Fog Index" and sometimes, when he is checking the draft of one of my reports, he plays it on me: You take a sample of text 100 words long and count two factors:

(A) The average number of words per sentence; and

(B) The polysyllable percentage, which is the number of words three syllables or longer in each 100 words. Don't count verbs where the third syllable is "ed" or "es" (like "adjusted" or "balanced"), combinations of short easy (like "pipefitter"), or capitalized words.

The Fog Index is 0.4 X (A + B).

For that article it came to 17.8, which is the number of years of school you are supposed to need to understand that paragraph easily. In other words it was written for readers with a Master's degree.

Ken complains if I go over 12. He says. "If you use a lot of big words and long sentences, people have trouble understanding what you are talking about. Then someone makes a mistake that costs a lot of money."

Hey! That's what that article was trying to say!

Dave Elovitz

May 1982

BOILER EGGS

Once a biology professor needed some fertilized eggs for an experiment. He sent a requisition to Purchasing for one dozen eggs priced at $5.98. The requisition came back with a hand scrawled note: "Eggs are 98 cents a dozen." The dauntless professor tried again: "One dozen eggs - $0.98, Services of rooster - $5."

Could a similar misunderstanding have made the building superintendent pick a bargain priced "outdoor weather control" for his boilers? The control turned out to be an outdoor temperature sensor and a time clock which shut the boiler off if the outdoor temperature was above 60 during occupied hours and above 55 during unoccupied periods.

Sure enough, the control saved about 10% of previous annual fuel costs. The building superintendent did not recognize the difference between the bargain control and the more expensive temperature-time type of outdoor weather control that runs the boilers only part of each hour, even during occupied periods. This control automatically varies on-time with outdoor temperature. The lower the outdoor temperature (the greater the need for heat), the larger the percentage of on-time.

During unoccupied periods the control only starts the boiler if heat is needed to maintain the selected setback temperature in the building. An energy audit determined that the additional savings from a temperature time control would be a further 20% reduction in energy costs: A few months' savings would be more than enough to pay for replacing the bargain control with the better type.

I guess you never know eggsactly how fertile a device can be until it weathers an analysis.

Ken Elovitz

June 1982

MURPHY'S LAWS
(and other wisdom)

Murphy's Law
Anything that can go wrong will, (And at the worst possible moment.)

Murphy's Other Law
It is impossible to make anything foolproof, because fools are so ingenious.

Park's Pronouncement
Good judgment comes from experience; experience comes from poor judgment.

Woody Allen's Distinction
The lion and the calf may lie down together, but the calf won't get much sleep.

Einstein's Illumination
Everything should be made as simple as possible, but no simpler.

90-10 Rule of Scheduling
The first 90% of any ten month project takes the first nine months. The last 10% takes the next nine months.

Segal's Law
A man with one watch knows exactly what time it is. A man with two watches is never sure.

Farber's Genealogy
Necessity is the mother of strange bedfellows.

Wethern's Geneaology
Assumption is the mother of all screw-ups.

The Golden Rule
Whoever has the gold makes the rules.

Chisholm's Second Law
Any time things appear to be going well, you have overlooked something.

Pipefitter's Postulate
Architects are geniuses. They are the only people capable of taking one acre of land, with the sky as the limit, and conceiving a structure in which there is not adequate space for a two inch plumbing trap.

First Law of Bio-engineering
Under the most rigorously controlled conditions of pressure, temperature, volume, humidity and other variables, the organism will do as it damn well pleases.

Churchill's Commentary
Man will occasionally stumble over the truth, but most of the time he will pick himself up and continue on.

Ken's Corollary
We'd worry a lot less about what people think of us if we recognized how rarely they do.

Ritchie's Rubric
Fool me once, shame on you. Fool me twice, shame on me.

Dave Elovitz

July 1982

PIRATED PINS

"We've got these air handlers in ship shape!" the building engineer boasted. Our new time clocks will save us 43% by shutting down all the fans from 7 PM to 7 AM."

"Damn clocks!" the maintenance superintendent growled. "Someone keeps taking out the pins. Even breaks off the lock! I give up. We're going back to 24 hours a day."

And the custodian announced triumphantly: "About time they quit putting those pins back on the clocks. I've got enough pins here to sink a ship. Floors don't dry at night when the fans are off. Gotta keep 'em running."

Would the building engineer have to let the air handlers run all night just to dry the floors? Or could he implement his idea without disrupting the rest of the operation? Did he need a computer to do it? Those were some of the questions he had for the energy consultant.

During his survey, the consultant talked with the custodian. He found out that it takes about an hour for the floors to dry when the fans are on. And he came up with a solution that everyone could agree on. A new "minute-minder" type twist timer would override the time clock and run the fan while the night shift washed the floor. By locating the new time switch in a Janitor's Closet, it would be handy for the cleaning crew to use. And best of all, the twist timer would shut the fan off automatically after it was no longer needed to dry the floors.

Adding the twist timers would cost only a few dollars for each time clock. The building engineer reported that the time clocks with twist timers could save a handsome 37%, the maintenance superintendent no longer had to lock up the time clocks, and the custodian stopped taking the pins out of the clocks!

Ken Elovitz

August 1982

655-2556

That's our new primary telephone number in Natick, equipped with various bits of electronic wizardry which are supposed to make it easier for your call to get through. It's an easy number to remember: 655-2556 - the same backwards as forwards - but don't rely on memory. Jot it down now so you can use it next time you call.

LABOR RELATIONS

"You better call this woman back right away! She really sounded frantic!"

The anxious caller turned out to be the administrator for a prominent congregation. It seemed the clergy had been complaining of a choking, biting odor in the air in the sanctuary. This noon they had collectively announced that if the odor wasn't gone before the weekend, they weren't going to conduct services! Here it was Wednesday afternoon, and how could she explain that the clergy had gone on strike?

Thursday morning I stood at the lectern and didn't smell a thing as each of the clergymen described the gasping, burning sensations. No trace of odor in the air conditioning system, and the Administrator was giving me a "See, I knew it was all in their heads" look. For lack of any other idea, I dropped to my knees and sniffed at the rug, then reared back choking, tears running down my cheeks. Whatever it was, it wasn't imaginary.

No, the rug was not new. It had been down for some years and there had never been a problem before this fall. It had even been extra thoroughly cleaned during the summer, the building superintendent explained, adding proudly how much he had saved by shampooing the carpet himself.

"Could I have a bucket of hot water and a scrub brush?" I asked and proceeded to scrub a small area vigorously with clear water - which quickly turned to thick suds.

"There's your culprit! You haven't rinsed all the soap out, and it has supported a growth of fungus! Scrub the carpet thoroughly with hot clear water until you don't get suds any more, and the strike will be over."

Dave Elovitz

September 1982

COMMUNICATION

As Architect Drew It

As Estimating Bid It

As Engineering Designed It

As Shop Fabricated It

As Field Installed It

What the Owner Wanted

Shamelessly plagiarized by Dave Elovitz

October 1982

"IMPOSSIBLE"

"If there is flow to any unit, there must be some flow, even if it is just a trickle, to every one," I thought when the Superintendent of the luxury high-rise condominiums told me that sometimes some faucets got no hot water flow at all. But that was before I traced out the complicated piping system and made the pressure drop calculations.

I could understand why there might not be enough *hot* water: Formulas indicated as much as 115 Gallons per Minute of hot water would be required for peak periods, and the heaters could only heat about 60 GPM to acceptable temperatures. It might not be hot, but how could there be *no* flow at some outlets?

Pumps recirculated a little flow from each riser, about 56 GPM to keep the 56 risers hot. Calculating the pressure to push that flow plus the 115 GPM peak- demand through the system and still have a little pressure at the faucets showed the pressure at the heater inlet would have to be a pipe-busting 190 pounds, not the 45 pounds the designer provided! That's why!

Further calculations showed that the pressure at some far outlets could actually drop below zero (because of recirculating pump suction) at flows as low as 20% of peak demand. Fortunately, the system could be corrected easily without adding a new booster pump and more heaters: A large insulated storage tank would provide hot water for the peak and allow the heaters to catch up later. The building's hot water would not flow through the hot water heaters, so the pressure drop would be much lower. A separate flow loop would connect the tank to the heaters, which would recharge the tanks with hot water as fast as they could make it, but without the extra flow for building recirculation.

So the impossible proved to be actual, but some basic engineering made it all satisfactual!

Dave Elovitz

November 1982

HOT TODAY, CHILLER TOMORROW

"We need a new chiller!"
The old one is fouled!
And get it by Monday!"
The manager growled.
    A central plant chiller is
    Not, I must say
    The kind of appliance
    You buy every day.

Yet this was a case
Where we had to be quick,
'Cause the worn out machine
Would no longer tick.
    They'd tried to repair it
    Without much success.
    The parts were all broke
    'Twas a terrible mess!

So I picked up the phone
And made call after call
Requesting machines
That were set to install.
    "Centrifugals, please,
    Though I'll take a Recip.
    And they need it today -
    So how soon can you ship?"

The first little challenge
Was not nuts and bolts.
It was finding machines
Made for 2-0-8 volts.
    And how about clearance
    Through doorways and hall?
    It looked like we'd need
    A big hole in the wall!

Meanwhile we thought
About what we could do
To make the place cooler
Until we got through.
    Remember those hoses
    We used once before?
    Would this be a day
    We could use them once more?

Pipe in chilled water
From just down the street.
Now there's a solution
That sounds pretty neat.
    The hoses, however,
    Could stay where they lay
    'Cause we found the right chiller
    The very next day.
Despite all those troubles
Things worked out quite nice,
So we settled right back
For some seltzer on ice.

Ken Elovitz

December 1982

DOUBLE TROUBLE

The manufacturer's representative wasn't argumentative or unfriendly, but he was very positive: "We've checked these units from stem to stern. There is nothing wrong with them."

"If there is nothing wrong with these units," the project manager answered, "why do we only get this terrible vibration on these two and not the other four just like them?"

The rep scratched his head. "I just don't know. But I do know that we have checked everything on these units and even had a factory man install some flow straightening baffles, but everything checks OK."

We started listing all the things that should be checked. "How about fan rotation?" I asked. "Yep. All turning the way they are marked," the factory mechanic answered. "Four clockwise and two counterclockwise, just like the arrows marked on the fans."

My ears perked up. "WHAT? Two of them have different fan rotations? The two that vibrate? Let's take another look inside. Wait a minute, what's this label on the inlet guide vanes? Clockwise! These inlet guide vanes are for a fan that rotates clockwise, but this fan turns counterclockwise!"

"The fan moves air by spinning it around," I explained. "The faster the fan turns, the more air pressure it makes. Inlet guide vanes reduce fan capacity by starting the air spinning in the same direction as the fan before it gets into the fan, so the air thinks the fan is going slower than it is. These "backwards" vanes spin the air in the other direction, and the air acts as if the fan were running faster. When the fan tries to produce a high pressure at low flow, the airflow becomes unstable, shaking the whole big unit! Get counterclockwise vanes into these units and let me know how they sound then."

"I1's unbelievable!" the project manager phoned later. "You can hardly tell the units are running, and they used to shake the whole building. I'm delighted. But I sure feel sorry for whoever has two units with clockwise fans and the counterclockwise vanes that belonged in ours!"

Dave Elovitz

January 1983

to the tune of
JINGLE BELLS

Cellulose or batt
Stuffed in ev'ry wall
Give yourself a pat
For answering the call

Calculating "U"
Checking building codes
Looking for the smartest way to
Cut down monstrous loads! ...

    (Refrain) Energy Engineers,
    Saving energy
    Oh what fun it is to be
    Saving energy

Switch to VA V
Shut off reheat coils
Cut down outside air
We'll save lots of oil.
Variable speed
'Stead of inlet vanes
Moving all the air you need
For bills that aren't insane ....

    (Refrain)

Decorative lights
Gleam against the snow
Cutting down the volts
Makes consumption low!
Electronic clocks
Count and calculate
Never run those costly lights
When you're not open late....

    (Refrain)

Measure C-O-Two
Cut back excess air
These things will bring you
Savings you can't spare.
Use recovered heat
Keep those filters clean
These are ways that you can beat
The high bills we have seen! ...

    (Refrain)

Sometimes we see jobs
That just will not work
And we hear the sobs
Of people gone beserk.
That's when it is time
To find out what went wrong
Fix it up and think of rhymes
For calendars in song!

    (Refrain)

Fran and Dave and Ken
Sara, Gary, too
Hope the coming year
Will be good to you!
To our many friends
All across the land
Goes our wish for peace and joy
And a New Year that is grand!

    (Refrain)

Dave and Ken Elovitz

February 1983

AIN'T IT SO!

And the Lord said unto Noah: "Where is the ark which I have commanded thee to build?

And Noah said unto the Lord: Verily, I have had three carpenters off ill. The gopher-wood supplier hath let me down - yea, even though the gopher wood hath been on order nigh upon 12 months. What can I do, 0 Lord?"

And God said unto Noah: "I want that ark finished even after seven days and seven nights."

And Noah said: "It will be so." And it was not so. And the Lord said unto Noah: "What seemeth to be the trouble this time?"

And Noah said unto the Lord: "Mine subcontractor hath gone bankrupt. The pitch which thou commandest me to put on the outside and on the inside of the ark hath not arrived. The plumber hath gone on strike. Shem, my son who helpeth me on the ark side of the business, hath formed a pop group with his brothers Ham and Japheth. Lord, I am undone."

And the Lord grew angry and said: "And what about the animals, the male and female of every sort of that I ordered to come unto thee to keep their seed alive upon the face of the earth?"

And Noah said: "They have been delivered unto the wrong address, but should arriveth on Friday."

And the Lord said: "How about the unicorns, and the fowls of the air by sevens?"

And Noah wrung his hands and wept, saying: "Lord, unicorns are a discontinued line; thou canst not get them for love nor money. And fowls of the air are sold only in half-dozens. Lord, Lord Thou knowest how it is."

And the Lord in his wisdom said: "Noah, my son, I knowest. Why else dost thou think I have caused a flood to descend upon the earth?"

Dave Elovitz

March 1983

OUT OF SIGHT...

Piping in the machine room had frozen during the cold spell and the designer had recommended putting a steam coil on the combustion air inlet. "Before we install a steam coil on a gravity combustion air intake, we better take a careful look. That just doesn't seem like a normal thing to do."

The offending combustion air inlet was temporarily sealed up with tape and insulation, and the outdoor air ventilation dampers were closed, but the room still seemed cold. The outside air dampers were the expensive, low-leakage kind, with gaskets around *every* blade, but there were streaks of daylight showing. Jets of cold air were streaming into the room through the closed dampers. Poor damper adjustment and unsealed joints provided the path for the daylight and the cold air, but why was so much air coming in through these small openings?

"Look at the gates on the barometric dampers swinging out, not in!" Cold air was pouring down through the flues. The room was under a terrific negative pressure. Why?

The exhaust hood over the chemicals was too small to have much effect, and we couldn't find any opening in an exhaust duct or a return fan inlet duct as it passed through the machine room. Cracking the door showed that airflow was into the machine room from the building. The source of the negative pressure had to be in the machine room, but where?

Hey! Is that daylight up there above those pipes? A big roof top exhaust fan - Out of sight - Out of mind. And out of control. A thermostat set at 80° was supposed to operate the outside air ventilation dampers and cycle the fan. The room was well below 80° and the dampers were closed, but the exhaust fan just ignored the thermostat and kept spinning away. Stop the fan, stop the outside air. No more freezeup problem.

Tough break for the steam coil salesman.

Dave Elovitz

April 1983

WHAT, A BARGAIN?

"For only $50,000 our microcomputers will save you $30,000 a year," the sales engineer crowed. His proposal ticked off a list of savings:

$4000 per year by resetting the baseboard radiation water temperature to match changes in outdoor temperature. Reset is a good idea, and produces savings along with better performance, but the building controls already included that feature.

Another $2000 saving from resetting chilled water temperature upward as outside temperature dropped. But the steam absorber data showed that raising the chilled water from 44F all the way to 50F reduces steam use by less than 3%. Besides, increased supply air temperature for the VAV systems would mean more airflow, thus more fan horsepower, to do the same cooling.

$8500 from duty-cycling? In a VAV system, the fan automatically reduces horsepower to just what's needed. When the fan comes back on after duty cycling, it has to move extra air to catch up, so duty-cycling actually increases VAV fan energy use.

Then there was $3000 "Optimized Start" savings by delaying the start-up of the air handler fans on mild days. The guy who calculated that should have recognized that all the heat is from perimeter baseboard, and the air handlers have nothing to do with warming the building back up in the morning. The only motor that could be optimized started would be a 5 HP pump that only uses $2400 of power all year. Hard to save $3000 by starting it at a different time every day!

$3200 would be saved by reducing outside air. You'd need some tools to accomplish that with the existing controls, but it wouldn't take more than an hour on each air handler.

There were more claims, and they all looked pretty much the same: Someone had written down a lot of savings promises without paying much attention to what was really going on in this particular building.

Anyone want to buy some aluminum siding?

Ken Elovitz

May 1983

WHAT, ME WORRY?

No need to worry, now! Gasoline prices were flirting with a buck a gallon 'til they added new taxes. Heating oil is al ready there. Fuel charges on the electric bills have been holding pretty steady since last summer, about a penny below what they were in 1981. OPEC's on the run, they aren't calling the shots anymore. Plenty of places to get oil, now, and I even had a gas station clean my windshield Tuesday!

Did you turn the thermostat back up a couple of degrees in the house this winter? Start looking at a bigger car, even if it would only get 16 miles to the gallon? A spectacular glass curtain wall for that new building? After all, energy isn't really that expensive any more.

Don't do it. The energy crisis isn't over. At best, it has just been rescheduled for 20 or 30 years from now. Newly discovered major new oil fields are not limitless, and they, too will peter out. Someday - probably be early in the 21st Century - we will have just plain used up all the easy to get at oil and natural gas. Even if OPEC can never again charge anything they want for oil, when all the fuel is from tertiary recovery or super-deep gas, higher costs of production will mean prices that seem unbelievable to us today.

At worst, the energy crisis is just waiting in the wings for the first disruption in supply from overseas: If you remember, it was the overthrow of the Shah of Iran that temporarily choked off a major source of oil, panicked buyers and sent prices doubling, then doubling again. Even with fuel switching, new finds and very effective conservation, we in the US still import about 25% of our oil, and oil still provides 40% of our energy needs. As energy prices fall as a result of the present oversupply, there is less incentive to do the very things that led to our current energy independence. And less incentive for us and our companies to invest in energy conservation measures - not the incentive to invest the capital dollars in new equipment or in retrofitting existing systems and buildings, and not the incentive for the emotional investment in maintaining good habits and good energy practices.

What, me worry? YOU BET!

Dave Elovitz

June 1983

CASE DISMISSED

"We need an expert witness to testify for the installer and the boiler manufacturer," the lawyer explained. "The plaintiff's man says the new boiler produced steam with too much water in it, which caused water hammer in their equipment, damaging it and shutting down production."

Water hammer is caused when flowing steam whips up a wave of liquid condensate ahead of it, just as the wind whips up waves on the ocean. When the waves of liquid hit an elbow, or the end of a heat exchanger they crash against it, just like waves crashing on the beach. That can set up serious vibration and damage piping and equipment. Steam from all boilers has moisture, and the standard for low pressure boilers like this one is 2%. The installer piped straight up from the boiler, over and down to a large header, connecting all the equipment with pipes coming off the top of the header. That way, steam traps on the header would drain away any moisture so only dry steam would go out.

"Our clients are only responsible for the boiler installation, not hooking up the process equipment." I remarked while we were sitting outside the courtroom, waiting for our case to be called, "It's too bad someone didn't do a simple throttling calorimeter test when this happened 10 years ago. We would have evidence as to how much moisture actually was in the steam leaving the boiler at the time. But it doesn't really matter. You see, moisture in the supply steam is nothing compared to the moisture created when the steam condenses in the equipment. You know, don't you, that steam does heating by turning from vapor to liquid and that all of it turns to liquid in the heater? If there was too much liquid in the equipment, it was because it was not being removed quickly enough, not because the steam supply was wet."

"Oh?" Defense counsel perked up. "Excuse me a minute. I need to talk with opposing counsel." Half an hour later, he returned. "We won't need you after all. The case is settled."

Dave Elovitz

Page 67

July 1983

MORE FROM MURPHY

Murphy's Conclusions
(1) Every solution breeds new problems.
(2) Whenever you set out to do something, something else must be done first.
(3) Matter will be damaged in direct proportion to its value.

Rasiej's Rule of Rueful Reality
The other line always moves faster.

Chisholm's Checkpoint
Just when things just can't get any worse, they will.

Cahn's Convention
When all else fails, read the instructions.

Koningisor's Criterion
The man who can smile when things go wrong has thought of someone he can blame it on.

Peter's Placebo
An ounce of image is worth a pound of performance.

Paul's Palliative
Opportunity always looks bigger going than coming.

Mary's Maxim
There's never time to do it right, but there's always time to do it over.

Milt's Mandate
Nothing is impossible for the man who doesn't have to do it himself.

Conway's Contradiction
Anything can be made to work if you fiddle with it long enough. If you fiddle with anything long enough, it will break.

Fayva's Formula
If the shoe fits, it's the wrong color.

Danville's Dictum
One good scare is worth more than three hours of good advice.

Dave Elovitz

August 1983

BUDGET MENDER

"That third floor tenant is bustin' my budget," the building manager groaned. "His lease says we gotta give him cooling when they work late, and they're always workin' late. With one big air handler for all ten floors, I have to cool the whole building just for their floor! I need to put in a separate system."

"I don't think so," I grinned after looking over the drawings. "See these single inlet induction boxes that control the temperature in each zone? With a couple of relays on each, we can make them shut right off wherever you don't want cooling."

"Neat! But what'll it cost to wire every one of those relays all the way back to the office?"

"I think we can beat that, too, with one of those remote controls that sends a signal over the regular building power wiring. You just pick which zones you want to shut off on whatever schedule you set up, and the control does it all automatically!

Now, if we can figure out how to cut fan capacity at low flow, you can save on fan operation, as well as cooling." We could motorize the existing inlet guide vanes pretty cheap, but that wouldn't save much either. Adjustable speed drives are really efficient, but with only about 28 hours a week of low CFM operation, it would take 8 years of savings to pay back.

Hey, we don't need variable speed! Just two speeds. A two speed motor costs a lot less, and saves almost as much. The darn fancy two speed starter costs plenty, though. Less than five years' savings to buy that whole rig, but can't we do better?

Shame to have to buy a whole new 75 HP motor and starter when we already have one. But we can't use that one for two speeds. Or can we?

We'll just add a separate 15 HP motor with its own belts and sheaves to run the fan for low CFM. Then the energy savings will pay off the investment in less than 18 months, and that little motor and the remote control gizmo will mend our building manager friend's budget better than new!

Ken Elovitz

September 1983

SHOEMAKERS' CHILDREN

Every client thinks he is the only one who has trouble getting an HVAC installation done right and on time. Here's what happened when I decided to air condition my house:

The contractor arrived on the job a week late, well into the warm weather. That's OK, I thought, I'll be nice and cool in just another couple of days. After all he has all the equipment here and only three short duct runs to install.

"Where's the drain pan on that fan coil unit?" I asked the installer. "You have to order it special on horizontal units. Have you checked the installation manual?"

"Don't worry," was the smug response. "It's there. I've put dozens of these in before, I know what to do."

"I'd still like to look at the book," I answered. "Try and find it for me."

By Friday, still no book, but the crew returned to wire up the unit, charge it with refrigerant and start it up. "Cool at last," I thought after a solid week in the 90's. By late afternoon the electrician announced, "Just a few more connections" and I could practically hear the compressor humming merrily away.

Nine PM, ten, eleven, then finally midnight before those few connections were complete.

"Before you button it up," I called to the mechanic in the attic, "show me how you have connected the drain pan." I groped up the ladder and squirmed into the attic. Why was there insulation on the bottom of the fan compartment, and a puzzled look on the mechanic's face?

"Could this be the drain pan?" I asked pointing to the galvanized pan wrapped neatly around the TOP of the coil.

"Oh, ----," he groaned wearily. "You're right. We installed the fan coil unit UPSIDE DOWN! We'll have to come back next week to turn it over."

Back they came, everything now works, and the house is nice and cool. But each time I turn on the system, I remember stumbling into bed that night and mumbling to myself over the roar of my window unit, "When all else fails, read the directions!"

Ken Elovitz

October 1983

WHEN I WAS A CLIENT

"I'm sure you guys are very busy," the client remarked, "but whenever I call with a question, I never feel rushed. Whether it is you or Ken or Gary, you always take the time to explain things carefully, as if my question was the only important thing in front of you."

I liked hearing that. When I was a client - and I spent 25 years at it before I became a consultant - there were two things that annoyed me most about consultants; things I vowed I would never do:

(1) "You're not the only client ... " or "I have other projects . . ." or "I have this big important project ... " I may not have been the only client that consultant had, or even one of his biggest clients, but he was the only consultant I had for that problem, and his assignment from me was the most important one to me. Why didn't he tell me how unimportant my job was when he was being interviewed for the assignment?

(2) Speaking of interviews, what about the heavy hitters who show up with all the impressive credentials at the interview, and then you never see them again during the job? The guy who actually runs your project is a junior apprentice, and another guy who is studying for his PE exam occasionally shows up at meetings, but he knows less about the progress of your job than you do.

Well, now I'm a consultant, not a client any more. But those things still bother me, and Ken and Gary share my attitude: Every client should feel we will be there when he needs us, that no other job is more important than his, and that his job deserves the full attention of a fully qualified professional. Because, as far as he's concerned, his job is the most important one we have.

We know how important it is, too. If you ever catch one of us forgetting that, remind us of this calendar!

Dave Elovitz

November 1983

CRACKING UP

Bruce Long, VP of Construction Services for Prudential, sent me the following story. His source swears it is an actual, on-the-job accident report submitted by an accident victim in Alberta, Canada:

"When I got to the building, I found that the storm had knocked some bricks off the top. So, I rigged up a beam, with a pulley at the top of the building, and hoisted up a couple of barrels full of bricks. When I had repaired the building there were a lot of bricks left over, so I filled a barrel with these extra bricks. Next, I went to the bottom of the building and cast off the line."

"Unfortunately, the barrel of bricks was heavier than I am. Before I knew what was happening, the barrel started down, jerking me off the ground. I decided to hang on. Halfway up, I met the barrel coming down, and received a severe blow on the shoulder. I continued to the top of the building, where I banged my head against the beam and jammed my finger on the pulley."

"When the barrel hit the ground, it burst its bottom. This allowed all the bricks to spill out. I now was heavier than the empty barrel, so I started down at high speed. Halfway down, I met the barrel coming up, and severely injured my shin. When I hit the bottom, I landed on the bricks and got several painful cuts from the sharp edges."

"At this point I must have lost my presence of mind. I let go of the line. The barrel came down and gave me another heavy blow on the head. This put me in the hospital."

Thanks, Bruce.

CRACKED ICE

For those who enjoyed our first little booklet of reprints, Cracked Ice, we have compiled some more recent calendar messages under the title, More Cracked Ice. A post card or a phone call will bring you a copy.

Dave Elovitz

December 1983

BUSY AS A BEE

The building was historic, but recent renovations included a modern variable air volume HVAC system with perimeter baseboard heat. Unfortunately, there were no drawings, but the new superintendent had been thoroughly briefed on the HVAC system and its controls by his predecessor. And he had been on the run ever since with HVAC complaints.

The first room we looked at was freezing, but cold air was whistling out of the diffuser. "Where's the thermostat for this area?" I asked.

"That one's for heat," he replied. "It's set way down now so we don't heat the room in summer."

"Strange," I thought, "a heating thermostat in an interior zone." Indulging my curiosity, I turned it to its highest setting. The air noise faded and the airflow dropped to nothing. "That's a cooling thermostat. This room may be cold, but the controls are doing just what they're being asked to do, cool it off. Set the stat for 75, and I think this room will be OK."

The fan on top of a filing cabinet in the next room indicated not enough cooling. The balancing damper in the diffuser was tightly closed. Cool air from the VAV box couldn't get into the room.

Our tour found many more rooms that were overcooled or undercooled, almost all with thermostats set well outside the comfort zone. In the few where changing the thermostat setting did not change the airflow, we found disconnected pneumatic control tubing or broken damper linkages.

By now, the building superintendent was enthusiastic about being able to solve his HVAC problems himself, and very distrustful of the "thorough briefing" by his predecessor. After I explained the system and each of its components, he was ready to check each and every VAV box and thermostat in the building. Which he did, adjusting and repairing as he went along, as the complaints evaporated.

Sometimes life is less complicated than we think it is. In this case, just understanding the design, plus a few simple repairs, turned a real hornet's nest into a honey of a building.

Ken Elovitz

January 1984

to the tune of
DECK THE HALLS

Stuff the walls with insulation,
Fa La La La La, La La La La.

Now's the time for conservation,
Fa La La La La, La La La La.

Pay attention to this carol,
Fa La La, La La La, La La La.

Use your wits and save a barrel.
Fa La La La La, La La La La.

Use reflective film on windows
Fa La La La La, La La La La.

Weatherstrip where in the wind blows
Fa La La La La, La La La La.

Tune up burners, check the chimney,
Fa La La, La La La, La La La.

You will save big dough, by Jim'ney
Fa La La La La, La La La La.

Use economizer cooling
Fa La La La La, La La La La.

You will see that we aren't fooling,
Fa La La La La, La La La La.

If you use all power wiser,
Fa La La, La La La, La La La

'Lectric bills won't be so high, Sir
Fa La La La La, La La La La.

Check controls for outdoor reset,
Fa La La La La, La La La La.

Set them right, you'll always be set,
Fa La La La La, La La La La.

Calibrate, check operation,
Fa La La, La La La, La La La

Good controls fight cost inflation,
Fa La La La La, La La La La.

If you've still got ducts with reheat,
Fa La La La La, La La La La.

V AV will make them run sweet.
Fa La La La La, La La La La.

Vary speed instead of guide vanes,
Fa La La, La La La, La La La

Maximize resulting cost gains.
Fa La La La La, La La La La.

Wasting energy is folly,
Fa La La La La, La La La La.

Doesn't make your budget jolly,
Fa La La La La, La La La La.

Change your lamps to lower wattage,
Fa La La, La La La, La La La.

You'll still have a nice, bright cottage,
Fa La La La La, La La La La.

David, Sara, Fran, Ken, Gary,
Fa La La La La, La La La La.

Hope your Holidays are merry,
Fa La La La La, La La La La.

Wishing you a jolly new year,
Fa La La, La La La, La La La.

Full of joy, and full of good cheer,
Fa La La La La, La La La La.

Dave and Ken Elovitz

February 1984

MOTOR MATTER

"The problem we're having
Is driving me nuts.
Control air compressors
Just don't have the guts.

"The motors are running
Beyond full load amps,
We can't find the place
Where gremlins made camp.

"Just listen to one of
The fixes we tried
As soon as the first
Of the motors had died:

"We lowered the pressure
By ten p-s-i
Thinking of one place
The answer might lie.

"When that didn't help us
The puzzle remained.
That foolish compressor
'Bout drove me insane.

"On top of it all
Is another small pest.
That poor little motor
Seems never to rest."

Now, too high a current's
A problem that's tricky.
Perhaps that compressor
Had parts that were sticky.

Before dis-assembling,
I looked for some dat-er
To see what performance
Was claimed by the rater.

I also decided
To measure the speed
To see how much power
That motor would need.

And that's all I needed.
I picked up the phone.
And spoke to the client.
I let it be known

That I had the answer
For airflow a-plenty;
His fifteen horse motor
Was s'posed to be twenty!

So don't chintz on motors
When buying compressors.
Else you'll leave a problem
For me to un-mess, sirs!

Ken Elovitz

DONKEY PLANNING

At the beginning of time, when the world was new, the donkey was regarded as the wisest of all the beasts. Other sheiks came from the far corners of the desert to marvel at the wisdom of the herd owned by one Sheik El Prin-Ci-Pal, as the donkeys in his herd could not only talk and recite proverbs, they could also make beautiful drawings, and do complex calculations.

One such group of visitors included The Prophet himself, that most learned and wisest of all men. El Prin-Ci-Pal led out his herd and invited, "Behold, 0 Prophet, these wise and talented asses. Examine them. Test them and see for yourself if they are not verily wiser than 40 trees of owls."

So The Prophet addressed the donkeys. "Let me test your wisdom. How long would an ass require to design me a new palace?"

One ass replied, "To design a fine palace, o Prophet, an ass should spend six months on planning and two years making working drawings." "That soundeth reasonable," quoth The Prophet, but he continued, "I have for one of you a commission to design a palace, but I will not wait so long. Let him who will do it more quickly stand forth.

All the donkeys started to talk at once, each offering to design the palace in less time than the one before until one especially long eared ass said he would design the palace in but one single week.

"Fool!" quoth The Prophet, "you cannot even make all the drawings in one week, much less do any of the calculations."

"True," replied the long-eared one, "but I wanted to get the order."

And from that day to this, asses have been known as fools, and those who promise to deliver work too quickly have become known as asses. And donkeys have quit talking.

Dave Elovitz

April 1984

SENTENCE SENSE

Like many businesses, clear communication is the essence of our work. Sometimes, the way that you say something (or at least the order you say it in) can change the whole meaning. We've collected a few examples. The right words are all there, but in the wrong order the result comes out very different from the intent:

The man ran down the street shouting fire in his pants.

The speaker urged the people to vote frequently.

The man was seen by a forest ranger who was thought to be lost.

The police found the man who had been hit by a car in a rooming house on Main Street.

The star pole vaulter raised the pole over his head which was made of bamboo.

The coach reprimanded the unsportsmanlike conduct of the player with the kindness of a father.

The scouts came upon an unknown lake hiking in the woods.

The man was thrown from the saddle when the horse stumbled and broke his arm.

Mrs. Thomas went to North Dakota after her husband entered prison to live with relatives.

The farmer wanted to hire a helper to milk and drive the tractor.

He confessed that he murdered the woman whom he had secretly married in self defense.

Mr. Roberts died shortly after the accident in the hospital.

Jackson was fined $1000 under a charge of transporting liquor made by Derne County authorities.

Ken Elovitz

May 1984

A ROSE BY ANY OTHER NAME

At the February AGA Architect/Engineers Advisory Council meeting, a number of us met with the Department Heads of 7 of the 9 Architectural Engineering programs in the United States. It was tremendously exciting to learn that colleges were training young people precisely for our industry - Energy Systems Engineers with career goals in consulting offices and facilities engineering departments, graduates so familiar with the technology we use every day that they can be productive in a matter of weeks rather than years. ArchE, we learned, is that field of engineering that trains young men and women who work with and for architects to design the engineered systems of building.

The only disappointment was learning that all these fine programs turn out only about 50 - 60 graduates each year. But this year's crop is not committed yet, and you can still interview some of them by contacting:

Professor David Claridge
Civil. Environ. & Arch Engineering
University of Colorado, Boulder
Campus Box 428
Boulder, CO 80309

Professor Robert Dahl
Dept of Architectural Engineering
Kansas State University
Manhattan, KS 66506

Professor Ronald Helms
Dept of Architectural Engineering
University of Kansas
Lawrence, KS 66045

Professor Franklin Johnson
Dept of Civil Engineering
ECJ 51210
University of Texas at Austin
Austin, TX 78712

Professor Howard Harrenstien
Dept of Architectural Engineering
University of Miami
P.O. Box 248294
Coral Gables, FL 33124

Professor David Hateher
Dept of Architectural Engineering
Cal Polytechnic State University
San Luis Obispo, CA 93407

Professor Howard Kingsbury
Dept of Architectural Engineering
Pennsylvania State University
University Park, PA 16802

Professor William Streat, Jr.
Dept of Architectural Engineering
North Carolina A & T State
1601 E. Market Street
Greensboro, NC 27411

If you're one of the many friends who have asked Ken or me where you can find people trained for building energy systems, at last we have an answer. And Professor Helms promises that after you have hired your first ArchE, you'll never want to hire a graduate from any other discipline.

Dave Elovitz

June 1984

WHOOMP!

WHOOMP! The control technician scrambled to shut down the air handling system. What was going on? The system had been beautiful for two days - the fan tracking control worked like a charm: As the VAV boxes varied, the supply CFM went up and down, and the return fan always handled a steady 9600 CFM less than the supply so there would always be 9600 CFM of outside air. Now, in the last step of the checkout, the control technician had put the system into its warmup cycle, and the sides of a big 84" x 60" duct started to crumple in like an old Kleenex, but with a lot more noise.

The system normally starts out in its warmup cycle - VAV boxes open wide and outside air dampers close, to recirculate air so temperatures even out before people started coming into the building.

Everything started smoothly. The supply and return fans came up to speed, the outside air damper stayed closed, and suddenly WHOOMP! The duct was starting to collapse again, and the technician lunged for the stop button. Four more tries and four more WHOOMPs. Then he called the engineer, certain that the system would never run right in warmup.

"Try it once more while I watch." Sure enough, everything started smoothly, just as the control man described. Supply and return air CFMs climbed steadily on the gages, then WHOOMP, the big duct started to buckle. But something caught my eye just before the WHOOMP. The return air CFM had started to fall rapidly. "Here, let me look at your control drawings. There it is! Of course the 'duct collapses! The fan tracking system controls the return fan so it delivers 9600 CFM less air than the supply fan, bringing in 9600 CFM of outside air, regardless of total airflow.

But during warmup, the outside air damper is closed, so both fans are connected in series, and they have to move the same amount of air. The tracking control measures the supply air flow and tries to get the return fan to deliver 9600 CFM less. Of course it can't, no matter how much the control cuts it back, because all the air for the supply fan has to come through the return fan during warmup. When the return fan cuts back enough, the suction of the supply fan just sucks the ductwork inside out. If you simply disable the tracking control during warmup, your WHOOMP won't even whimper!"

Dave Elovitz

July 1984

CHASING YOUR TAIL

The new control system had worked beautifully all summer long, and the new economizer controls had saved money and improved comfort during the fall. But as soon as the bitter cold weather arrived, the freezestats began shutting down the fans unexpectedly.

"Let's check a few temperatures," I told the building superintendent. "Outside air is 17F, mixed air is only 30F, and preheat coil discharge is 59F. That's strange. Why is the mixed air so low? And why is the preheat higher than 55F?"

A look at the control diagrams and a few quick calculations revealed the answer to the problem. "To get 30F mixed air your outside air damper must be 70% open," I started to explain, "but it only needs to be open enough to make the 55F air you need for cooling - about 30% when it is 17F outside.

"These controls remind me of a dog chasing his tail. The faster he goes, the faster his tail gets away from him. And that's exactly what's happening here. You see, the discharge air temperature controller positions the outside air dampers to make 55F air for cooling. But the preheat coil is between the dampers and the discharge air control sensor. When the discharge controller positions the dampers for 55F air, the preheat coil heats it up to 59F. Then the discharge air controller calls for more cold outside air, trying to get 55F supply air even though the preheat coil is calling for 59F. More cold air causes the preheat coil to open wider to satisfy its control.

"Eventually, the outside air damper opens so wide that the preheat coil can no longer handle the load, and the preheat coil discharge temperature drops. When the air reaches 40F, your freezestat does its job and shuts the fan down to protect against freezeups. If you reset the preheat coil control to 55F and add a low limit control so the mixed air temperature can't go any lower than 53F, these controls will stop chasing their tails!"

Ken Elovitz

August 1984

SERVICE

One of our clients insists that he recently sent the following letter to the President of an air conditioning service company with which he was very unhappy:

"Dear President,

"I haven't always lived in the city. In fact, I am a country boy, really, raised on a farm.

"One of the things I remember from my younger years growing up on that little dairy farm is that from time to time my father would make all of us kids stay in the house all afternoon. Of course, we never did like that, but he explained that it was absolutely necessary: We had to stay in whenever he took the bull down to service one of the cows.

"Well, Mr. President, that was a long, long time ago, but I have always remembered it. And would you believe that it wasn't until just recently, after all these years, when I started doing business with your company, that I learned what "service" meant?"

Well, maybe our friend never really sent such a letter, but then again, maybe he should have.

N.A.F.E.

The National Academy of Forensic Engineers is a chartered affinity group of the National Society of Professional Engineers devoted to recognizing those engineers qualified through background, expertise and experience to provide expert testimony when matters of engineering content are at issue within the jurisprudence system. Ken and I are pleased that we have recently been elected to NAFE.

Dave Elovitz

September 1984

COMMISSIONING

Several sessions at the recent ASHRAE annual meeting were devoted to the topic of commissioning new HVAC systems and equipment. We were not surprised at the spirited discussion among designers, installers, and users, as we are being called on more and more frequently to help owners determine if they "got what we bought" in new systems.

It is difficult to conduct a thorough checkout of all the complex system components, but some simple, common sense tests that you can run yourself can determine if the system is generally performing the way it should:

- Check the performance 01 zone thermostats by setting them to their lowest temperature. Cooling should increase - airflow should go up on a VAV system, and air temperature should go down on other system types. Then turn the thermostats to their highest temperature. Cooling should decrease.

- Check supply air temperature by sticking a pocket thermometer through a canvas duct connection or up into the neck 01 the diffuser. For a VA V system and for other systems on lull cooling, the supply air temperature is usually in the middle to upper 50's. Look at the discharge air temperature control setpoint on each air handler and compare it to the measured supply air temperature. Raise the control setpoint, and the air temperature should climb (within a few minutes). Lower the setpoint, and the air temperature should fall.

- Check the freezestat by spraying a small amount of refrigerant on the sensor. You can buy a one pound can of refrigerant and a tapa-can valve at most auto parts stores for under $10. Install the valve, hold the can up side down (for liquid flow) and spray refrigerant over a 12-18" length of the freezestat bulb. The fan should stop, and the outside air dampers should close.

- Standby equipment (like pumps) with automatic changeover can be tested by shutting off the lead motor. The standby motor should start right up.

These and similar simple tests cannot check every detail, but they do give you a quick look at overall system performance and check some important safety functions that protect your system from damage. If gross problems exist in your new system, these tests give you a way to spot them. In fact, it would be worthwhile to check the system with tests like these from time to time even after it has been accepted, just to be sure it still works properly.

Ken Elovitz

October 1984

TRUE STORY

Pamela told me this story while she was straightening up the office the other day.

Seems a friend's father sold his campervan to an excitable little guy with a thick accent and a thin skin: Every attempt to explain how anything worked was cut off indignantly.

"This is the head. You pump this handle.

"I know all about it. Do you think I am stupid?"

"This is the cruise control. Shall I show you how ... "

"I know about cruise control, everybody does. I'm not stupid."

"This is the stove. Before you ... "

"That's simple. You don't have to show me."

And so on. So Jane's father was taken aback when he answered the door a few days later to find the little guy, all messed up, and flanked by two state troopers. The troopers explained that the little guy had wrapped his camper around some trees, but between the accent and his being so excited and upset they couldn't find out what happened. All they could get out of him was that he had to get to this address right away. So they brought him.

The little guy had been bouncing up and down impatiently through that explanation. Finally, he could contain himself no longer and burst out, in his thick accent, that the van was no good. The cruise control was lousy, didn't work at all. Why, he set it and just went in back to make "a good cup of Columbian coffee ... "

'Struth! At least Pamela said so. I don't know what it has to do with energy or engineering, but it was too good a story not to share.

Dave Elovitz

November 1984

STILL MORE MURPHY

Law of Probable Dispersal
Whatever hits the fan **will** not be evenly distributed. (Also known as the **How Come It All Landed On Me Law**)

Sam's Syllogism
L'Occhia dei padrone ingrassa el cavallo. (The eye of the owner fattens the horse.)

Callahan's Consensus
If a problem causes many meetings, the meetings eventually become more important than the problem.

Meade's Maxim
Always remember that you are absolutely unique. Just like everybody else.

Goethe's Growl
When ideas fail, words come in very handy.

Paul's Principle
When you are up to your ass in alligators, it is hard to remember that you came here to drain the swamp.

Tyger's Truism
The most important thing in solving a problem is to begin.

Bergson's Belief
Think like a man of action. Act like a man of thought.

Broder's Law
Anyone who wants to be elected so much that he'll spend two years organizing and campaigning for it is not to be trusted with the office.

Grier's Law
No good deed goes unpunished.

Leon's Law
If all you have is a hammer, everything looks like a nail.

Cole's Law
Thinly sliced cabbage, a little vinegar, oil, and mayonnaise.

Dave Elovitz

December 1984

HOW MUCH?

Last year one of our clients decided he wanted to know what the real costs were for different systems on a new building, so we suggested he get proposals on more than one system for his project. We had studied the building and either rooftop VAV units or water-source heat pumps would work well, with reasonable operating cost. The heat loss was low enough that the operating cost savings with gas heat would not justify a large extra investment compared to electric heat.

We asked six contractors to submit proposals on three different alternatives:

> VAV with gas hot water heat
> VAV with electric heat, and
> Water source heat pumps.

Not all the contractors bid all three systems, but two guys wanted just about the same money whether we took electric or gas heat, and two said electric heat would cost much less. The heat pump system was less expensive than VAV - not much less from some, but a lot less from others. Comparing the quotes on a percentage basis, with the lowest proposal as 100%:

Bidder	VAV/ Gas System	VAV/ Electric System	Heat Pump
A	136%	132%	124%
B	121%	112%	100%
C	134%	134%	
D	130%	112%	
E	141%	125%	123%
F			122%

What did we learn from all this? Two things.

First (and I know I'll hear from our contractor friends for saying this) was that mechanical contractors don't seem to know what it costs them to do work!

Second, that it didn't make any difference, because we didn't buy any of those systems for this project! Contractor B said he could put in a dual duct VAV system (which we had dismissed as being too expensive) for 10% less than the lowest heat pump bid!

Dave Elovitz

January 1985

to the tune of
FROSTY THE SNOWMAN

Energy Savings
Is a very worthy goal.
You can light for less
With H-P-S,
And convert from oil to coal!

Energy savings
Is a good way to invest.
You must check the rate
And calculate
To find out which way is best.

There must be no mystery
To saving B-T-Us.
Weatherstrip and time controls
Are the tools that you must use.

Oh, Energy Savings
Is an awful lot of fun.
You'll surprise yourself
And build a wealth
Of ideas to get done!

Energy savings
Come from solar heating, too.
You can use the sun to save a ton
And have fun with something new.

Energy Savings
Can come from hydro pow'r.
With dams so tall
And waterfalls
Saving money by the hour.

And of course you must be sure
To insulate your walls.
Then sit back and laugh with glee
As your heating fuel bill falls!

Oh, Energy savings
From a brand new shower head,
With flow so low
That's the way to go
So your budget won't see red.

Energy savings
Come from insulating sash.
Be it two or three
Take a tip from me
Extra panes are worth the cash.

Energy savings
Is our theme for the new year.
You can hum this song
Or sing along
In a voice that's loud and clear.

Once a year we send these rhymes
To bring a smile or two.
And to let you know we care
For our friends - that's all of you!

Oh, energy savings
Was our message number one.
And for message two
We say to you
Have a Happy New Year, too!

Dave and Ken Elovitz

February 1985

TROLL CONTROL

"It's as if the machines were inhabited by trolls!" the shopping center's chief engineer proclaimed as we climbed toward the mechanical penthouses. Sure enough, the hot water and chilled water valves were pumping up and down like a calliope, and the outside air and return air dampers were flapping like a startled pheasant. No wonder he asked if the whole system would wear itself out before the warranty expired!

The large two story mall was heated and cooled by air handlers located in penthouses on opposite sides of the mall. Distribution ductwork with sidewall registers ran along the length of the mall on each side, so each system delivered air from one wall toward the middle of the mall. Unlike a VAV system, airflow was constant, and thermostats in the mall would change the supply air temperature to provide heating or cooling, depending on the space temperature, by operating the hot water and chilled water valves.

"I have an idea what's happening," I told him. "And I think I know how we can tone down your trolls. Let's shut off the air handler on the other side of the mall, and watch this one."

Sure enough, the valves settled right down, and the supply air temperature evened out nicely.

While walking through the Mall, I had felt the supply air temperature changing rapidly from warm to cool, even when I walked along the far wall. If I could fee l those temperature changes, then so could the thermostats! The controls were working perfectly: The throws on the registers were so long that the airstream from one unit blew directly onto the thermostat for the other unit. When the blast of air hitting the thermostat was cold, it responded with a call for full heat. Then the blast turned warm, and it called for full cooling. Simply adjusting the registers and relocating the thermostats gave the Chief Engineer complete con-Troll of the situation.

Ken Elovitz

March 1985

APOLOGIES TO DR. SEUSS

This is a story
'Bout restaurant eating
That couldn't begin
Due to no water heating.

The heater was modern
And fired by gas,
But the owner's opinion
Was "Modern, my ass!"

So lend me your ear for some
Good information
And you'll get the word
On the whole situation.

In order to keep the thing
Out of the way
A closet was built
Where the heater would stay.

The clearances
All were exactly as needed.
So how come the water
Just would not get heated?

The chimney was straight
And went up through the roof
But the pilot was out
And the flame had gone poof.

The heater was rated
At 5 therms per hour.
By all calculations
That's plenty of power.

And 65 gallons
Is lots of hot water
Provided the burner
Gomes on when it oughta.

Each time that the serviceman
Came to inspect

He found all the workings
Exactly as spec'd.

In fact, while the serviceman
Worked on his chore
The dishwashing crew had
Hot water galore!

And there lay the key
To the "question mystere".
The poor water heater
Had not enough air.

For while the technician
Was peekin' and pokin'
The door to the closet
Was always wide open.

The answer of course
Was to change out the door
And install one with louvers
From ceiling to floor.

Ken Elovitz

April 1985

97.8% SOLUTION

Want to sav* a ton of tim* on typing? Us* this n*w word proc*ssing syst*m. It go*s a lot fast*r than any oth*r syst*m, and it works v*ry w*ll *xc*pt for on* k*y. Th* oth*r forty fiv* k*ys work p*rf*ctly, and th* syst*m r*ally is sp**dy. Aft*r all, on* out of forty six is only 2.2%, but som*how that 2.2% s**ms to mak* quit* a diff*r*nc*.

Som*tim*s w* ar* call*d in on a "probl*m" to h*lp th* own*r g*t a syst*m to op*rat* prop*rly aft*r h* has giv*n up on th* original install*r and d*sign*r. W*'r* probably not any smart*r than thos* guys, and w* probably don't hav* any r*f*r*nc* books or gaug*s th*y don't hav*, but w* go in th*r* pati*ntly, and s*ttl* down and ch*ck *v*ry asp*ct of how that syst*m is working and what th* own*r n**ds it to do, and that t*lls us what's wrong.

Som*how, th*r*'s tim* to do that aft*r all th* argum*nts, but *v*ryon* was in a r*d rush during d*sign and installation, and it was too h*ctic during start-up to tak* th* tim* to ch*ck *v*rything compl*t*ly. "Hav* to m**t th* sch*dul*. No tim*, no tim*."

Th*n, wh*n things didn't work quit* right, th* d*sign*r was pr*occupi*d with som* oth*r n*w proj*ct, and th* m*chanics w*r* all off on oth*r sit*s, putting in oth*r syst*ms, and th* contractor had start*d som* oth*r structur*, and *v*ryon* just want*d to g*t back to thos* n*w proj*cts.

So, inst*ad of ch*cking thoroughly, th* air balanc* man says it's th* controls, th* control man says it's th* unit, th* manufactur*r says it's th* installation, th* contractor says it's th* d*sign, and th* *ngin**r says it just n**ds to b* air balanc*d, *tc., *tc., *tc.

Som*tim*s small goofs tak* a lot of *ffort to corr*ct onc* th* installation is compl*t*. Most of th* tim*, som*on* took a shortcut. H* did most of th* calculations, or th* surv*y, or th* calibration, or th* adjustm*nts, but h* didn't tak* tim* to do th*m all. H* assum*d, but didn't v*rify. H* mad * most of th* *ffort, but not that last littl* bit to mak* sur* th* r*st was corr*ct.

Th* last coupl* of p*rc*nt in th* way w* do our jobs *v*ry day do*s mak* a diff*r*nc*, just as th* 2.2% mad* such a hug* diff*r*nc* in this word proc*ssing syst*m.

Dave Elovitz

May 1985

THREE GREAT TRUTHS

Once upon a time a hard working farmer lay in bed after a long day's work. Tired as he was, he could not go to sleep because he was enchanted by the melodious singing of a nightingale. The bird sang all through the night, and the farmer was so delighted that he decided he would set a trap and capture the bird.

The next night the farmer set his trap and succeeded in capturing the bird. Once the bird was securely in the cage the farmer said to her, "Ah, my beauty, now that I have captured you, you shall perch in this cage and sing me to sleep every night."

The nightingale quickly replied, "We nightingales never sing in a cage. If you imprison me, I shall sicken and die, and you shall never hear my song again."

The farmer, not to be fooled so easily, retorted, "Then I shall put you in a pie and eat you, for I have heard that nightingale pie is a dainty morsel."

"Please do not kill me," the nightingale begged desperately, fearing for her very life. "If you will set me free, I will tell you three great truths that will be worth far more than my tiny body."

So the farmer, realizing that wisdom has far more value than food or song, set her free.

As the nightingale flew up to a branch of a nearby tree the farmer cried, "Wait! What are those three great truths you promised to tell me?"

The nightingale trilled a few happy notes and sang out:

> "You don't have to own the ballpark to enjoy the game."
>
> "It's easy to fleece a sheep in wolfs clothing."
>
> "A forced promise binds only as long as there is force."

Ken Elovitz

June 1985

BACK TO BASICS

The Chief Engineer at one of the big banks told me this story. His chilled water mains, like those in many high rise buildings, have lubricated plug cocks to isolate the chillers for service. To close those valves, you slowly pump special grease into a fitting while gradually turning the plug. The grease helps the plug turn and seals between the plug and the *valve* body. Without proper grease, the *valve* will leak. Problem is, the new grease pushes out the old grease, which gets into the chillers, fouling the tubes.

A common problem, but no one John talked to had found a way to prevent it. So John asked his water treatment company to put their chemists to work finding something that would lubricate his valves without fouling his chillers.

After a few weeks with no results, John was getting frustrated. There just had to be an answer, but none of the experts could seem to fine it.

Now, you should know that John didn't just happen to become an operating engineer. It is in his genes. His father had been the chief operating engineer in a clothing factory for many years until he retired, so John heard engineering talk at the supper table even as a kid. And even though that little clothing 'plant was a far cry from the sophisticated systems that John runs in a modern high-rise office building, they still talk engineering over the supper table when they get together. So it wasn't surprising that one Sunday afternoon found John telling his father about the problem. The older man leaned back into his chair with a far-away look in his eyes, as if he could still hear the stitchers chattering to each other over the clack of the sewing machines, and said, "You know, we used to have the same sort of problem lubricating the steam engines in the old days, 'til we started using Ivory Soap. Just soaked it in hot water 'til it was all squishy, then loaded it into a grease gun. Any extra just dissolved and never gave us any of the trouble that regular grease did."

Ivory worked on John's plug cocks, too, of course. Which ought to remind all of us that the first step in solving any problem is getting "Back to Basics!"

Dave Elovitz

July 1985

THE NOISE THAT ANNOYS

"The President is wild!" moaned the Facilities Manager. "Brand new headquarters building, fancy modern open-office system, plush Board Room, and he says no-one can hear him at the end of the table because the air conditioning is so noisy. The contractor has been back four times, and then the design engineer a couple of times. They've fiddled with this and that, and now they say we have to spend $8000 to rebuild this whole part of the system and put the Board Room VAV box out across the hall. We're getting pretty annoyed."

It really was pretty noisy, that hissy sound like air leaking through a damper, or out of duct seams, that makes it hard to hear what someone is saying. To check the VAV box damper, turn the thermostat for full cooling, then to full heating. There was no change in sound either way, so it could not be the damper. That left the fan in the VAV box or leaky ductwork. One of the maintenance men opened the concealed spline ceiling near the VAV box so I could take a look. As soon as I got my head above the ceiling, it was obvious that the sound was coming right from that VAV box. I could see it, but I couldn't get an arm out to feel around it for leaks. Then I pulled my head back down and asked the mechanic to shut off the power to the VAV box fan box. I could hear the breaker click, but the noise didn't change.

"That fan must still be on. Let's open up the whole ceiling around the VAV box and I'll shut it off right there," the mechanic offered, and slid out another dozen ceiling tiles to flip off the service switch. But the noise continued. This time I climbed the ladder and squeezed my way up in the narrow space between the VAV box and the wall, getting closer and closer to the sound. I could see the fan was off, and discovered that the noise wasn't coming from the VAV box at all, but from somewhere above it!

Although I couldn't see up there, I could touch something metal, and something else, something that felt like heavy paper. And two thin wires. Of course! I eased down the ladder with a big grin.

"You've been annoyed at the wrong guys! There's a loudspeaker up there that belongs in the open plan office area, for the "white noise" sound masking system!

Dave Elovitz

August 1985

CONVERSION FACTORS

Usually, we try to bring you a little story every month, something that will bring a smile, or an "Aha!" We thought we'd do something a little different this month and give you a handy pocket guide of conversion factors. You know, like 1 mile = 5,280 feet. Only we tried to include the ones you wouldn't find in all the handbooks:

1 Boiler Horsepower = 33,472 BTU/Hour

1 Liter = 0.0353 Cubic Feet

10 Millipedes = 1 Centipede

1 Barrel (oil) = 42 Gallons

1 Decadent = 3.3333 Tridents

7.48 Gallons = 1 Cubic Foot

10^{21} Picolos = 1 Gigolo

7000 Grains = 1 pound

10 Monologues = 5 Dialogues

5 Dialogues = 1 Decalogue

2150.42 Cubic Inches = 1 Bushel

10^{12} Microphones = 1 Megaphone

10 Rations = 1 Decoration

1.85325 Miles = 1 Kilometer

1 Kilometer = 109 Micrometers

4 Seminaries = 1 binary

1 Terapin = 10^{12} pins

1 Millicent = 10^{-5} Dollars

1 Megacycle = 1000 Kilocycles

10^6 Bicycles = 2 Megacycles

And, finally, for Greek Mythology buffs, the quantity of beauty which it takes to launch one ship is, of course, 1 Milli-Helen.

Dave Elovitz

September 1985

SAVVY SOLUTION

Once upon a time, in a far off Island, two engineers were arguing over how to control a gas fired boiler to maintain the minimum temperature in a heat pump water source loop. "You must allow the glycol to flow through the boiler at all times and cycle the burner on and off to prevent the loop temperature from falling below 60F," the First Engineer announced.

"Nay, nay," the Second Engineer retorted, "For if you circulate that cold glycol through the boiler, the boiler and flue will be below the dewpoint. Moisture will condense and form acid which will dissolve the boiler in no time. You must keep the boiler no less than 140F to prevent condensation, and you must use a control valve to pass small amounts of cool glycol from the loop through the boiler for heating.

The First Engineer recoiled in horror: "Alas, my good friend, at high loads the control valve will admit large quantities of cool glycol from the loop to the boiler, and there is danger of thermal shock."

So the two Engineers argued and argued, until the Project Manager called a meeting. The First Engineer presented his theory, and everyone nodded in agreement. Then the Second Engineer presented his theory, everyone nodded in agreement once again, and much discussion ensued.

Finally the Third Engineer, who had many grey hairs, and thus was known to be wiser and more clever than the other Engineers, rose from his seat and cleared his throat. "The First Engineer and the Second Engineer are both correct," he began, "and we have discussed this subject at length. I have a suggestion which may be unorthodox but which answers all the concerns. We must buy a boiler with a tankless domestic hot water heater, and we shall circulate the fluid in the loop through the tankless heater when we need to raise its temperature. The boiler water will always be hot enough to prevent condensation, and there will be no risk of thermal shock because cold glycol from the loop will never enter the boiler directly.

So the meeting was adjourned, and the First Engineer returned to his desk to call the boiler manufacturer and ask if he really could circulate a glycol solution through a tankless domestic hot water heater. Whereupon the boiler manufacturer replied, "No problem at all. We do it all the time. What's it for, water source heat pumps?"

Ken Elovitz

October 1985

ALTERNATIVE ENERGY

My engineering professors always said to consider new techniques and innovative solutions, so two recent projects got me thinking. One was a brand new building with well-insulated walls and the newest high performance glass. The other was a laboratory animal facility, the first time I had come across the data on heat generated by various animals.

Think about a 10 foot wide exterior office in that modern building with, say, 30 square feet of glass. In October, when the temperature rarely drops below 30 F, the heat loss is 432 Btuh. Now, a normally active golden retriever, like Ken's, gives off about 390 Btuh. This is close to what we need, but not quite. Also, Joshua isn't always so active. When he lies down and goes to sleep, he produces only about 183 Btuh. The solution to this is to get a cat. A normally active 6.6 lb. cat will produce 45.6 Btuh of heat. The cat not only brings us to the required heating but will also keep the dog active enough to keep us warm all through October.

But in November the temperature can dip to 20 F, and we'll need another 108 Btuh of heat. Now, a lamb will produce 1.56 Btuh per pound of body weight. So if we get a 60 lb. lamb on November 1, we'll get 93.6 Btuh. Not quite enough for the coldest days of late November. But my research also indicates that our lamb will grow by about ½ lb. per day, so that by late November, our lamb will be big enough to meet the greater heat loss.

In December we'll need an extra 108 Btuh, because it can get down to 10F outside. Our lamb's normal growth rate will provide only about a third of this increase, so we still need another 75 Btuh. A 12 lb. monkey will provide about 71 Btuh, which is enough for all but the coldest days. Certainly you could buy a chimpanzee and be safe, but that seems silly. If it gets really cold we could make a modest investment in a supplementary guinea pig.

So you see, you could get through the whole winter without buying any expensive fuel. I haven't yet solved summer cooling, but with all those animals in your office, you'll probably want to keep the windows open anyway.

Gary Elovitz

November 1985

DAMPER SCAMPER

The building was new
But from what I'd been told
After opening day
It was terribly cold.

The engineers came
And the engineers went
But the problem remained
Though much money was spent.

The manager thus became
Very frustrated
For the lack of good heating
Had not been abated.

"The cooling is great,"
He said, stamping his feet.
"But the problem we have here
Is we need some heat!"

The contractor told him,
"Please don't get upset.
We'll work on your system
And you musn't fret."

Now all of that day
And all through the night
The thermostat settings
Were exactly right.

And then the next day
It beg an to unfold
When even interior
Rooms had grown cold!

The Manager came
And extolled with great glee,
"At last all you people
Will listen to me!

We all have cold hands
And we all have cold feet!
Forget about cooling
Let's turn on the heat"

So I went around
To inspect a few things
And to see what a change
In the T-stat would bring.

When I raised the setting
To ninety degrees
The place I was sitting
Continued to freeze.

The thermostat hissed
And the motor had stroked
But the cooling supply air
Was not fully choked.

A check up above
At the damper position
Revealed it was not
In a tight closed condition.

Rework the linkage,
Adjust and align
And the system would then
Begin working just fine.

And so I now leave you
A few words of gold:
If the damper's aren't closing
Then you will be cold!

Ken Elovitz

December 1985

THERMO-TRIVIA

The basic concept behind the modern thermostat was discovered in the mid-1700's, when "natural philosophers" (the scientists that we call physicists today) discovered that different metals have different coefficients of expansion. When different metals were bonded in layers to form a two-ply strip, the strip would curl as the temperature changed because one metal expanded more than the other. The first practical use of the principle came in the 1770's to compensate a timepiece for changes in temperature.

The next recorded use was the *thermoscope,* a French device (circa 1820) which indicated that temperature was changing but didn't measure by how much. The actual word "thermostat" was first mentioned in 1831, by Scottish chemist Andrew Ure, and the first patent for a temperature control dates from 1870: Julien Bradford of Portland, Maine developed an "electric heat and vapor governor" to regulate temperature and humidity levels for textile weaving and spinning rooms. He used a bimetal thermostat as a sensor controlling an electric circuit and gear train to open and close steam valves.

Modern use of bimetals for temperature control began when Minneapolis inventor Albert Butz developed his "damper flapper" temperature regulator for coal furnaces in 1883. Apparently drawing on ideas in Bradford's patent, Butz used a bimetal thermostat and an electric circuit that responded to temperature changes to operate a spring motor to open and close dampers on a coal furnace. As the furnace draft was changed by the dampers, the coal burned more or less rapidly. Butz completed the development of his device in 1885, the year he formed the Butz Thermo-Electric Regulator Company (which eventually became Honeywell, Inc.). His patent for the damper flapper was granted in May 1886, and the automatic temperature control industry was born.

Even today most thermostats still rely on the same bimetal strip principle that was discovered more than two hundred years ago.

Dave Elovitz

January 1986

to the tune of
SANTA CLAUS IS COMIN' TO TOWN

You better watch out
You better not cry
Better not pout
I'm tellin' you why:
Energy is what we're about.

We're making a list
And checking it twice.
We want you to know
That waste is not nice.
Energy is what we're about.

We check your insulation.
We calculate your loads.
We turn on all your pumps and fans
And decide what that forebodes.

Oh, You better watch out ... (refrain)

We measure your air
And see if it's right,
Gonna find out
Whose building is tight.
Energy is what we're about.

We try cogeneration,
We look at solar heat,
We check out thermal storage tanks
For a system you can't beat.

Oh, you better watch out ... (refrain)

Reduce the exhaust
And use V-A-V.
You might not know
What fun it can be.
Energy is what we're about.

You weatherstrip your windows.
You tune your burner flame.
And before too long you come to learn
That's the way to play the game.

Oh, you better watch out ... (refrain)

Check all your traps
'Cause strange as it seems
One leaking trap
Wastes mountains of steam.
Energy is what we're about.

And use fluorescent lighting
Plus setback thermostats.
Control them automatically
And you'll find out where it's at.

Oh, you better watch out ... (refrain)

Gary and Ken,
David and Fran,
Sara joins in
Right where we began:
Sending you our annual wish.

We wish you all good fortune
In everything you do.
We hope you have a joyous Yule
And a happy year, too.

Oh, you better watch out ... (refrain)

Dave, Ken, and Gary Elovitz

February 1986

WHO ARE THE ELOVITZS?

David graduated WPI in 1953, worked as a plant and facilities engineer for several technology-based companies, then for a major HVAC design-builder, before he started MASCO in 1972. He started consulting full-time in 1980. He is registered in MA, and one of the authors of the ASHRAE Systems Volume.

Ken is a Lehigh grad, 1975, with a JD from Suffolk in 1984. He was a process engineer at Bethlehem Steel, then Texas Instruments, joining Energy Economics in 1980. Ken is registered in MA and PA, President of the local PE Chapter, a member of the Massachusetts Bar, and teaches at Northeastern.

Gary got a BA from Dartmouth (1980), an MA from Harvard, then a BS from Northeastern in 1985. While he was getting that education, he also worked here at EEI, and as a manufacturing engineer at Ein Zurim Metals, an Israeli manufacturer of very high pressure valves and fittings. Gary has passed the first half of his PE Exams, and will be eligible for the second half in 1989.

Fran keeps the books, runs the office, manages the business side of things and makes lunch, which was Dave's best subject in school.

WHAT DO THEY DO?

About half of what we do involves existing energy systems (HVAC, power plants, power distribution) that don't work properly or efficiently. We diagnose what is wrong and devise corrective measures. Often this is related to a potential lawsuit, and involves expert testimony. (Ken and Dave have been elected to the National Academy of Forensic Engineers.)

Clients we have helped with a problem job often ask us to review system design documents before the systems are installed, to try and avoid the problems instead of solving them later, or to suggest ways to improve the system.

Clients who want the advantages of both design-build and competitive bidding have had us prepare a Scope of Work that they can use to get competitive proposals, and to measure the performance of the completed system.

Of course, as the company name suggests, we analyze energy use in existing buildings to find ways of reducing operating costs, and some clients use us as an extension of their own staffs, to provide specialized expertise on energy systems issues of all kinds.

Dave Elovitz

March 1986

TO SHADE OR NOT TO SHADE

"Of course you need interior shading! Otherwise the cooling load will be astronomical!"

"That may be, but at the price my tenants pay for office space, I can't tell them to close the blinds in order to keep their rooms cool."

Actually, the debate over shades and blinds generates more heat than light. According to the current ASHRAE load calculation method, windows *with* interior blinds often have a *higher* peak heat gain than windows without. The following table is based on tinted insulating glass (SC=.57) and medium colored vertical blinds (SC=.39 with blinds):

Orient-ation	Month	Solar Gain BTUH/sf	
		w/blinds	without
N	June	17	22
NE	June	50	44
E	June	67	63
SE	Sept.	71	72
S	Sept	65	66
SW	Sept.	73	76
W	June	69	69
NW	June	55	53

How could blinds, which keep direct sunlight out, *increase* the peak heat gain through some windows?

Direct sunlight which enters the space does not become part of the cooling load right away. The sun's rays must strike objects that absorb the solar energy. That absorbed energy heats the objects which in turn transfer heat to the air in the space.

If the windows have blinds, the rays of the sun strike them directly. Blinds have very little thermal mass, so they heat up quickly. They also have a large surface area compared to their thermal mass, so they release heat to the room rapidly. In other words, blinds are much more efficient at converting the sun's radiative energy to heat than other furnishings. Blinds make the window work like a passive solar collector with no storage.

Of course, blinds reflect some of the sun's rays back out through the glass, reducing the total amount of solar heat gain over the course of the day. By reducing the total number of BTUs which enter the space, blinds reduce operating costs. However, blinds do not significantly reduce the number of BTUs entering the space during the peak hour, and that is the important factor for determining cooling capacity.

So, designers, next time you think you have it made in the shade, make sure you were not just left in the dark.

Ken Elovitz

April 1986

I REMEMBER MURPHY

Murphy's Law of Thermodynamics
Things get worse under pressure.

Re's Restatement of the Laws of Thermodynamics
(1) You can't win. (Energy can neither be created nor destroyed.)

(2) You can't break even. (No engine can be 100% efficient.)

(3) You can't even quit the game. (Law of Entropy)

Maryann's Law
You can always find what you are not looking for.

Rowe's Rule of Order
He who shouts the loudest has the floor.

Lunstead's Last Law of General Engineering
Any wire cut to length will be too short.

Commoner's Second Law of Ecology
Nothing ever goes away.

Wilgoren's Worry
The hidden flaw never remains hidden.

Rubin's Rumination
If you're feeling good, don't worry. You'll get over it.

Cronin's Corollary
If you explain so clearly that nobody can misunderstand, somebody will.

Python's Principle of TV Morality
There is nothing wrong with sex on television, just as long as you don't fall off.

D'Agostino's Dilemma
No matter what goes wrong, there is always someone who knew it would.

Prachniak's Principle
There is nothing so small that it can't be blown out of proportion.

Hammond's Harangue on Life's Highway
If everything is coming your way, you must be in the wrong lane.

Gallagher's Gambit
Opportunity always knocks at the least opportune moment.

Quinlan's Question on Queues
Why is it that the longer you wait in line, the greater the likelihood that you are standing in the wrong line?

Murray's Musing
You may never know who is right, but you always know who is in charge.

Dave Elovitz

May 1986

THE ENGINEER AND THE DEVIL

There once was a man who had three sons, and he wanted each of them to prosper. The man did not have enough money to send his sons to school, so he sold his soul to the Devil to raise money for the boys' education. The Devil gave the man the money, and one son became a priest, one became a lawyer, and the third became an engineer.

Years went by, and each of the sons was successful. Just as the man was beginning to think the Devil had forgotten about the deal, the Devil showed up, demanded the man's soul, and prepared to take him to Hell.

The three sons were visiting their father when the Devil showed up to collect his debt and were determined to use their skills to prevent the Devil from taking their father's soul.

The first son, who had become a priest, began to pray and beg and appeal for mercy from the Devil. His efforts paid off, and he got a few more years of good life for his father.

When the Devil appeared a second time, the lawyer used his negotiating skills and appealed for mercy. He pleaded inequality of bargaining power and was able to convince the Devil to grant the father an extension of a few more years of good life.

The third time the Devil showed up to claim the old man the engineer was in the room. The engineer spoke to the Devil, "You have been exceedingly kind to my father and have spared him twice already. I am just an engineer and fear I will not be able to convince you to spare my father again. But, as a last request, would you spare my father while that candle remains?"

There was a butt of a candle burning on the table. The Devil noticed that it would not burn much longer and granted the engineer's request.

At that the engineer picked up the butt of the candle, blew it out, and put it in his pocket! The Devil was shocked, but he had to keep his bargain and left without the old man. The engineer held on to the candle, and the old man lived for many years.

Trust an engineer to outfigure the Devil.

Ken Elovitz

June 1986

ASHRAE

Ken and I will be attending the ASHRAE Annual Meeting for most of the fourth week in June. The American Society of Heating, Refrigerating and Air-conditioning Engineers is the technical society that sponsors most of the significant research on better calculation methods for the design of HVAC/R systems and determining human comfort; writes and publishes the ASHRAE Handbooks and Manuals, which are the definitive references on HVAC/R design and practice; and provides several forms of continuing education for engineers concerned with heating, ventilating, air conditioning and refrigeration systems.

Why do we attend local ASHRAE chapter meetings every month, and the two national meetings every year? When we pinpoint a system problem, evaluate the operating costs of different systems, or advise on the design for some special application, chances are we can do those things because of the extensive information developed and provided through ASHRAE. When we come across an unusual issue, we can call on experts in that field from all over the world, experts we have met through ASHRAE.

There is more to life than getting. Just as we rely on ASHRAE for information, ASHRAE relies on us to share our experience. This June Ken will be presenting a paper on commissioning large building HVAC systems. I will be meeting with Technical Committee 9.1 to put the final touches on the 1987 Systems Volume of the ASHRAE Handbook, for which I wrote Chapter One on System Selection, and helped update Chapter Six on Multiple Packaged Unitary Equipment Systems. I will also be following up, with other TC members, on a research project we initiated on perimeter heating systems, and setting up a potential new research project on control of outside air quality and building pressurization with VAV systems. Ken and I will each hear between 30 and 40 different papers presented, and we'll bring home and study copies of a couple of dozen others that we didn't get to hear in person.

So, if you call the fourth week in June and have to wait a couple of days for us to get back to you, we hope you'll understand. Active participation in ASHRAE benefits our whole industry.

Dave Elovitz

July 1986

C'EST LA VIE

The dispute was over a 6300 CFM return fan that would only deliver 5200 CFM. Careful measurements showed that the pressure drop through the ductwork was higher than the fan's design pressure, even though the flow was less.

"Changing those fittings won't make a bit of difference," the sheet metal foreman growled. "That SMACNA stuff is OK for you book engineers, but out here in the real world we have to make things fit together in the space we have. I was putting in ductwork when you were still in kindergarten, and I can tell you that fan's just not big enough. Why don't you just admit it, and stop wasting everybody's time?"

The young engineer's neck turned red, but he didn't raise his voice. "Look, short radius elbows need turning vanes to work properly, and that's what the spec calls for. And just plugging a round duct into the side of a rectangular duct doesn't comply with the spec, either. You owe us what the spec says, so make it the way the spec says. And this time, don't forget to seal all the joints according to the specs."

After the sheet metal contractor had reluctantly replaced the abrupt square to round with a proper pyramid tapered connection and had installed the turning vanes, they all gathered at the job-site again, to watch the air-balancer take new measurements. He had already measured a much lower pressure drop through the ductwork than before. Now, as the air-balancer man moved his pitot tube to each point on an imaginary grid in the duct cross-section and called out the velocity, both the engineer and the sheet metal man kept careful score. After the last reading, each rushed to average the velocities and compute the CFM before the other: "Over 6900 CFM! 10% over design," the engineer exulted, "and 30% more than before the fittings were corrected."

"Well, we pulled the fat out of the fire for you again, I guess." the sheet metal foreman responded, as he packed up his tools. "There's no substitute for practical experience. I hope you engineers learned something, so you'll lay the ductwork out right on the next job and we won't have to show you how to make it work."

C'est la vie!

Ken Elovitz

August 1986

7 WAYS TO WASTE MONEY

(1) Blow down boilers at least daily. This will insure a TDS concentration several times lower than necessary and will waste 1250 BTUs for every gallon of extra blowdown. As a bonus, you will increase water and chemical costs.

(2) Replace fluorescent lighting with HID in offices and retail stores. When ballast losses are considered in total system efficacy, 8-foot instant start lamps are more efficient than all but 400 watt and 1000 watt high pressure sodium. Popular 4-foot rapid start fluorescent system with energy saving lamps and ballasts is more efficient than all but 1000 watt metal halide (or high pressure sodium over 250 watts), and are more readily available than HID, cost less, and last just as long.

(3) Duty cycle fans to save energy. A fan that is off for 15 minutes out of the hour moves 75% as much air and uses 75% as much energy. But a fan than runs at 75% speed still moves 75% as much air but uses only 42% as much energy and doesn't shorten bearing and motor life by frequent starts.

(4) Use cheap, low efficiency filters. Since they hold less dirt, you will have to change them more often and can be sure of higher labor and disposal costs (and the material cost for cheap filters is probably higher per gram of dirt collected). As an added attraction, the cheap filters pass more dirt, allowing more buildup on coil surfaces, making it necessary to wash the coil more often.

(5) Raise the supply air temperature on VAV systems, especially if someone complains about being too cold. By raising the supply air temperature, everyone else will need more air to meet their loads, and fan operating cost will increase.

(6) Lower the supply air temperature on reheat systems to satisfy areas that need additional cooling capacity. By lowering the system supply air temperature instead of increasing the airflow to the problem room, all the other spaces will need more reheat. In one step, you can increase your costs for cooling *and* for reheat energy.

(7) Shut off reheat coils in summer. This will insure that all areas with minimum airflows or fluctuating loads will overcool. Some occupants will drive the building superintendent nuts, but the more resourceful ones will bring an electric heater from home. Energy that used to come from a fossil-fired boiler will be replaced with electricity costing 3 times as much per BTU.

Ken Elovitz

September 1986

FAST TRACK
With Apologies to Longfellow's
The Village Blacksmith

Over a cluttered drawing board
    The AC draftsman stands;
With frown upon his furrowed brow,
    His head held in hands;
Determining the C-F-M's;
    And specifying brands.

The Architect had worked and drawn
    for days, from morn 'til night;
The Owner told the Architect
    "At last the layout's right!"
And now the hapless engineer
    Must rush with all his might.

Fitting ducts between the beams
    That regularly loom;
No time to calculate the load
    For every tiny room;
The builder's putting up' the steel;
    Delay means certain doom.

"This is so like the job before,
    I need not calculate;
"They all come out about the same
    And I must meet the date.
"No time for aught but specs and plans,
    The contract to create."

"The shop has got the ductwork done;
    It's hanging on the floors.
"Delivery time is fourteen weeks,"
    The AC builder roars.
"I've got to order units now!"
    Sweat flows from all his pores.

"Never mind the Tenant's lists
    Of occupants and gear;
"We've got to get him moved in now,
    The lease says so right here.
"So put up grilles and just ignore
    The Tenant's engineer."

Before too long the Tenant calls
    His PC's are too hot.
His engineer comes to the site
    "In spec, this job is not.
"These calculations show what's wrong,
    What's needed spot by spot."

"This duct's too small and that too long,
    These CFMs too low.
"To rip part out and put it back,
    It costs a lot of dough;
"But it will never work this way;
    Enough air it must blow.

"Why is there never time enough,"
    The Owner, he did roar,
"To do it right this first time 'round,"
    Before the cost will soar?
But always time, for much more dough,
    To later do it o'er?"

ON A PERSONAL NOTE

Those of you that have had an opportunity to work with Gary will be pleased to note that he will wed Miss Dena Bach on September 14.

Dave Elovitz

October 1986

SHAKE 'N' BAKE

On frequent occasions
Without any warning
Same heat pumps shut off
In the midst of the morning.

With the temperature rising
And air growing stale
It did not take long
For the people to wail.

"Your heat pump's not running:'
Mechanics would say
Then they'd reset a switch
And go far, far away.

The owner was angry
And foresaw great doom.
So he called the installer
To come to his room.

"Please take a seat,"
The good owner did roar:.
Then he rose from his desk
And he slammed closed the door.

By venting his anger,
The wall he did jounce.
Made the thermostat shake
And its mercury bounce.

The temperature rose
And their brows they did mop
Before they had noticed
The heat pump had slopped.

h took a few minutes
But soon they had proved
The problem occurred
When the thermostat moved.

If the heat pump is running
It switches to off.
But then turn back on,
As quick as a cough.

The current's so high
When the motor restarts
That the relay locks out
And the power departs.

All safety controls
Go through one common switch.
There's no way of telling
Which one is the glitch.

So various people
Would try out their skill
And fiddle with breakers
And switches until

The unit restarted
They didn't know yet:
When cut off from power,
The relays reset.

And as for the problem
That caused all this mess,
We moved all the stats
So they'll vibrate much less.

Ken Elovitz

November 1986

EXPERIENCE

The new shopping center had a problem. A big food tenant needed space for his refrigeration condensing units, and it looked like the only place they could go was a pocket cut into the roof for air conditioning fresh air intakes. Would heat from the condensing units warm up the outside air enough to cause problems? Calculations proved fruitless - too many variables: How many condensers, how much outside air, which way the wind blows, and a few other things. Clearly this project needed some "gray hair" advice.

"Why worry about all that when you can eliminate the problem altogether?" my father asked, "Divide the roof pit in half with a partition, and arrange the condensers so that they blow through the partition. Put the fresh air intakes on the same side as the condensers; you'll have all the fresh air on one side and warm air from the condensers on the other." That sounded pretty clever, but I saw a complication: "What happens when one of the condensers cycles off? Warm air from the other units will get back through the idle condenser's opening and into the air intakes, won't it?"

"How about a motorized damper on each fan, to close when the fans are off?"

"That sounds too expensive and complicated. Why not a gravity damper like on an attic fan? "Good idea, but condenser fans are usually pretty small and may not have the oomph to open one of those." Then he paused, stared into space, and dove into one of our mysterious file cabinets, filled to the brim with apparently unrelated catalogs, and held up a little pamphlet. "I knew I remembered seeing one! Here's a catalog for a backdraft damper made of overlapping strips of flexible plastic instead of hinged metal blades. It doesn't take much pressure to open it, and will close tight enough for this." A couple of photocopies and a short letter later, and our client was happy. But I had one more question: "How did you come up with that idea?"

"I wish I could tell you I was smart enough to think it up by myself, but I can't. Supermarket designers used to build refrigeration machinery rooms that way all the time. I can't remember the last one I saw, but it worked fine then, and it should work now."

Gary Elovitz

December 1986

DEBT COLLECTION

A mason was hired to build a fireplace and chimney for a very wealthy man. On his way to the job, the mason encountered his friend, a plumber, who asked him where he was working.

"That old miser?" the plumber scoffed. "Wait 'til you try to get your money!"

Knowing that the plumber was a cynic, but keeping the advice in mind, the mason asked the wealthy man to pay for the bricks and mortar before he started work. The wealthy man refused, saying he would pay in full as soon as the job was done.

When the mason was almost to the top of the chimney, he began to worry about getting paid. So he climbed down the ladder and told the wealthy man he was almost done and asked the wealthy man to get the money ready. The wealthy man looked into his wallet and then put on a long face saying he could not pay just then as he didn't have the correct change.

"That disappoints me greatly;' the mason told the wealthy man. "But if I have to wait, then so do you. You must not build a fire in the new chimney until you have paid me." With that the mason went back to work, finished the job, and went home.

Shortly after the mason left, the wealthy man decided to build a fire anyway. After all, he thought, why shouldn't he enjoy the fireplace just because he still had the mason's money?

So the wealthy man tested the mortar with a stick and found it had set up dry and hard. Then he bent over, stuck his head into the throat of the fireplace. Seeing daylight coming down a very straight flue, he determined that the fireplace was ready for use.

The wealthy man carefully placed twigs and kindling on the hearth and tried time and again to light the fire. The flames would flicker and die, and smoke puffed back into his face, filling the house. Enraged, the wealthy man stormed over to the mason, cursing him and accusing him of slipshod work.

"I told you that you could not use the chimney until you paid me," the mason said very calmly. "Pay me and I'll fix your chimney."

So the wealthy man reached into his wallet, which contained the right change after all, and paid the mason. The mason returned to the wealthy man's house carrying some bricks and climbed a ladder to the top of the chimney. Then he dropped a brick down the chimney, smashing out the pane of glass he had mortared across the flue.

Ken Elovitz

January 1987

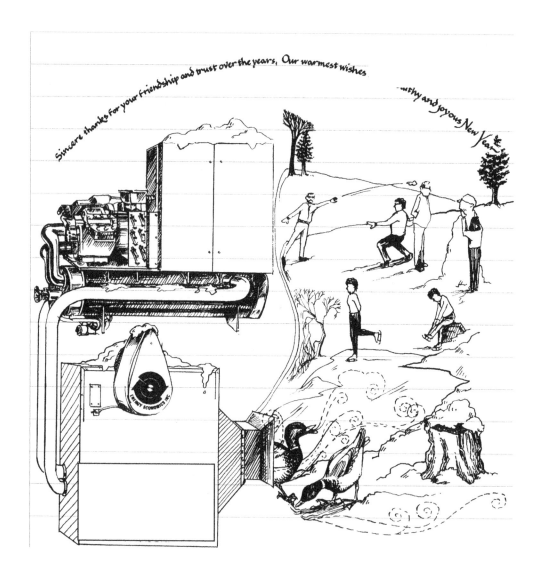

Dave, Ken, and Gary Elovitz

February 1987

LESS CAN BE MORE

"This preheat coil hasn't worked since day one", the operations supervisor complained. "Our freezestat trips every time we get any cold weather. And no wonder. What jerk sized this coil to heat the air only 2 degrees?"

Sure enough, the design calculations were based on mixing -10F outside air and 72F return air which would make 53F air at full flow. Thus, it would only need to be heated two degrees to the 55F supply air temperature. But this was a variable air volume system, and at the times when the outside air temperature dropped down below zero, the cooling load drops, too, and the system would deliver much less than design airflow. Fan tracking controls kept the outside air ventilation quantity constant, so at 10 below, the mixed air temperature would be so low the preheat coil didn't warm it enough to keep the freezestat from tripping. It sure looked like the preheat coil was too small.

Could we increase the coil capacity just enough? "Can't be done," the manufacturer insisted. "At the air flow rate and temperature rise you're talking about, the water would be in laminar flow. That coil just won't work. Buy a new one."

That didn't seem right. After all, for any given set of air and water flow conditions, the coil must deliver some performance. The only question was what that performance would be. So I set the computer crunching on coil performance data.

The answer quickly became obvious, but surprising. The preheat coil actually had too much capacity for the job, not too little! With the available water temperature, only a trickle of water flow would provide more than enough BTUs, but it would provide them all at the beginning of the coil. Then the rest of the coil face would let cold air through, tripping the freezestat.

But I knew that reducing water flow is not the only way to reduce a coil's heating capacity. You can also lower entering water temperature. All we need do is replace the 2-way valve with a 3-way mixing valve and install an inexpensive recirculating pump to provide 120F mixed water entering the coil. At a lower entering water temperature, the flow would be high enough for the coil to heat evenly without overheating the air.

Ken Elovitz

March 1987

STANDARD PROGRESS REPORT
(for those with no progress to report)

During the report period which ends 28 February 1987, considerable progress has been made in the preliminary work directed toward establishing the initial activities.[1] The background information has been surveyed and the functional structure of the component part of the cognizant organization has been clarified.[2]

Considerable difficulty has been encountered in the selection of optimum materials and experimental methods, but this problem is being attacked vigorously, and we expect that the development phase will proceed at a satisfactory rate.[3] In order to prevent unnecessary duplication of previous efforts in the same field, it was necessary to establish a survey team which has conducted a rather extensive tour through various facilities in the immediate vicinity of manufacturers.[4]

The Steering Committee held its regular meeting and considered rather important policy matters pertaining to the overall organization levels of the line and staff responsibilities that devolve on the personnel associated with the specific assignments resulting from the broad functional specifications.[5] It is believed that the rate of progress will continue to accelerate as necessary personnel are recruited to fill vacant billets.[6]

NOTES:

1. We are getting ready to start, but we haven't done anything yet.

2. We looked at the assignment and decided that George would do it.

3. George is looking through the handbook.

4. George and Harry had a nice time in New York.

5. Untranslatable - sorry.

6. We'll get some work done as soon as we find someone who knows something.

(from More Effective Writing for Business and Industry *by Robert Gunning)*

Ken Elovitz

April 1987

THE DAMPERS WERE HAMPERED

"These units never worked right," the Operations Supervisor insisted. Damfool design engineers blocked up half our outside air intake for a duct, and we can't get the air we need on these systems. If you run them one at a time, they're fine, but when they all run together they just starve." Sure enough, about half the intake area was blocked, but my calculations showed that there was still enough free area for the desired airflow.

Four air handling units backed up to a long "outside air room" about 5 feet deep. The outside air louver formed one wall of that outside air room, and the outside air dampers for the individual units formed the other. Two units mixed return and outside air, and the other two were 100% outside air systems.

"How about if we measure each fan with all of them operating together and then check each one operating by itself. That way, we can see how the systems influence each other," I suggested. "Before we start our measurements, let's take a look at the louver and the outside air intakes."

As soon as I stepped into the outside air room I saw the problem, but how could I get this crusty supervisor to see it? I decided to go ahead and take all the measurements and then sit down with him to go over the results.

"AC-1 looks OK," I began. "The data match the balancing report from the original installation. AC-2 is a little low on CFM and high on static. Did your mechanic tell you that the outside air dampers were open about 10% and the return air dampers were open only about 50%? Those two dampers should work together so their percentages always add up to 100% for minimum pressure drop. If you adjust the linkages and controls I think you'll see that high static pressure converted into the CFM you're looking for.

"Your 100% outside air units, though, are real problems, just as you said. The statics we measured are so high those fans must be stalled."

"I told you that before you wasted all this time taking readings," the supervisor growled.

"Come with me and take a look. See the outside air dampers on both these outside air units? They are not only closed, they've been that way so long the pressure from the fan has actually bent them. Fix the damper motors and I'll bet you'll find your missing CFM."

Ken Elovitz

May 1987

from the PUNABRIDGED DICTIONARY

allege - (a-LEJ) - a high rock shelf.

brotherhood - (BRUTH-er-hud) - your brother, the crook.

bulkhead - (BULK-hed) - hat size larger than 7½.

carpet - (KAR-pet) - a dog or cat who enjoys riding in an automobile.

deduce - (de-DOOS) - de lowest card in de deck.

dogma - (DOG-ma) - a mother dog.

eclipse - (e-KLIPS) - what a gardener does to the hedge.

falsehood - (FALS-hud) - someone who pretends to be a gangster.

fission - (FISH-un) - where Huck Finn went when he played hookey.

grateful - (GRAT-tul) - what it takes to build a good tire.

handicap - (HAN-de-kap) - a ready to wear hat.

intense - (in.:rENTS) - where campers sleep.

ketchup - (KECH-up) - what the last runner in a race wants to do.

locate - (LO-Kate) - nick name for a short girl named Catherine.

minimum - (MIN-a-mum) - a very small mother.

nitrate - (NI-trate) - cheapest price tor calling long distance.

paradise - (PAR-a-dis) - two ivory cubes with dots all over them.

paralyze - (PAR-a-liz) - two untruths.

rattan - (ra.:ran) - what a rat gets vacationing in Florida.

stirrup - (STER-up) - what you do with cake batter.

tenure - (TEN-yer) - the number following nineure.

vermillion - (ver-MIL-yen) - the number following vernine hundred ninety nine thousand nine hundred ninety nine.

warehouse - (WAR-haus) - what you say when you are lost.

zealotry - (ZEL-a-tre) - what a tree salesman wants to do.

Shamelessly plagiarized by Dave Elovitz

June 1987

'COS WAS THE CAUSE

We needed to project how much electricity patient rooms in the hospital's new building would use, so we measured usage in an existing block of patient rooms. I set up our recording electric meter in an inconspicuous place and came back a week later to collect 14 feet of meter tape showing how much electricity was used in the patient rooms every hour of the week. I transcribed the data onto a graph, and found a fairly consistent daily pattern. It was interesting to see just how closely the meter tape mirrored the hospital routine:

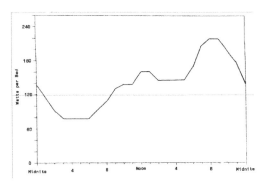

Electricity usage is lowest overnight, when most of the patients are asleep and most of the lights are out. Usage begins to pick up about 7 AM as the patients start waking up and doctors make their rounds, and levels off around 9 AM. There's a little bump about noon when lunch is served, but then the usage returns to a fairly steady rate throughout the afternoon. Many patients leave their rooms during the day to sit in the sun room, and quite a few others are gone to doctor's appointments or therapy sessions. Those who stay in their rooms seem to be satisfied by daylight coming in the windows.

Later in the afternoon, usage starts to pick up. Visitors come by in the afternoon and early evening, so more patients tend to be in their rooms. As evening sets in, more lights come on. Lots of patients watch TV in the evening, and that showed up too. Usage starts to taper off about 10 PM, settling down to its overnight level by about 1 AM. This pattern repeated itself day after day.

But then there was one point where dots on the tape suddenly deviated from the normal pattern. "Something went wrong with this meter," I thought as I looked at the blip. There it was, 8 PM on Thursday night, and usage suddenly jumped over 30%. Could Maintenance have been running some kind of test? No tests on Thursday nights, they assured me. Must be something else. Wait a minute! Thursday at 8 PM. Isn't that when "The Cosby Show" is on?

Gary Elovitz

A Personal Note

We are pleased to announce that Sara Elovitz and Dave Adelizzi will be married on June 7.

July 1987

TOO EFFICIENT?

"The leaks are so bad, the walls are getting moldy - It's just awful!" The manager of the apartment complex wailed. "We never had problems like this until we spent all that money to have the building renovated. New roof, storm windows, and new insulated stucco panels over the old decaying wall panels. "They did a lousy job, though. The building is leaking like crazy!" He showed us photos of wet and moldy strapping inside the walls. "Find the leaks, and get me evidence against the contractor."

The first apartment was hot and humid. Mold grew in thick patches where the walls and ceiling met. There was a puddle of water at the base of each window, and water dripped down the inside panes. In places the wall board was like a sponge. A huge kettle boiled on the stove, and six adults and three children watched us with interest.

We visited several other apartments and found similar conditions. The problem apartments seemed to be spread randomly throughout the building, many on lower floors below apartments with no mold at all. If the leaks were from the roof, the mold should be worst on the top floor.

Temperature and humidity readings showed that the problem apartments were so humid that condensation would form on any surface colder than 65F. The architectural drawings showed no insulation on the edges of the concrete floor slabs, and we calculated that the surface temperature of the ceiling near the wall would be below 65F whenever the outside temperature was below about 50F.

Water wasn't leaking in from outside! It was condensation. But why did it only show up only after the renovations? The old windows and decaying wall panels let so much dry outside air leak in that the humidity levels in the apartments were kept down. The renovations made the building so airtight that moisture couldn't escape. Not the contractor's fault. After all, he was supposed to seal up the leaks.

The renovation tightened up a leaky building and provided spectacular energy savings. For most of the tenants, there was also a big improvement in comfort. But tenants who keep a kettle boiling all day need more ventilation because of the humidity. Now those few have to open a window.

Gary Elovitz

August 1987

A PROMIS IS A PROMISE

The counterman looked familiar as he slid a steaming mug of tea across the counter. I was puzzling over where I knew him from when I was startled by what sounded like several blasts from a diesel horn. That clamor jarred my memory: He was an electrician on a client's new building, and he had stuck in my memory because he seemed to blow his nose every couple of minutes.

"Didn't I see you working as an electrician down the street?" I asked. In a thick accent he replied that he had indeed. "What are you doing here, working in a lunch counter? This can't be anywhere near as good as a job as being a construction electrician."

He shrugged and explained in his thick accent that he had been fired, ending the sentence with another thundering honk into his handkerchief.

"But why?" I asked. "I remember you as always being very busy, working very hard, and doing very neat work."

It seems the other electricians complained about the resounding blasts when he blew his nose over and over, and he admitted that he had just plain refused to drive any screws up into the underside of the floor slab above. Several of the journeymen had threatened to leave the job because of his repeated' and unnerving trumpeting. And the foreman had warned him repeatedly, to no avail, that he would lose his job if he didn't drive screws up into the overhead. The contractor had finally laid him off, and he hadn't been able to get another job as an electrician once word about his peculiarities got around.

"Couldn't you just break the habit of blowing your nose? Or at least learn to blow it quietly?" I asked earnestly. "And for gosh sakes, it doesn't seem so hard to put screws into the slab overhead. What's the big deal?"

"I just couldn't do that," he shrugged. "I promised my father when left the old country. Just before I got on the boat, he said to me, 'You are going to be representing the whole family when you get to America. Keep your nose clean and don't screw up!'"

Dave Elovitz

THE CONSULTANT ON A DESIGN-BUILD PROJECT

Many owners see significant advantages in the design-build approach to HVAC systems, where the contractor is responsible for design, construction, and (usually) a satisfactory end result. Some also see disadvantages and risks and retain a consultant to help reduce them, but the consultant's role is not always well understood by all parties. This is how we see the Owner's consultant contributing to the design-build process.

SCOPE OF WORK: The consultant prepares a document that defines the performance the design-build contractor (DBC) undertakes to deliver (in terms of specific items that can be measured) and the design conditions the system must meet, including occupancy and internal heat gains. The Scope should also, establish the level of quality and identify any constraints or special requirements like operating efficiency, space limitations, or esthetic requirements.

DESIGN REVIEW: The consultant reviews the DBC's design and advises the Owner if he believes the system shown by those documents meets the criteria presented in the Scope and is generally consistent with the Scope. The review may raise questions or flag misunderstandings before they become difficult to correct. Since the DBC is solely responsible for meeting the performance requirements, the DBC's engineer is the engineer of record, he stamps the drawings, and he makes the final decision in case of any disagreement.

SUBMITTALS: The consultant checks submittals for the major equipment to advise the Owner if they are consistent with the Scope and the DBC's design. The consultant does not approve submittals and does not have the right to reject them -- only the engineer of record can do that. However, the consultant's review informs the Owner about the major equipment, and alerts him if there is a deviation.

ACCEPTANCE: Once the DBC says the system is complete and operating properly, the consultant can evaluate the system operation and performance under simulated operating modes and advise the Owner if the observed performance is gene rally consistent with the Scope of Work. The consultant might be assisted by the DBC or by an independent air balance technician for the performance checkout.

Most DBCs welcome the Owner's consultant when his role is defined in this manner because the consultant helps both the Owner and the DBC: Both want to catch errors or problems before they become difficult to correct. But if the consultant is set up as an adversary, the spirit of team effort toward a common goal will be severely damaged, defeating one of the major advantages of design-build.

Dave Elovitz

October 1987

CAN'T WIN 'EM ALL

I read recently that half of all commercial tenants are unhappy about their space temperatures, and that 56% of those who complain about temperature at least three times per year won't renew their leases! A 1984 survey for the Building Owners and Managers Association identified some of the factors that give rise to tenant complaints:

Workers complain about cold floors. If a building with an overhang does not have heat supplied beneath the floor, workers' feet will be cold. Just adding insulation is not enough to provide a comfortable floor surface temperature. The cavity below the floor must be heated, not just insulated.

Workers near windows complain about uneven radiation and drafts. Even if the average temperature in the room is at the proper level, people will feel cold unless the heating system blankets the entire window. Heat covering the whole window warms air infiltrating through cracks before it reaches the occupants. Heat covering the whole window keeps the inside surface of the glass warm. Occupants don't lose heat by radiation to a cold glass surface and there is no layer of heavy, cold air next to the glass to slide down and make a cold draft along the floor.

Workers complain about cold airstreams. As little as 50 FPM (which is barely perceptible) moving across the neck will bring complaints from ¼ of the people even if the air is only 2 degrees colder than the room. Drafts can come from inadequately heated vestibules or from elevator shafts, but they can also come from the air conditioning system itself. If diffusers don't mix supply air with room air thoroughly before the airstream reaches the occupants, the cool supply air can cause draft complaints.

Surprisingly, the survey found that just as many men complained about cold temperatures as did women. The myth that women are always cold is just that - a myth. Finally, the study revealed that even under the best circumstances, about 5% of any group will still be uncomfortable. You just can't win 'em all.

Ken Elovitz

Calendar message for: ~~September 1987~~
~~October 1987~~

November 1987

PROCRASTINATION - THE THIEF OF TIME

"Never do today what you can put off until tomorrow." If that's your motto, then you're a procrastinator. We all procrastinate from time to time, and we all know procrastination is no good, so why do we do it?

Believe it or not, perfectionism is one cause of procrastination. Perfectionists can be their own harshest critics, relentlessly demanding better performance. Perfectionists procrastinate because they feel they are not up to the task yet or that their initial ideas just aren't good enough. Waiting for the right "inspiration" is just an excuse for putting the job off. Perfectionist procrastinators can get themselves going by realizing that everyone makes mistakes, and if they can turn their internal critics to evaluating the first draft, chances are they will catch and correct their mistakes.

Procrastination is sometimes a response to frustration. That frustration might come from a feeling that the task to be done is unimportant. But if it has to be done, it doesn't matter how menial it is, and putting it off won't make it go away. Looking for little things to interfere with doing the job is another excuse for procrastinating. Most of the time, those little things don't have anything to do with whether or not you can get started.

Work best under pressure? That's another sign of procrastination. We wouldn't always be responding to crises if we didn't wait so long to get started that everything had to be done in a rush. Remember that in the long run, slow and steady wins the race, and most crises can be avoided by getting started on time.

Procrastinators need to begin mending their ways immediately - or at least get started first thing tomorrow!

Ken Elovitz

December 1987

CIRCUMSTANTIAL EVIDENCE

Both Ken and I have occasion to testify as experts, and the lawyer for our client always warns us not to try to outsmart opposing counsel during cross-examination. "Just answer the questions directly, and don't add anything. If he twists things around by asking a limited question, let me straighten it out on redirect" is the usual advice. Recently heard this story that shows what sound advice that is:

A farmer, going for a load of grain with his horse and wagon, was struck head on by an inexperienced driver. The farmer suffered severe and permanent injuries. The insurance company wouldn't pay, so the farmer sued.

During the trial, the insurance company's lawyer asked the farmer, "Isn't it true that immediately after the accident a person came over to you and asked you how you felt?"

"Yes, I remember that," the farmer replied.

"And isn't it also true that you told him you never felt better in your life?"

"I guess I did."

"No further questions," the defense lawyer said.

The farmer's lawyer walked over to the witness box. "Would you tell the jury the circumstances in which you made that statement?"

"Sure," the farmer said. "Immediately after the accident, my horse had two broken legs and was neighing and kicking. The deputy sheriff came over, put his gun to the horse's ear and shot him dead. Then he went over to the dog, which had a broken back and was yelping. He put his gun to the dog's ear and shot him dead.

"When he came over to me and asked me how I felt, I said I never felt better in my life."

What you say is greatly influenced by the circumstances in which you say it!

Dave Elovitz

January 1988

TO THE TUNE OF ....

O come all ye wasteful
Tune up all your systems
O come ye, O come ye, save energy
Replace aging burners
Keep your furnace well in tune
O come let us adjust them
O come let us adjust them
O come let us adjust them, for
    e-fficency

Check insulation
Weatherstrip your windows
Fuel costs won't stay this low forever,
    you know
Slow down your fan speeds
Trim your pump impellers
O come ye, optimize them
O come ye, optimize them
O come ye, optimize them, for
    e-fficiency

Call your control man
Have him calibrate things
Check all your thermostats and
pneumatics, too

Control air compressor
Should run only half the time
O come fix all those air leaks
O come fix all those air leaks
O come fix all those air leaks, for
    e-fficiency

If you've got reheat
Fans on constant volume
You'll save a pile if VAV is installed
Retrofit dampers
Inlet vanes or speed control
O come let us convert them,
O come let us convert them,
O come let us convert them, for
    e-fficiency

Clean heat exchangers
Insulate the piping
O come ye, O come ye, reduce
    wasted heat
Use heat recovery
Capture heat that you can use
O come let us improve them
O come let us improve them
O come let us improve them, for
    e-fficiency

Fix leaky steam traps
Change your dirty filters
Clean all your coils and get a low
    delta-P
Blow down your boilers
Lubricate all moving parts
O come let us maintain them
O come let us maintain them
O come let us maintain them, for
    e-fficiency.

Franny & David
Ken & Dave & Sara,
Gary and Dena send warm wishes to
    you
Happ'iest of New Years
Luck and joy and happiness
O come and let's be thankful
O come and let's be thankful
O come and let's be thankful for good
    friends like you

Dave, Ken, and Gary Elovitz

February 1988

Rx

News headline: "Sick" Building Evacuated!

Workers in a large office building complain of coughing, wheezing, tightness of the chest, headaches, muscle aches and fatigue. They are diagnosed as having Hypersensitivity Pneumotitis. The cause of the illness is determined: high levels of toxins in the air-"Sick Building Syndrome".

People who work in large buildings are beginning to suspect that the indoor environment is no longer safe. And in fact, people really do get sick because of indoor air contamination.

Sound ominous? Well, it is a serious problem, but in most cases, the causes and the solution are quite mundane: good old common sense maintenance.

Almost always, we find the causes of a Sick Building complaint to be either insufficient fresh air ventilation or growth of fungus and mold spores in the building. Both of these can result from a lack of close attention to maintenance.

A fresh air damper seizes in the closed position. Or the controls malfunction and the damper doesn't open. Perhaps a building manager eager to reduce operating costs locks the damper in the closed position. Without sufficient fresh air, the building gets stuffy, and people suffer the result.

Sometimes the fresh air intake doesn't get really fresh air at all. It may draw in the discharge from a nearby exhaust system, or even actually draw in toxins: Birds roosting in the fresh air intake leave excrement behind; fumes from a nearby truck dock or mist from a contaminated cooling tower can be drawn in instead of fresh air.

What if the rooftop unit has a little leak, and rain gets in and soaks the casing insulation and the filters? The condensate drain on the cooling coil backs up, and there is a puddle of water in the unit. Or the filters are not changed often enough, and dirt builds up. Perhaps the filters are torn, and they let some dirt through to build up on the coils. A waste pipe with a tiny leak drips on a ceiling tile in a return air plenum just enough to keep the tile moist. These are all hospitable environments for mold and fungus growth. When allowed to grow unchecked, they will spread throughout the building.

When a building gets sick, people panic and begin to lose faith in the safety of the indoor environment. But 99 times out of 100, the prescription for a healthy building is conscientious preventive maintenance.

Gary Elovitz

March 1988

GETTING OUT OF HOT WATER

"The tenants are driving me crazy! We don't have enough hot water in Building 2, especially at shower time," the building manager grouched. "The plumber says we need a storage tank just like the ones in the other two buildings."

"Only thing is," I replied, "the other buildings don't have storage tanks at all. They just have external heat exchangers. Building 2 has water heaters built into the boilers. Better let me check my files for some history."

Earlier temperature measurements had indicated dirty heat exchanger surfaces. The city water side had been cleaned, but the boiler water side had not. The measurements also suggested the tempering valve was stuck in one position instead of mixing in a varying amount of cold water to maintain a constant temperature at the tap. It took some convincing, but the building manager reluctantly had the plumber clean the boiler water side of the heat exchangers and replace the tempering valve instead of installing a storage tank. Exasperated and yet triumphant, the plumber reported that the building still ran out of hot water and insisted we needed a storage tank. The building manager agreed indignantly.

All the symptoms pointed to an overloaded heat exchanger. This was puzzling because calculations confirmed there was plenty of water heating capacity. On top of that, Building 3 had no complaints, and it had only half as much domestic water heating capacity as Building 2. Could the flow sometimes exceed the heat exchanger's rating? Double the flow, and the same rate of heat transfer makes only half the temperature rise.

"We need to limit the flow through the heat exchangers to their rated value," I told the building manager. "That way, the water leaving the heaters will always be hot enough that the tempering valve mixes in some cold water. It may take a few minutes longer to fill a tub, but more people will be able to use hot water at the same time."

A few days later, a sheepish building manager reported the results: plenty of hot water. "I'm glad you wouldn't let me buy the storage tank. That little flow limiter valve takes up a lot less space than a big old tank, and costs a lot less, too!

Ken Elovitz

April 1988

PLASTERED

"We've had heating problems ever since we finished renovating this building," the developer explained. "This unit is the worst. The contractor looked at it, the engineer looked at it. They even had the factory rep come out and take all kinds of measurements. Everyone agrees there isn't enough airflow, but no one agrees on why. The contractor says the unit is no good, and the distributor blames it on the installation. In the meantime, these poor people are freezing."

The airflow we measured was only about half the 1400 CFM the drawings called for, and the static pressure was well below design. The catalog said the unit should deliver 1500 CFM at that pressure. The pressure drop across the heating coil was very low, confirming the low airflow. There was plenty of area for airflow on the inlet, and the filter was clean. I checked the fan rotation; it wasn't running backwards. The fan was just not performing up to the catalog ratings, so I called the local factory service manager. After I outlined my tests and conclusions he agreed to come out and check the unit himself."

George repeated my tests and got the same results. He agreed that even though there were some poorly made duct fittings, the fan definitely should have been putting out a lot more air than it did. "You're right, Dave. It just doesn't make any sense, these units always perform OK. The only thing I can figure," George mused, "is that maybe a piece of insulation fell down and is blocking airflow through the heating coil."

"I don't think so," I responded. "We checked the pressure drop across the heating coil, but how hard could it be to pull the fan out so we can see up inside?" We started disassembling the unit in the cramped equipment closet, chatting as George removed pieces and passed them up to me. As he pulled out the fan rotor, he suddenly collapsed back against the closet wall and shook with laughter. I twisted down to peer in over his shoulder. There, almost filling the bottom of the fan housing, were four or five pounds of plaster in big chunks that blocked off much of the airflow before it could ever get out of the fan!

"Did these units come in early and get stored here during demolition?" George asked. "Stored, but obviously not covered," I answered, "So chunks of plaster from the demolition could fall in and slide unnoticed to the bottom of the fan housing to drive us all crazy."

Dave Elovitz

May 1988

OFCI

Ken presented a paper on the method he developed for calculating the Optimum Filter Change Interval (OFCI) at a Symposium on Building Operation Dynamics at the ASHRAE national meeting in Dallas in February. There was so much interest from people at the conference, I thought you might want to know about it, too.

The method projects the total cost of fan energy, electric demand for the fan, new filters, and labor and disposal of dirty filters. The pressure drop across dirty filters depends on the type of filter, the filter face velocity and the amount of dirt allowed to accumulate between filter changes. If filters are changed more frequently, the cost for filters increases. If filters are loaded to their maximum dirt holding capacity, the pressure drop increases, increasing both fan energy and fan KW demand.

Ken's paper points out the difference in dust content of outdoor air and return air, and explains how to calculate the effect on filter dust loading of the changing proportions of outside air with an outdoor air economizer cycle. It also recognizes that the varying air flow in a VAV system brings different amounts of dust into the filters under different cooling load conditions, and shows how to calculate that effect on filter loading.

Examples in the paper show how the method is applied to calculate total annual filter costs for a VAV system in a Boston office building and a constant volume system in a New Hampshire hospital. Results for any given installation depend on the operating factors and energy costs for that specific installation; the method provides a way to determine the costs for any specific situation. But it was interesting to note that, in both examples, it was less expensive overall, to use high quality filters and change them at maximum pressure drop than to follow the traditional approach of routinely changing filters, four times a year.

With Ken's method, you can calculate the Optimum Filter Change Interval for your system in a few hours with a personal computer spreadsheet program, standard weather data, electric rates, and the published data from the filter manufacturer. If you'd like to try it, drop us a line or give us a call, and we'll send you a copy of the paper that explains how.

Dave Elovitz

June 1988

THE IDEAL CONSULTANT

Doctor John Walsh was a prominent obstetrician and gynecologist in Providence, RI. Over the years his patients included my grandmother, Sarah Elovitz, some of my aunts, and several of my cousins. It was my cousin Charlotte who related what Doctor Walsh told her about Bouba, which is what we called our grandmother. I remember Bouba holding court for all her children and grandchildren in her kitchen on Sundays. She was a beautiful woman with snow white hair and limitless energy and kindness. But that is not part of Dr. Walsh's story.

"Whenever I had a deeply troubling problem," Dr. Walsh told Charlotte, "I would mull it over and over, and eventually I would call Bouba," for that is what he, too, called her, "and ask if I could come by to see her. She would always say she would be happy if I would do so, and I would soon find myself ushered to the table and a nice cup of tea and freshly baked cookies.

"I would tell Bouba about whatever problem had been troubling me, and she would listen carefully and ask questions. Then we would talk about the situation, and she would say I could do this .and if I did, then I could expect such and such would happen. On the other hand, I could do that, and if I did, so and so would probably happen. And so on, until she has described all the choices and outlined the probable outcome from each.

"She never told me what to do, but by the time the tea and cookies were gone, I was no longer troubled about whatever it was that brought me to see her, and I always left her knowing what to do."

That was Dr. John Walsh's story, as my cousin Charlotte told it to Bouba's great grandchildren.

I never met Dr. Walsh, and I first heard the story only a few months ago. It was a moving story for me, of course, because it was about my grandmother and her great wisdom and her great sensitivity for other people. As I mused on the story, I realized that Ken and Gary and I were not the first consultants in the Elovitz family. Two generations back, in Providence, an immigrant woman from Poland, had mastered what it takes to be the ideal consultant: Not to tell the client what to do, but to identify what the options are, and to outline the implications of each option. I only hope that is what we do for you.

Dave Elovitz

July 1988

BEST OF MURPHY

Koningisor's Criterion
The man who can smile when things go wrong is the man who has thought of someone he can blame it on.

Einstein's Illumination
Everything should be made as simple as possible, but no simpler.

Carli's Conclusion
There's never enough time to do it right, but there's always time to do it over.

Westley's Worry
Just when things can't get any worse, they will.

Prachniak's Predicament
The first 90% of any project takes the first 90% of the time. The last 10% takes the other 90% of the time.

The Golden Rule
He who has the gold makes the rules.

Re's Rubric
The lion and the lamb may lie down together, but the lamb won't get much sleep.

Looney's Lament
Nothing is impossible for the man who doesn't have to do it himself.

Potter's Principle
Anyone who will do what it takes to get elected is not to be trusted with the office.

Kranz's Law
If all you have is a hammer, everything looks like a nail.

Cassin's Law
No good deed goes unpunished.

Cole's Law
Thinly sliced cabbage, a little vinegar, oil and mayonnaise.

Rowe's Rule of Order
He who shouts the loudest has the floor.

Frank's Formula
When ideas fail, words come in very handy.

Dave Elovitz

DON'T FORGET - We will have a new telephone area code starting July 16: Dial 617 until July 15, then start using 508 (even from Boston) on July 16. Toll charges remain unchanged.

August 1988

THE ENGINEER

Karen Gooze of Honigman Miller Schwartz and Cohn in Detroit sent us this story, and we thought it was worth sharing:

During the French Revolution, a priest, a lawyer, and an engineer were put on trial as enemies of the Revolution. The priest tried to defend himself by explaining how he had ministered to the poor, brought food to the hungry, and tended the sick, but to no avail. He was sentenced to the guillotine.

In those days, when a person felt he had been sentenced unjustly, he would demonstrate his feelings by making the supreme gesture - he would lie face up rather than face down in the guillotine. The priest did so. When the rope was tripped, it happened that the blade stuck in the channel about six inches above his neck. It was the rule that if ever the guillotine stuck, the victim was set free. So the priest was spared.

Then came the lawyer who also argued for his life. He reminded them of the many criminals he had defended, of how he had served the poor without fee, and had argued against unjust laws. But he too failed and was sentenced to die. Like the priest, he insisted on looking up at the blade as it descended. And in his case, too, the blade stuck in midair, and he was let free.

Then came the turn of the engineer. He pointed out that he had built the water works. There were public buildings and roads he had made for the people. He had never been engaged in politics, nor had he become rich at the people's expense. But these arguments fell on deaf ears, and he expressed his contempt by lying on his back looking up at the blade. Then, just as the executioner was about to pull the rope, the engineer let out a cry, "Look - I see your trouble. The rope has slipped off the pulley up there. No wonder the thing doesn't work."

Ken Elovitz

September 1988

ANGUISHED ENGLISH

Our business is Engineering, but language plays a crucial role in how we get things done. We always try to write clearly and concisely, so we can't be misunderstood. *Anguished English*, by Richard Lederer (Wyrick & Co., 1987), gives some marvelous examples of good English gone bad. (We hope you don't catch any like these in our writing!)

For sale: A quilted high chair that can be made into a table, pottie chair, rocking horse, refrigerator, spring coat size 8 and fur collar.

"We do not tear your clothes by machinery; we do it carefully by hand."

Dog for sale: Eats anything and is fond of children.

Sign in loan company office: Ask us about our plans for owning your home.

Letter to Welfare Office: In accordance to your instructions, I have given birth to twins in the enclosed envelope.

Welfare case report: Mrs. Jones has not had any clothes for a year and has been visited regularly by the clergy.

Newspaper report: Joining Wallace on stage were new School Committeewoman Elvira Pixie Palladino and Boston City Councilman Albert (Dapper) O'Neil, both active opponents of court ordered busing and Wallace's wife.

Sign in a Bangkok dry cleaners: Drop your trousers here for best results.

Migraines strike twice as many women as do men.

Correction: Our paper carried the notice last week that Mr. Oscar Hoff is a defective on the police force. This was a typographical error. Mr. Hoff is, of course, a detective on the police farce.

Headlines:

POLICE DISCOVER CRACK IN AUSTRALIA

TRAFFIC DEAD RISE SLOWLY

WILLIAM KELLY, 87, WAS FED SECRETARY

FARMER BILL DIES IN HOUSE

Mixed Metaphors:

It's time to swallow the bullet.
It's time to grab the bull by the tail and look it in the eye.

Gary Elovitz

October 1988

BUT, MR. LINCOLN ......

"No matter what I do, I can't seem to keep everyone comfortable," the building manager groaned. "Even when it's cold outside, some people swelter if our chiller is off. And if we turn our chiller on, other people freeze. Then in the summer, people on the shady side of the building want us to turn off the cooling. The building engineer gives me some gobbledygook about differential condenser temperatures and false loads. He says we have a two pipe system, so we have to cool all of the people all of the time if we want to cool any of the people any of the time."

As we toured the mechanical room, I asked the building engineer why there were so many pumps. "We have two for the chiller, two for the cooling tower, and one for each face of the building," he explained.

The drawings showed a constant temperature fixed air supply to each room. A room thermostat controlled a coil which could add heating or cooling depending on whether hot or cold water was available. It was a two pipe system, but the pump for each face of the building could be connected to hot or cold water independent of the others.

The air supply delivered a constant 2200 BTUH of cooling to each room. Some of the time, some of the rooms needed more cooling than that, even in winter. Much of the time, most of the rooms needed less, even in summer. At any given time, some coils needed to add cooling while others needed to add heat.

"That may be what you want," the building engineer countered, "but with a two pipe system, you can either have heating or cooling."

"Not really. You provide a constant amount of cooling to every room with the air system. That's important because it makes a certain amount of cooling available to all of the rooms all of the time. Then you use the water system to adjust the amount of cooling according to sun and outside temperature. Since the sun is the biggest effect on how much cooling each face of the building needs at any time, having a separate pump and changeover valves for each face of the building lets you select heating or cooling when and where you need it."

So, even though you have a two pipe system, it is true that you can cool all of the people some of the time; you can even cool some of the people all of the time; but you don't have to cool all of the people all of the time.

Ken Elovitz

November 1988

LONG ARM

It was not what you would call an elegant installation, but I didn't see why it shouldn't work. The old house apparently once had a gravity warm air furnace in the center of the basement with ducts radiating our in all directions like an octopus. The contractor had simply plunked down a new forced air lo-boy model with sheet metal supply and return plenums connecting it to the old ductwork. An A-coil sitting on top of the furnace outlet inside the supply plenum, with a 3 ton condensing unit outdoors, provided cooling. "All kinds of people have looked at it and say it's OK, the homeowner complained, "but it is still so hot you can't sleep."

I slid back the insulation on a round branch duct and unhooked it from the plenum. Now I could look into the plenum, see the coil, and take some temperature measurements. "That's odd," I mused. "The supply air is only about 60F, but the coil feels really cold. And every once in a while I get an even higher supply air temperature." The refrigerant line leaving the coil measured 33F. Why wasn't the air colder? All the other symptoms suggested low airflow, but that should give very cold supply air.

We measured the air flow at each supply outlet and each return register. About 850 CFM compared to the usual 1200 for a 3 ton system. "With that low air flow, the supply air ought to be in the forties, not 60," I thought. "And look at that, now there's ice on the suction line. I better look at that coil again." I slipped the branch duct back off and peered in to see the upper coil silvery white with heavy frost. But the air was not particularly cold! Twisting around to get my arm in through the duct collar, I felt around the lower part of the coil. It was frosted, too. and there was the edge of the condensate pan. I could reach down and feel the bottom of the drain pan and even get my finger tips all the way around the drain pan to the other lip ...

WAIT A MINUTE! I had reached all the way down around the coil into the furnace discharge! There was the problem that had eluded the installer, two service companies, and the factory rep. And if I wore a 32" sleeve instead of a 35", I wouldn't have spotted it either! When the mechanic installed the A-coil on top of the furnace, he had left out the baffle plates that force the air to go through the coil. No wonder the supply air was warm when the coil was so cold: Most of the air never went through the cooling coil at all, just went right around it into the supply ducts!

Dave Elovitz

December 1988

A MODEL SOLUTION

In 1966 they were pals. "Look, I'm building a new building for my own office anyway. Why don't I make it two stories, and you'll put your restaurant on the first floor. It's not worth spending the extra money to install separate systems," the landlord said. 'You pay the whole utility bill, and I'll reimburse you for my share. It's only a few dollars.

After two oil embargoes and spiraling fuel costs, the tenant had laid out almost $130,000 over the years, and the landlord never paid his share. "Look," he claimed, "you have all those freezers and big machines. My offices hardly use any electricity compared to all that."

"My kitchen doesn't use that much more power than a kitchen at home," and you should be paying the whole heating bill. Everyone knows most of the heat loss from a building is through the roof, and the roof is over your space."

As is so often the case, there was some truth on both sides. The refrigerated cases in the restaurant did use a lot of electricity, but they also reduced the air conditioning load. All the people in the restaurant reduced the amount of heat required downstairs, but they also increased the amount of cooling need. The roof is a big heat loss, but so is the storefront in the restaurant.

"Why not let us model the building as two separate single-story structures," we suggested to the lawyer, "a restaurant with exactly the same construction as the ground floor, and an office building with the same construction as the second floor."

Our computer made it easy to take different operating schedules and internal activities into account. Then we divided the actual energy use of the two-story building in proportion to the relative energy uses of the two imaginary separate buildings.

Calculations are no substitute for actual metering, but the amount of energy used was not in dispute. The argument was over how much of the bill each party should pay. After a simple explanation, and a little prodding from the judge, both parties understood the idea and accepted the results as the basis of an amicable settlement before the first witness was even sworn in.

Ken Elovitz

January 1989

to the tune of
THE TWELVE DAYS OF CHRISTMAS

For the FIRST way of saving, we recommend to thee:
    A BOILER THAT'S TUNED UP CARE FREE.

For the SECOND way of saving, we recommend to thee:
    ICE STORAGE MODULES and a boiler that's tuned up care free.

For the THIRD way of saving, we recommend to thee:
    BRONZE WEATHERSTRIPPING, ice storage modules, and a boiler that's tuned up care free.

For the FOURTH way of saving, we recommend to thee:
    FLUORESCENT LAMPS, bronze weatherstripping, ice storage modules, and a boiler ....

For the FIFTH way of saving, we recommend to thee:
    FIBERGLASS BATTS, fluorescent lamps, bronze weatherstripping ....

For the SIXTH way of saving, we recommend to thee:
    AUTOMATIC TIMERS, fiberglass batts, fluorescent lamps ....

For the SEVENTH way of saving, we recommend to thee:
    STEAM TRAP LEAK DETECTORS, automatic timers, fiberglass batts ....

For the EIGHTH way of saving, we recommend to thee:
    HEAT RECOV'RY CHILLERS, steam trap leak dectectors, automatic timers ....

For the NINTH way of saving, we recommend to thee:
    VAR'ABLE SPEED DRIVES, heat recov'ry chillers, steam trap leak detectors, automatic timers ....

For the TENTH way of saving, we recommend to thee:
    HID LIGHTING, var'able speed drives, heat recov'ry chillers, steam trap leak detectors ....

For the 'LEVENTH way of saving, we recommend to thee:
    NEW STORM WINDOWS, HID lighting, var'able speed drives ....

For the TWELFTH way of saving, we recommend to thee:
    WATER SAVING SHOWERHEADS, new storm windows, HID lighting ....

With warm wishes for a joyous, healthy new year, we thank you for your friendship and trust over the years.

Dave, Ken, and Gary Elovitz

February 1989

HOT AIR STORY

We heard this story, with a slightly different twist, from our accountant friends at Tofias Fleishman and Shapiro:

After scanning cloudless skies, two hot air balloon enthusiasts decided to ignore a threatening weather forecast and take off anyway. They drifted upward in the crisp, clear early morning light, enchanted by the beautiful views below and reassured by clear skies in every direction.

Suddenly they were surrounded by towering thunderheads that seemed to have materialized out of nowhere. Before they could descend, the balloon was enveloped by the clouds and they were being shaken by violent winds. They decided the safest course was to ride out the storm high up in the clouds. As the balloon rose, the power of the storm sent the gondola lurching and swinging, so that they lost all sense of direction. It was all they could do to hang on and keep from being pitched out as the storm whipped them through the sky.

After what seemed an eternity, the storm abated and the two balloonists found themselves drifting serenely once again, but this time far above an unfamiliar and deserted landscape. They reduced buoyancy and floated gently toward earth until they spotted a lone golfer lining up his putt and they headed the balloon toward him. Being courteous types, and sometime golfers themselves, they waited until he sunk the hole, and then called down, "Hullo down there! We're lost. Can you tell us where we are?"

The golfer looked up, startled, squinted up at the balloon, closed his eyes a moment in thought, then responded, "Yes, you are precisely 287 feet above the 14th green." Then he turned and without another word strode off out of sight.

"Just our luck!" one balloonist said to the other. "Here we are hopelessly lost and the only guy in sight is a building code consultant."

"How do you know he's a building code consultant?" puzzled the second balloonist. "Why from his answer," responded the first. "It was technically correct but totally useless."

Dave Elovitz

March 1989

INDOOR AIR QUALITY

"We have a terrible indoor air quality problem," the call started. "Our people complain the air is so bad it makes their contact lenses pop out. So we called an industrial hygienist who set up his Petri dishes, then said we have a sick building syndrome and our HVAC system is no good. I don't see how that could be: We never had any complaints all summer. As soon as the weather cooled off it's been terrible. We need real help!"

At the site, I got a quick tour. A warehouse building had been converted to offices. The building was served by rooftop units, so plenty of outside air was available. Outside air intakes were all located well away from truck docks and exhausts, so the problem wasn't contaminated air being drawn in. Reviewing the documents, I noted the system had been designed and installed by reputable people. They just shouldn't be having the problems the facilities manager described.

As we sat looking over the drawings, the south sun came streaming in through the window, and the office did seem stuffy. "Even on a cold day like this the sun can create quite a cooling load," I commented. "Maybe people are just too warm and attribute the stuffy feeling to poor air quality."

Just then, the cooling finally cycled on, with a roaring supply fan. "If this were my office, I think I'd complain more about that noise than indoor air quality," I started joking as we continued looking over the drawings. Then it struck me: If the fan runs only when the system calls for cooling, they don't get the design 5 CFM of outside air per person at all! They only get 5 CFM while the fan is on. When the fan is off they don't get any outside air. If the cooling runs only 20% of the time in cool weather, they actually have only 1 CFM per person.

"These thermostats are so-called smart stats," I explained, walking over to the wall and opening the cover. "You have them set to save fan energy by running the fan only when heating or cooling is required, but you don't get any outside air while the fan is off. Before you have us do a big study with lots of calculations, I think I can solve most of the problem right here. Just slide this switch up to run the fan continuously during occupied hours and have it cycle off during unoccupied hours except when heat is required. Running the fan continuously during occupied hours will provide all the ventilation the design calls for, and that should be enough anti-biotic for this sick building."

Ken Elovitz

April 1989

FURTHER ADVENTURES

So many people told us they liked the February story about the golfers who went up in the balloon, I decided to tell of a further adventure by one of them, even though it has nothing to do with engineering or construction.

Finding himself with a free afternoon one weekday, he hurried out to the club to play 18 holes. The course was busy, so he was joined to a threesome of nuns that were just getting ready to tee off as he came out of the locker room. They introduced themselves and stepped quietly up to the first tee. Each nun in turn teed up and got off a competent but unspectacular drive right down the middle. Our friend stepped up last and, much to his dismay, shanked the ball badly off into some woods.

"Dammit!" he growled.

"Sir," said the first nun, "I have to ask you to remember to watch your language. Remember you are playing with ladies." Flustered, he apologized and promised to be more careful.

The nuns hit straight down the middle again on the second tee, but this time he sliced into the rough. "Dammit!" he grunted again, and the second nun stepped forward. "Sir, I must insist that you refrain from blasphemy. We are not accustomed to such language. Remember, we are not just ladies, but also ladies of the cloth." Again, he apologized and promised to be careful.

But approaching the third green, his chip shot went right over the green into a sand trap, and out came a third involuntary "Dammit!"

The oldest nun stepped forward. "Look, mister! We've asked and asked, as nicely as we can. Now I'm warning you. Next time you use foul language in the presence of the servants of the Lord, you will be struck by lightning!"

The golfer took that warning to heart and was very careful about his language from then on. Everyone loosened up and soon all four were chatting pleasantly. Until, putting on the 17th green, his ball rolled right along the rim, then off the far side. "Dammit!" just burst out.

Suddenly the sky darkened, thunder rolled and a bolt of lightning streaked down ....

.... and struck the oldest nun, just as a majestic voice thundered through the clouds, "Dammit!"

Dave Elovitz

May 1989

HATFIELD COPY & McCOY CAFE

One of our real estate clients had sub-divided a single, large rental space and leased it to two tenants: a cafe and a copy center. Since the space had been built as a single store, the two tenants shared one rooftop heating/air conditioning unit. The thermostat was in the copy center, but the main electrical panel was in the cafe.

On busy days, the copy center has several large copy machines which run most of the afternoon until the store closes at six. The cafe usually has a brisk dinner trade and is often crowded after the nearby cinema lets out in the evening.

When the copy center is busy, the copy machines heat up the room, and the thermostat calls for cooling. Meanwhile, the cafe is in the midafternoon lull and doesn't need any cooling. The rooftop unit, responding to the conditions at the thermostat in the copy center, provides cooling to both. The copy center is comfortable, but the cafe freezes.

When this first happened, the cafe owner would walk over to the copy center and push up the thermostat. Soon the people in the copy center began to swelter, and their manager pushed it back down. Before too long, the manager of the copy center stopped letting the cafe owner in. It didn't take the cafe owner long to find out he could solve his problem by shutting off the power to the rooftop unit from the panel in his space!

Later in the evening, after the copy center closes, the cafe gets busy, and the cooking and the customers generate heat. Since the thermostat is in the copy center, it doesn't sense the heat build up; the rooftop unit doesn't come on, and the cafe overheats. One night it was so unbearable that the cafe owner actually took an axe and broke down the door to the copy center so he could get at the thermostat.

"You've got to find a cure for this problem before someone gets hurt," the building manager pleaded. "No problem!" I responded. "We know just what to do in a case like this. All you need to do is install thermostatically controlled dampers in the ductwork to each space. That way, each space can control its temperature independently by shifting cooling to the room that needs it and away from the room that doesn't.

"This scheme would require some simple ductwork and control changes to keep the rooftop unit coil from frosting, but there is a standard control package available for just this application. Sure, the controls will cost some money to install, but just think what you'll save on doors!"

Gary Elovitz

GAS PAINS

"We'd like to know if you can check our gas bills for us," the caller started out. "The gas company says our usage is up compared to last year because of colder weather, but we want to check their theory. They said this December had two degree-days per day more than last year."

"Send them on over," I suggested, "and I'll look at them and give you a call."

A few calculations and reference to data from previous years confirmed that colder weather accounted for some of the increase. "Besides colder weather, this December's bill covers 34 days while last December the bill covered 33 days. Looking at the overall month by month pattern, the December bill is not out of line.

"What stands out, though, is a sudden increase in Building #68. Until October, all four buildings used about the same amount of gas every month, but starting in October, Building #68 jumped about 300 therms a month ahead of the others. What could have changed last fall?"

We tossed around a few ideas, but none could account for 300 therms a month. A small gas leak had been repaired, but that was only a few therms a month. The manager thought one tenant was using the oven to heat her apartment, but even that did not come close to the amount of the mysterious increase.

"We've checked and checked," the manager reported, "and can't find anything. Can you take a ride up here and see if you can find something we can't see?"

At the site, I had the manager set the boiler controls so we could run each module individually while I tested the boilers. "These test show Module #1 could use a little tuning up, but Module #2 has a very low flue gas temperature and a very low carbon dioxide content. That tells me you have unburned gas going up the chimney!

"Your energy management panel runs only Module #1 in the summer, but when the cold weather arrives, it brings on Module #2. That explains why the low efficiency in Module #2 showed up so sharply last October. A good boiler tune up should get usage back in line."

A few weeks later the phone rang, and the manager was on the other end. "You hit it right on the head!" he applauded. "We tuned up the boilers, and when the gas bills came today, all four buildings were right in line -- within 1 therm."

Ken Elovitz

July 1989

CFCS AND OZONE

Papers and magazines are full of articles about the hole in the ozone layer over Antarctica that has been blamed on CFC gases. In 1986, the US and other nations signed an international agreement called the Montreal Protocol where they agreed to cut CFC production in half by 1998. That is of particular concern to our industry, because about 40% of the CFC's used in the US are used in air conditioning and refrigeration systems. A new law in Vermont will prevent anyone from selling or registering a car with CFC-based air conditioning in that state after 1993.

There is a good deal of confusion about what will be affected by these restrictions on CFC's. Most smaller air conditioning systems use R-22 as a refrigerant, and R-22 is apparently far less damaging to the ozone than R-11 and R-12 -- by a factor of 20. In the long term, it may be necessary to control R-22 as well, but for the foreseeable future, refrigerant should be available to maintain most air conditioning which uses R-22.

R-11 is used primarily in centrifugal chillers, which are common in large systems, but R-11 is also widely used as a solvent for cleaning up contaminated systems. R-12 is primarily used in low temperature refrigeration systems and is also used in a few high speed centrifugal chillers. One of the biggest HVAC/R uses for R-12, and a major source of CFC discharges to the atmosphere, is auto air conditioning.

All systems using R-11 and R-12 will be substantially affected by the reduced availability of these refrigerants and by strict controls that will be imposed on their release and disposal. Refrigerant manufacturers are hard at work trying to develop substitute refrigerants which will be as non-toxic and almost as efficient as the harmful ones. One new compound, R-123, looks promising as an ozone-safe direct replacement for R-11 in existing centrifugal chillers. It is expected to be available in 1991. No likely direct replacement for R-12 has yet emerged from the lab, but there is a great deal of research under way to find one, and a new refrigerant, R-134a, is being evaluated.

Certain CFC's may represent a real threat to the environment, but the majority of air conditioning systems use relatively harmless R-22, so most system operators need not be concerned about causing damage to the ozone layer.

Dave Elovitz

August 1989

FIELD REPLACEABLE UNITS

Gary had some problems with the mouse for his computer sticking and skipping. When he inquired from the computer manufacturer about getting a replacement mouse, they advised the mouse could be repaired and sent the following service tip bulletin:

"Mouse balls are now available as FRU's (field replaceable units).

"If a mouse fails to operate or performs erratically, it may be used in need of ball replacement. Because of the delicate nature of this procedure, replacement of mouse balls should be attempted by trained personnel only.

"Before ordering, determine type of mouse balls required by examining the underside of each mouse. Domestic balls will be larger and harder than foreign balls. Ball removal procedures differ, depending on manufacturer of the mouse. Foreign balls can be replaced using the pop-off method, and domestic balls replaced using the twist-off method. Mouse balls are not usually static sensitive, however, excessive handling can result in sudden discharge. Upon completion of ball replacement, the mouse may be used immediately.

"It is recommended that each servicer have a pair of balls for maintaining optimum customer satisfaction, and that any customer missing his balls should suspect local personnel of removing these necessary functional items.

"P/N 33F8462 -- Domestic mouse balls
P/N 33F8461 -- Foreign mouse balls"

So help me, I copied the service tip bulletin just as we got it.

Dave Elovitz

September 1989

DARK SUCKERS

Thanks to Crawley Cooper for calling to my attention some startling new research on illumination. After all these years thinking that electric bulbs gave off light, we now know that what they actually do is suck up dark!

Just look at the room you are in: You can clearly see that there is less dark near the bulbs than elsewhere and the larger the bulb the more dark it can suck in. Of course, bulbs don't last forever. Once they are full of dark, they can no longer suck any more in, and must be replaced.

You may find that hard to accept after all these years of misguided belief that bulbs give off light, but just look at a candle for further proof. A new candle has a white wick, but as soon as you start to use it you will notice that the wick turns black, having been stained by the dark being sucked in. (You see similar dark stains on old bulbs which are almost full of dark.) When dark is sucked in to a dark sucker, friction generates heat. Because a candle draws the dark in through a solid wick, there is a lot of friction, and a great amount of heat, so it is dangerous to touch an operating candle.

Portable dark suckers, called flashlights, must have very small bulbs, so they have auxiliary devices (called batteries) to hold all the dark sucked in by the bulb. When the battery is full of dark it must be replaced, or have the dark drained out by a device called a battery charger.

The researchers also tell us that dark is heavier than light. You can prove that yourself the next time you go swimming. If you swim right near the surface there is a lot of light, but as you dive deeper and deeper, there is less and less light. Eventually, if you have scuba gear, you can dive deep enough that you will be in total darkness. That is because the heavier dark sinks to the bottom and the lighter light floats to the top.

Because the dark is so heavy, when we let it out of the bottom of lakes or rivers through turbines to flow to the oceans (which can hold an immense amount of dark because they are so deep) we can use the force of the falling dark to generate electricity.

If you still have doubts about the new theory, you can prove it to yourself by standing in an illuminated room and opening the door to a dark closet. The dark will be sucked out so quickly that you won't even be able to see it go by.

PS: Thanks for the many calls and notes about the August story. We're glad you enjoy these messages. There are still some copies of a collection of earlier calendar messages. Just call or drop a line and we'll send you one.

Dave Elovitz

October 1989

RED HERRING

"You've got to help me!" the robust blond burst into my office, sobbing. "They're all after me because of the leaks." I could see it was serious, so I holstered my smoke gun, and started for the car. "You can explain on the way, sweetheart," I said.

"I can't take it any more," she wept. "They're all complaining that the windows leak, but I can't find any leaks at all. I had a few windows tested, and they outperform the spec! Then we had an infrared scanner in here, and that didn't turn up anything either. They just won't let up! They say water leaks in all winter, and the blinds sway in the breeze."

It turns out "they" were the condominium owners, I discovered as we pulled up in front of the building. I'd seen the type before: a converted pre-war schoolhouse, with old high ceilings that had allowed the developer to squeeze in a couple extra floors between classroom floors. Most rooms had floor-to-ceiling windows.

I grilled the windows, but they wouldn't squeal. The sash was all weather-stripped, and the frame was caulked solid in the masonry opening. I ran a few blower door tests, but they came up empty.

"I think you're being framed, sweetheart", I told her as we went through the building doing smoke tests. "These old masonry walls are up to two feet thick, and the windows are set deep in the wall openings, forming 'wells' set off from the main room volume by the blinds. I'll bet it's supply air from the overhead diffusers swaying these blinds, not air leaking in." That was pretty easy to confirm, by flipping the fan on.

Then we walked into the laundry room. "This jazz about water leaking in has me stumped ... Hey, what's this?" I grabbed a metal box hanging in the dryer exhaust hose. 'The mechanic built these 'filter boxes' for the dryers to keep lint out of the ductwork," she offered.

"That's a great idea," I said as I turned on the dryer, "but he missed one thing. These filter boxes don't close tight. You see how much air comes into the room instead of going down the duct? This air is full of moisture from the wet clothes, so it makes the apartments humid. With the windows set back in the window wells, the glass will be quite cool, and the moisture will condense on the cold glass. A little duct tape over the openings in the filter boxes should wrap up your 'leaks' in no time!"

Gary Elovitz

November 1989

to the tune of
I'VE BEEN WORKING ON THE RAILROAD

I've been working on my chiller
All the livelong day
I've been working on my chiller
To improve efficien-say
Can't ya hear the water flowin'
Keepin' me cool all day?
Can't ya hear compressors whirrin'?
Pumpin' heat away

Pumpin' heat away, pumpin' heat away,
Pumpin' all the heat away, away

Pumpin' heat away, pumpin' heat away,
Pumpin' all the heat away, away

Someone's gotta clean the condenser,
Someone's gotta clean, I kno-o-o-ow
Someone's gotta clean the tower,
Gotta keep head pressure low.

Fee fi fiddle-y-i-o, fee fi fiddle-y-i-o-o-o-o
Fee fi fiddle-y-i-o,
Gotta keep head pressure low.

I've been working on my boiler
All the livelong day
I've been working on my boiler
To improve efficien-say
Can't ya see the flames a'dancin'
Keepin' me warm all day?
Can't ya hear the burner hummin'?
Heat is on the way

Heat is on the way, heat is on the way,
Heat is on the way, the way, the way.

Heat is on the way, heat is on the way,
Heat is on the way, the way.

Someone's gotta clean the chimney
Someone's gotta tune for good C-O-o-oh
Someone's gotta tune the burner
Make combustion losses low.
Fee fi fiddle-y-i-o,
Gotta keep the losses low.

I've been working on air handlers
All the livelong day
I've been working on air handlers
To improve efficien-say
Can't ya feel the breezes blowin'
Keepin' airflow up to snuff?
Can't ya see the people smilin'
When they're cool enough?

Airflow up to snuff, airflow up to snuff
Not too much, there's just enough, enough

Airflow up to snuff, airflow up to snuff
Not too much, there's just enough.

Someone's gotta clean the filters
Someone's gotta set the fan just so-o
Someone's gotta balance the outlets,
We won't overcool no mo'!

Fee fi fiddle-y-i-o, fee fi fiddle-y-i-o-o-o-o
Fee fi fiddle-y-i-o,
We won" o,ercool no mo'

Ken Elovitz

December 1989

A LETTER FROM MURRAY

"Dear Dave,

"Maybe reading all your calendars has done some good, or maybe I found something in my genes, but I must have learned from you.

"I had a group working in a proposal building, actually integrated trailers. One day the big open bay would be very comfortable, then terribly cold, then the next day the same sort of thing. I thought about thermal load, as people came and went; I thought about sun load because it was only a trailer building. The thermostat was sealed without even a thermometer showing. Just grin and bear it.

"We were working under heavy schedule pressure to meet the proposal deadline so I never gave it much attention for about two weeks. Finally, one especially cold afternoon I went over to see if I could pry the thermostat cover open. First I reached over the PC on the movable table (everything in these buildings is temporarily in place) then I shoved it to the side, but I couldn't get the cover open neatly.

"Then I said, 'Dave would have known the minute someone complained.' Sure! The PC and Laserjet were pumping out a lot of watts right under the thermostat. When the PC was on we froze. The AM's were fine until someone used that machine.

"Well, I shoved that PC another six feet down the wall (the room is about 40' x 40'), and we've been comfortable ever since! I knew I had to write and let you know how much good you've done -- and without ever even knowing about the problem!

"Love,
Murray"

We know from your comments that many of you get a kick out of some of these calendars, but we often wonder if anyone ever gets any practical benefit from them.

This letter we just got from my cousin Murray, an electronic engineer at a big west coast aerospace outfit, answers that question.

Do you think I should send them a bill?

Dave Elovitz

January 1990

to the tune of
ANGELS WE HAVE HEARD ON HIGH

Use efficient lighting source;
Fluorescent is good of course.
Switch to variable speed;
Motors use just what they need.
Sa-a-ving energy.
We're conserving power.
Sa-a-ving energy.
We're conserving power.

Insulate your walls and floors;
Keep all that cold air outdoors.
Burne-r's clean by Jim'ney;
No smoke up the chim'ney.
Sa-a-ving energy.
We're conserving fuel.
Sa-a-ving energy.
We're conserving fuel.

Thermostats must regulate,
Or else you'll refrigerate.
Shut off fans and pumps at night;
Start them when the time is right.
Sa-a-ving energy.
We're conserving power.
Sa-a-ving energy.
We're conserving power.

Water heater's turned down low;
Showerheads are all low flow.
Don't use BTUs in vain;
Don't let them go down the drain.
Sa-a-ving energy.
We're conserving fuel.
Sa-a-ving energy.
We're conserving fuel.

Keep your coils and filters clean;
You know what that dirt can mean.
Cooling towers need it, too;
Do as much as you can do.
Sa-a-ving energy.
We're conserving power.
Sa-a-ving energy.
We're conserving power.

Insulate your ducts and pipes;
Be energy saving types.
Fix those steam traps if they leak;
Keep plants running at their peak.
Sa-a-ving energy.
We're conserving fuel.
Sa-a-ving energy.
We're conserving fuel.

Dave and Ken and Gary, too;
We all wish the best for you.
Health and joy and much good cheer;
Have a happy New Year.
Sa-a-ving energy.
We're conserving power.
Sa-a-ving energy.
We're conserving power

Dave, Ken, and Gary Elovitz

February 1990

TOO MUCH OF A GOOD THING

"We have a big problem with our hot water heaters," the maintenance manager complained. "Actually, it's the cold water - at night it comes out hot. I thought we had a cross connection, but that can't be because the water in the cold water lines gets even hotter than the water in the hot water lines!"

A few years ago, the hospital had installed several heat exchangers to capture waste heat from steam condensate. I knew that during periods of very low water flow, the preheated domestic hot water temperature would approach the temperature of the condensate - 180°F or so - which is much to hot to use. But how could hot water get into the cold water system?

"The way this system works," I started to explain, "is that condensate flows through one side of the heat exchanger continuously. When anyone uses hot water, the incoming cold makeup water passes through the other side of the heat exchanger, picking up any available heat. There's also a small storage tank and a little pump that recirculates preheated water back to the heat exchanger to pick up additional heat if the temperature in the tank is less than 120°F."

"Your theory may have been OK when the system was installed," the manager protested. "But that's definitely not what's happening right now. And don't tell me we need another heat exchanger to capture waste heat from the cold water lines." "You don't need another heat exchanger," I continued, "but you do need to do a little piping.

A check valve in the cold water makeup line prevents the pump from pushing recirculated water back into the cold water system. I bet that check valve is leaking, letting hot water flow right up the cold water makeup line. You see, the cold water line rises straight up, providing a perfect route for hot water to rise by gravity through the whole cold water system. That phenomenon was used on purpose in early hot water heating systems, and there's even a name for it - it's called a thermal siphon.

"I think you can solve the problem by fixing the check valve. And while the system is drained down for repairs, you could move the cold water makeup connection to the top of the main line - that will break the thermal siphon so you won't have the problem if the check valve leaks again."

Ken Elovitz

March 1990

DID YOU KNOW ...

Only 20% of the Americans surveyed eat corn on the cob in rows (typewriter style) as opposed to in circles.

When eating a sandwich cookie, 41% of women pull it apart and eat it in pieces; only 16% of men do.

90% of men surveyed but only 72% of women would marry their same spouse again.

76% of us prefer to sleep on our sides, 14% on our stomachs, and 10% on our backs.

81 % put on their shoes and socks in sock-sock shoe-shoe order. 19% put them on sock-shoe sock-shoe.

61% of those surveyed admitted they would wear torn underwear. 48% of the blue-collar workers would not wear torn underwear while 74% of the white-collar workers would.

78% of women wear colored underwear; 76% of men wear only white.

Most men (51%) put their left leg into their trousers first, but most women (65%) put their right leg in first.

60% of those who earn $50,000 or more prefer that the toilet paper unwind over the top of the roll. 73% of those who earn less than $20,000 prefer the opposite.

82% of men and 63% of women squeeze the toothpaste tube from the top.

When asked if they liked the way they look in the nude, 68% of men said yes, and 78% of women said no.

42% of men and 31% of women clean their belly buttons daily; 28% of men and 15% of women never do.

In the 21 to 34 age group, only 4% of men can whistle by inserting their fingers into their mouths while 50% of women admit to having that talent.

More people have visited Disneyland or Disney World (70%) than Washington, DC (60%).

These and dozens of other equally insignificant data, broken down by age and gender, are reported in "The First Really Important Survey of American Habits" by Mel Poretz and Barry Sinrod (Price Stern Sloan, 1989, $4.95)

Ken Elovitz

April 1990

QUIZ

The following clues are for construction industry (mostly HVAC) terms or brand names. For example, "no cost, not off" is "free, on" or "Freon".

Answers next month, but for those who can't wait, send us your guesses, and we'll send the winners a totally useless, inexpensive but not overly ostentatious prize.

Have fun!

1. Helix of water vapor _____
2. Scale for descendants _____
3. Execute electrical power _____
4. Faucet from Prague _____
5. Arid partition _____
6. Top brass with farm machine _____
7. Tire out a groupie _____
8. Haircutting camp stove _____
9. Canard labor _____
10. Arctic 52 _____
11. Menu of party drinks _____
12. Refrigerated skyscraper _____
13. Condensation peninsula _____
14. Deep hole full of sweets _____
15. World hydrant _____
16. Brother of small Oriental cart _____
17. Rooster on a tightrope _____
18. Gaelic trainer _____
19. TV for Laplanders _____
20. Methane motor _____
21. Maharaja's hat _____

Answers:

1. Steam coil
2. Air balance
3. Kill a watt
4. Check valve
5. Drywall
6. General contractor
7. Exhaust fan
8. Barber Colman
9. "Duck" work (ductwork)
10. Cold deck
11. Punch list
12. Cooling tower
13. Dew point
14. Honeywell
15. Globe valve
16. Robertshaw
17. Balancing cock
18. Air Handler
19. Fintube
20. Gas engine
21. Turbine

Ken Elovitz

LAST THING YOU NEED

When a problem arises on a construction project and the parties can't work it out between them, often the next step is to call for outside help from a lawyer. Most people know about the role of experts in medical malpractice cases, but many, including some lawyers, don't know that experts can be essential in any case that hinges on technical issues. After all, lawyers are experts in the law; you can't expect lawyers to be expert in the technical aspects of every case they handle.

As forensic engineers, we're sometimes called on to testify as experts, and we know that a case does not need an expert witness unless all else fails, and the issues are going to be resolved in court or an arbitration hearing. But involving a technical expert from the very beginning can often lead to resolution of the differences without court proceedings.

You sometimes hear about expert witnesses who are branded as "hired guns": Advocates whose testimony conjures up technical justification for whatever position a lawyer wants taken. That's not the role of a forensic engineer. Rather, forensic engineers figure out what happened and why, then translate it into layman's terms. Being an expert witness -testifying in court or at a hearing -- is only a small part of the forensic engineer's role.

Consulting a forensic engineer who investigates and explains the technical truths of the matter before the positions on both sides have hardened makes the logic of technical analysis available as an additional bargaining tool to convince the other side to settle. If settlement cannot be reached, the attorney and his client will understand both sides of the technical issues in the case, not just those that are favorable to their position, before they proceed any further. If an expert doesn't show his client the whole story, chances are they will learn about the other side from the opponent's expert -- in court, where it can be most damaging.

Clear cut cases settle early on, leaving only the close calls or complex ones to go to trial. Of those cases that go to trial, ones that turn on technical matters are especially risky because the judge and jury may not understand the underlying issues. If the parties and their lawyers can understand the technical issues when a problem first arises, most disputes need never escalate to those last stages.

Dave Elovitz

June 1990

THE GREAT AIR HUNT

"Design error," the mechanical contractor announced, "and I can prove it! Total airflow at the fan is way above spec."

"Just speed up the fan 25%," said the engineer. "That'll give 25% more exhaust."

We already sped up the fan," the balancer replied, "but the exhaust at the inlets hardly changed. Besides, that fan's maxed out now. Must be duct leaks."

"No way. Every seam in that ductwork was sealed twice, inside and out," the sheet metal man defended. But the fact remained the exhaust flow at each inlet was well below design.

"Look," I said, "the measurements at the fan show high airflow and low static, so there must be a big opening somewhere. We know the traverse at each branch matches the sum of the inlets. But the traverse at the fan is much higher than the sum of the branches. The opening has to be inside shat shaft."

"Oh, no you don't," the GC jumped in. "I saw those ducts before we closed in the shaft, and there were no holes. That's a masonry shaft, with ceramic tile and drywall ceilings all around it. You're not chopping that all up."

"Yeah, and it'll be cheaper to just put in the next larger fan," the engineer and mechanical contractor chimed in.

"Putting in a bigger fan won't solve the problem," I explained. "The airflow into each inlet depends on the negative pressure at that point. With a big leak near the fan, the pressure available to draw air into the inlets is no more than the pressure drop through the leak. To overcome the effect of the leak, you'd need ten or twelve times as much air at the fan as now. Tell you what. If I don't find a big leak in the shaft, I'll pay for opening up and repairing the shaft myself. If I find the leak, you do the work for free."

"You're on, wiseguy. Where do I send the bill?"

Three days later, I wriggled through a small opening into the dim void, squeezing past pipes and other ducts. My heart sank as my flashlight beam showed ample duct sealant -- on three sides out of four -- but the unsealed top seams didn't account for the size of the leak I know had to be there. Then, squeeze in a little further to where the duct heads off toward the shaft. There! A big dent in the top of the duct looked suspiciously like it would match a telephone man's derriere. The adjacent seam broke wide open when the duct got crushed, and there was our leak. A little sheet metal repair, and we got the design airflows at the fan and at the inlets.

Ken Elovitz

July 1990

EYE TO EYE

My friend Paul Pritzker, PE, tells this story about his late father, Morris. Morris Pritzker was an electrical contractor and a very good one. His prices were quite competitive, so his fledgling business grew even during the depression.

Like the owners of many growing businesses, Morris found that he needed some financing to support his company's continuing growth. He approached the manager of his local branch bank and eventually found himself in the downtown office, seated across the table from the bank's Loan Committee.

"Mr. Pritzker," the committee chairman explained, "these are difficult times. We have many more applications for loans than we have funds available. The records of your account with us, and your business records do show that you have a healthy growing business. Our branch manager, Mr. Bradford, tells us that you have an excellent reputation for honesty and reliability. Still, the same could be said for many other loan applicants."

The committee members eyed Morris stonily for a silent moment or two while he tried to figure out what was going on, then another member spoke up: "I'll tell you what, Mr. Pritzker. I think your application is as good a risk as most of these in this pile, but it is probably no better. I'll give you a sporting chance. One of the five committee members has a glass eye. If you can tell us which is the glass eye, the loan is yours."

Morris looked slowly from one face to the next, then smiled and said, "Start counting out the money. It's the right eye of the second gentleman on the right."

The committee members were amazed, but a deal is a deal, and as the Chairman rose to end the meeting he promised, "The money will be in your account tomorrow morning. Thank you for bringing Mr. Pritzker in to meet us, Mr. Bradford." Morris and Bradford shook hands with the committee members and left the room.

As they were standing waiting for the elevator, Bradford said in wonderment, "What a lucky guess! But I am certainly glad you were able to get your loan."

"That was no guess. I just looked at each eye in turn and that was the only one that showed any hint of compassion!"

Dave Elovitz

August 1990

HAVE AN ICE DAY

As we sit and "quaff a few cold ones" on a sweltering August afternoon, most of us probably don't give a thought to how our drinks get cold.

Back in Scarlet O'Hara's day, people probably didn't give it much thought, either, but there's a decided difference between modern day and the Civil War era. We can have cold drinks any time we want because we have mechanical refrigeration. Scarlet needed ice cut from frozen northern ponds the previous winter to cool her mint juleps.

With the current chatter about banning halocarbon based refrigerants, some of us might take comfort in knowing that laws regulating ice ponds are still on the books, just waiting for a once thriving business to revive.

While ice can be cut from any existing pond that freezes over, Massachusetts law allows people to dam streams to create ice ponds. None of the regulatory thicket surrounding dams for hydroelectric power applies to dams for creating ice ponds. Any non-navigable stream may be dammed to make an ice pond, as long as the dam does not block water flow before November 1st or after March 1st of each year.

Since 1674, certain large bodies of water ("great ponds") have been dedicated to public use. Persons cutting ice on great ponds must not interfere with other members of the public like skaters and ice fishermen who have an equal right to use the great pond. However, the right to cut ice does take priority over some uses of lakes and ponds. Even a town using a great pond for its public water supply cannot enjoin ice cutting unless the cutting would interfere with the quantity or potable quality of the water. And the quality of ice cut from ponds was of just as much concern to our early legislators as the quality of drinking water. For that reason, the law forbade driving any animal onto an ice pond except those used directly for cutting or hauling ice.

The ice taken from the pond is protected personal property. A fine of up to $100 may be imposed for damaging harvestable ice or for cutting ice without permission from the owner of the pond.

We modernists think many old laws on our books are arcane, and perhaps they are. But the next time you sit back and enjoy a cold drink to help you cool off on a hot day, be reassured knowing that while today's lawmakers debate whether to ban halocarbon refrigerants, the old laws protecting ice harvesting are still in place.

Ken Elovitz

September 1990

POSITION IS EVERYTHING IN LIFE

They just couldn't get the meeting rooms comfortable. People complained constantly, and the hotel was beginning to lose business.

"What I can't understand," puzzled the chief engineer, "is that the temperatures don't seem that bad. We get a complaint that a room is too cold, so the first thing I do is check space temperature on the computer. Usually it reads about 74°F, within a degree of setpoint. So I send an engineer out with a thermometer anyway. Every time, the engineer confirms the computer's reading to within a degree. Of course, I raise the setpoint, too, but I can't understand why everyone is always so cold!"

"That does seem strange," I agreed. "Let's go take a look."

We set up several thermometers all over the room to check for uneven temperatures, and they all read the same. "I have my little smoke tracer bottle to check air patterns," I said. "Let's see if the diffusers are making cold drafts. If you have Engineering lower the setpoint to 65°F, we'll be able to work for a while before the compressor cycles off. We need the compressor to stay on so changes in air temperature won't affect the flow patterns."

Done. After tracing air patterns for a while, looking at the return path above the ceiling, and general mucking around, I began to feel cold.

I checked my thermometer, and it read 62°F.

"That's funny," I mused, "I would expect the compressor to cycle off by now. It's 62°F in here!"

We phoned down to Engineering: The computer said the room was 66°F, still calling for cooling.

"That's strange. This is the first time the computer doesn't agree with a thermometer." I walked over to the temperature sensor. "When your engineer puts his thermometer in the room, where does he put it?"

"He hangs it right from the sensor. I told him to make sure we're comparing two sensors in the same location."

"Right, and that's the answer: See the dimmer switches right below the sensor? The heat these dimmers give off throws off the room sensor, making it think the room is warmer than it really is. When your engineer puts his thermometer below the sensor, he gets the same wrong reading as the sensor. Move the sensor over a foot and your computer will read the right temperature.

Now as for the cold drafts, that's a story for a future calendar!"

Gary Elovitz

October 1990

YOU CAN'T TAKE IT WITH YOU

A wealthy real estate developer called his three closest personal advisers -- his accountant, his architect, and his contractor -- to his hospital bedside. Knowing he was near death, he handed each of them an envelope containing $40,000 in cash. The dying man instructed his friends to throw the envelopes into his grave after the funeral so he could "take it with him".

After the funeral service, the three men rode to the cemetery together in a limousine. As the casket was being lowered into the ground, each of the dead man's friends threw an envelope into the grave, just as the deceased had instructed.

During the ride home, the accountant confessed to the other two, "I only threw in $30,000 because the deceased could use some charitable deductions on his tax return. I knew no one would ever miss the $10,000."

The accountant's candor prompted the architect to confide to the other two, "Well, I didn't throw in the whole $40,000 either. I kept $20,000 to design a new wing for the hospital. That money would be better used by the living than the dead."

Those two confessions prompted the contractor to speak up, too. "I am really ashamed at the two of you," he lectured. "You are both professionals, and you were fiduciaries. You owed our friend the duty to keep your promise to him. Unlike you, my conscience is absolutely, clear. I threw in a check for the full amount!"

NOW WE ARE THREE

Three Registered Professional Engineers, that is. Gary has just completed the registration process which consists of an 8-hour "Fundamentals" exam, four years of qualified practical experience, and an 8-hour practical exam. Those of you who have worked with Gary already know he is well qualified to have the title "PE" behind his name and will be as glad as we are that the Commonwealth of Massachusetts has officially recognized that fact. Congratulations, Gary!

Ken Elovitz

November 1990

DON'T BELIEVE EVERYTHING YOU HEAR

"And another thing," the hotel's chief engineer complained; "That lounge at the top of the Atrium is unbearable! The heat rises up through the atrium and it all ends up in the lounge. The rooftop unit up there runs so long without cycling off that the refrigeration circuit occasionally cuts out on some safety, and my guy has to go up on the roof to reset it! That unit is way undersized."

I calculated the cooling load for the atrium lounge. Surprisingly, according to the calculations, the unit was not undersized, but more than doubly oversized. Plenty of capacity to handle heat rising through the atrium.

Up in the lounge it was definitely warm, even though it was not a hot day. As I began to measure the airflow from the supply outlets, I noticed that the supply air was up in the mid 80's. "Maybe the thermostat is wired wrong," I thought, "and the heating coil is on." To check this theory, I switched the thermostat to heat and turned it up. Then the supply air really got hot; I guess it wasn't wired wrong. So I flipped the thermostat back to cooling, and the supply air started to get cold. Real cold, down to the low 40's. I knew immediately what was wrong.

"The problem in the lounge," I explained to the chief, "is a restriction to the airflow. That 7½ ton unit is only moving about a third of the air it should. When the compressor tries to do 7½ tons of cooling on 2½ tons worth of air, it makes the air very cold. With air that cold, the coil surface must be in the 20's, which will freeze any moisture in the air. The frost builds up, clogs the coil, and the airflow drops even further, until finally it is so cold the compressor shuts down on the low pressure safety. With the compressor out, the unit simply recirculates air, and the frost eventually melts. The supply air then gets warm because of heat from the fan motor, and also because of outside air mixed with the return air. Shutting the unit off at the thermostat resets the safety, and when it is turned on again, the cycle repeats. Find out why the airflow is so low, and your problem is solved."

The next morning the chief gave me a call. "I sent a guy out to the unit; he reported, "and he found something interesting: The return air damper in the unit was stuck closed! As soon as we got that damper open, it all worked beautifully.

Gary Elovitz

December 1990

SONS OF MURPHY

Several earlier calendars cataloged some of those truisms known as Murphy's Laws. Here are some new additions to our collection:

S&L Law of Scandals
When wrongdoing is exposed, the real scandal is what was legal.

Hellman's Rule for Mayonnaise and Management. (From their label)
Keep cool, but do not freeze.

Ethan and Lora's Enigma
Marriage is the only union that can't be organized: Both sides think they are management.

Mycofsky's Maxim
Almost anything is easier to get into than out of.

Lashman's Lament
You can't make a fact out of an opinion by raising your voice.

Koningisor's Komplaint
When you're hot, you're hot. And when you're not, everybody is watching.

Dena's Dictum
The two most common elements in the universe are hydrogen and stupidity.

Perry's Parameter
Greasy hands make your nose itch.

Griffin's Grievance
There is always more going on than you think. And it is always worse than you imagine.

Litman's Law
When a person with experience meets a person with money, the person with experience will get the money and the person with the money will get the experience.

Pandora's Principle
Never open a box you didn't close.

Sweeney's Syllogism
You can fool some of the people all of the time, and all of the people some of the time, but if you work it right, that's all you need to make a comfortable living.

Ryzhkov's Realization
The difference between communism and capitalism is this: In capitalism, man exploits man; in communism it's the other way around.

Anderson's Law of 2%
When the government is involved, it takes twice as long, costs twice as much, and only half as much gets done.

Dave Elovitz

January 1991

SEASON'S GREETINGS

One of us is David,
One of us is Ken,
One of us is Gary;
The books are kept by Fran.

We're not always hard at work,
We like to have some fun.
To bring our friends a laugh or smile,
We'll even make a pun.

We're consulting engineers,
Computers at the ready;
When you call on us for help,
We try hard -- veddy, veddy.

We'll calculate your cooling load,
Account for BTU's;
Evaluate your lighting needs,
Efficient lamps to choose.

When you contemplate your budget,
And can't afford the fuel,
We'll help you use it wisely
When you heat and cool.

And if you're having trouble
Controlling your controls;
We think you'll find none better
To help you meet your goals.

Sometimes we do design reviews
On projects not yet bid.
To try and spot the gremlins
That are in the drawings hid.

On other jobs you call us in
To tell you what you need;
To put down in a Scope of Work
What contractors should heed.

And other times, alas, alack,
You call on us too late.
The job's all in; it doesn't work;
So we investigate.

If all else fails and you're in court
We'll take the witness stand.
Explain it all in simple terms
That lawyers understand.

But this is not the time of year
To speak of what we do.
This is the time to wish the best
To everyone of you.

Good cheer, our friends, a happy
    year;
Prosperity and more;
A year of peace, a year of health;
From us to you and yours!

Dave, Ken, and Gary Elovitz

February 1991

GOOD CONNECTIONS

"The most important thing about the new boiler is hot water," the client kept repeating. "We don't care about heat -- what we want is hot showers." With that admonition, I sized the system to heat several times as much water as sizing guides recommended and even added 200 gallons of storage to ride through the peaks. The hot water circulating pumps were sized to recharge the system fully in less than 10 minutes.

"Morning peak is fine, but we have no hot water in the middle of the day," the client complained. "How can a system that supports 5 or 6 showers at once conk out for a single kitchen sink?"

At the site, my own tests showed that the system worked fine at full load. The mixing valve which regulated the temperature going to the units held its setpoint beautifully. But then, at very low flow, the water temperature fell to only 80°F even though the storage tanks were at 160°F.

"You can't expect the valve to control at low flows," was the manufacturer's first response. "We're not asking it to control," I retorted, "We just want it to stop adding cold water when the send out is below setpoint. We have 80°F water right at the mixing valve with plenty of hot water available from storage." "I can't tell you why it does, but I'll get our factory expert up here."

The expert arrived and repeated the tests with contractor Roger Litman of North Shore Fuel. "Nothing wrong with the valve," he said, "but we did find one thing. The problem occurs only when this transfer pump is on, but I can't figure why that is."

'That's it!" Roger announced triumphantly. "See how the makeup water is connected at the pump discharge? That makes the pump discharge the same pressure as the cold water, which makes every other point in the hot water system a little lower. With low hot water draw, there is hardly any pressure drop through either port of the mixing valve. The small pressure difference between the hot and cold parts of the system gives the cold water a slight advantage. Too much cold water flows because the cold water side of the mixing valve can't close quite all the way. If I just move the makeup water connection to the pump inlet, the pump will add pressure to the hot water side, not subtract it, and hot water will always flow as it should."

Ken Elovitz

March 1991

to the tune of
OVER THE RAINBOW

Somewhere, in my computer,
My files lurk.
Locked up where I can't get them,
Ho-urs and hours of work.

Somewhere, in my computer,
Spreadsheets hide,
Filled with Lotus and Quattro,
Just when my hard disk died.

> When I boot the
> Directory,
> And files names aren't
> Where they should be.
>   I panic!

> Someday I'll learn
> To work this thing.
> 'Til then I guess
> I'll have to wing
>   It, 'till I'm manic!

Someday, my new computer,
Will work right.
Some say I'll never learn how,
E' en if I spend all night.

Somewhere, in my computer,
Your report,
Is there in my computer.
Or did I press abort?

Somehow this darn computer,
Hid that file.
I know it's there in mem'ry
If I just knew which dial.

> I typed it in
> And then pressed save.
> I can't make this
> Machine behave.

>   I'm frantic!
> The file has gone,
> I know not where.
> I typed it in,
> It must be there.
>   Such antics!

Somehow will my computer,
Spit it out?
Send it to the printer?
Ending my hours of doubt.

OUR NEW LOOK:

After 15 years, we decided it was time for a lighter, more modern look. You may have noticed that our calendars have a new look this year and we have new stationery to match. Our thanks to Peggy Brier of Studio Twenty Seven, who designed a new logo that is still legible after fax and photocopy.

Dave Elovitz

April 1991

to the tune of
IF I ONLY HAD A BRAIN

If the rent checks they're retainin'
And tenants are complainin'
Cold drafts are on their feet
While their heads are all sweatin'
And death threats you keep gettin'
'Cause you can't control the heat.

If you face a thorny riddle,
For any in·di-vid-le
Or for the place complete
You must fix up what you bought to
Make the thing run like it ought to
One that does control the heat.

Oh, we could tell you why the boiler makes that roar,
We could show you things don't want to ignore
You'll make it right, not like before.
You would then become a hero.
To tenants you'd be dear. Oh,
How they would think you neat.
And perhaps you'd deserve it,
And be even worthy erv it,
'Cause you now controlled the heat.

Life is said, believe me, Ryan,
When fuel costs go a flyin'
The boss, he gets so cruel.
And the Owner starts a callin'
Cause the profits are a fallin'
You just gotta save some fuel.

Well I don't think we'd be lyin'
To say, "Us you can rely on."
We'll teach you all the rules .
We can help with your mission
Make you system more efficient,
When you gotta save some fuel.

Change incandescent lamps, they're just like dinosaurs
Put in fluorescent lamps like ne'er before.
And H. I. D. to save some more!

When you clean out your condenser
There's ample recompense, sir:
Less pow'r to keep it cool.
Outside air economizer
Will make you seem much wiser,
'Cause it lets you save some fuel.

When you're coolin' system's whacky,
'Cause the installation's tacky,
And you gotta find a cure.
That is not time to gamble,
Not the time to wildly scramble,
It's a time you must be sure.

We must calculate the heat gain,
So you don't just have more pain,
After the cure de jour.
We must check the ducts and piping,
Find a way to stop the griping,
Make things better than they were.

So first we figure out just how the thing should run,
Then report to you on how to get it done
We list the tasks, done one by one.

When you finish up re-hab'in'
You'll have a comfy cabin,
Your efforts well repaid.
All the man'gers now respect you,
You have made a magic rescue,
Lemon now is lemonade!

Dave Elovitz

May 1991

PART TWO

"Moving those thermostats will clear up our control problem in these meeting rooms, but what about the cold drafts?" the hotel's chief said, reminding me of the promise I had made at the end of September's calendar. "The drafts are still so bad that we can't keep the rooms rented!"

My load calculations showed that the heat pumps in the meeting rooms were twice as big as they needed to be. (Maybe they shouldn't have let the equipment vendor size the equipment.) The oversized units made the airflow to each room very high: Air left the ceiling diffusers at such a high velocity that it rolled across the ceiling, down the walls, and across the floor (and occupants' ankles) before becoming fully mixed with the room air.

Slowing down the fan wouldn't help much: That would reduce the air velocity, but also the supply air temperature - the drafts would be slower, but also colder. Besides, even the minimum airflow necessary to keep the units from icing up was simply too much air to deliver to those small rooms without causing drafts. We pored over diffuser catalogs to try to find an outlet that would help, but no luck. In order to eliminate drafts, we would either have to make the entire ceiling the outlet, like a clean room, or reduce the airflow.

Could we reduce the airflow to the room? Bypassing some supply air directly to the return would only make the supply air colder, and that would make any drafts more uncomfortable. Adding hot gas bypass around the compressor would work, but it would double the energy cost. Replacing the entire unit with a smaller one would certainly work, but that seemed like a waste of a good unit - and besides, the existing units were so tightly shoehorned into the equipment closets that it would be very expensive to get them out.

We finally engineered an unusual approach: keep the existing units, but replace the compressors with smaller compressors, and slow down the fan. The smaller compressors would reduce the cooling capacity, so the coil would not ice with low airflow. The evaporator coil and the condenser would be oversized, but that is not a problem. In fact, the oversized evaporator and condenser would increase energy efficiency, and slowing down the fan would make the units very quiet.

Changing the compressor in one room as a demonstration worked perfectly. No drafts, even temperatures, very quiet operation, and happy occupants. With new compressors in the rest of the meeting room units, the hotel's meeting rooms have become much more popular.

Gary Elovitz

June 1991

BACK IN CONTROL

"Help! A tenant in my building hired a fancy consultant because their conference room was too hot. Now he's telling them the whole system is no good just because his temperature recorder showed that the supply air temperature was too high." The consultant's temperature charts showed that the supply air temperature usually ran about 70F except for occasional brief dips to the design temperature of 55F. Clearly, something was out of control.

"First we need to trace out the wiring and make sure none of this tangled spaghetti was inadvertently connected to a wrong terminal," I told the building engineer. "Then we need to figure out what all these relays are supposed to do."

After several hours of tracing wires and studying diagrams, I began to understand the system. "This factory mounted controller in the HVAC unit is supposed to provide the constant 55F discharge air temperature you need. This return air sensor gives you a morning warmup cycle by locking out the cooling during startup until the return air temperature reaches 70F. So far, so good. The problem is that nothing prevents the system from jumping back into the warmup mode if the return air temperature falls. These temperature charts tell me the unit spends most of the day in the morning warmup cycle."

"It has to do that," the building engineer protested. "Otherwise the offices get too cold." "Hold on," I pointed out, "If you're getting 70F return air, you must be overcooling much of the space. The way to avoid overcooling on a VAV system is to make sure the cooling to a zone shuts off completely when it's not needed. Raising the supply air temperature, either manually or automatically, because some room is too cold causes trouble in places like the conference room that need cooling. Besides, average return air temperature could be 70F while half the space is shivering at 60F and the conference room is sweltering at 80F. A simple wiring change will let the factory installed controller do its job and keep the unit from reverting to warmup once the normal cooling mode starts. But most important, you need to make sure the primary air minimum is set to zero on all zones that don't have reheat coils. That will stop the overcooling, and you'll be back in control."

Ken Elovitz

July 1991

DAFFYNITIONS

An architect friend sent us the following explanations of terms commonly used in a construction project:

Contractor:
A gambler who never gets to shuffle, cut or deal

Bid Opening:
A poker game in which the losing hand Contractor

Bid:
A wild guess carried out to two decimal places

Low Bidder:
A contractor who is wondering what he left out wins

Bid Documents:
Sworn to by the designer and sworn at by the contractor

Designer's Estimate:
The cost of construction in heaven

Project Manager:
The conductor of an orchestra in which every musician is in a different union

Critical Path Method:
A management technique for losing your shirt under perfect control

OSHA:
A protective coating made by half-baking a mixture of fine print, split hairs, red tape and baloney .... usually applied at random with a shotgun

Strike:
An effort to increase egg production by strangling the chicken

Delayed Payment:
A tourniquet applied at the pockets

Completion Date:
The point when liquidated damages begin

Liquidated Damages:
The penalty for failing to achieve the impossible

Auditor:
People who go in after the war is lost and bayonet the wounded

Lawyer:
People who go in after the auditors and strip the body

Shamelessly plagiarized by
Dave Elovitz

August 1991

NEW ELEMENT

We just received word that a new element, the heaviest ever discovered, was recently identified by the National Institute of Science. The element, tentatively named Governmentium, has no protons or electrons, and thus has an atomic number of 0. However, it does have 1 neutron, 125 assistant neutrons, 75 deputy neutrons, and 111 deputy chief underneutrons, for an atomic weight of 312. These 312 particles are held together in a nucleus by a force that involves the continuous exchange of meson-like particles called morons.

Since it has no electrons, Governmentium is inert. However, it can be detected chemically as it impedes every reaction with which it comes in contact. According to the discoverers, a minute amount of Governmentium caused one reaction to take over four days to complete when it would normally occur in less than one second. Before actually entering the reaction, Governmentium appeared to act exactly like a catalyst, but as soon as it became involved, its nature changed dramatically.

Governmentium has a normal half life of approximately two years, at which time it does not decay but, instead, forms fact finding commissions which result in assistant neutrons, deputy neutrons, and deputy chief underneutrons exchanging places. Some studies have already shown that atomic weight actually increases as a result.

Research at other laboratories indicates that Governmentium occurs naturally in the atmosphere. Although it tends to concentrate in government offices, it can also be found in large corporations, universities, and other administrative locations.

Scientists point out that Governmentium is known to be toxic at any level of concentration and can easily destroy any productive reactions where it is allowed to accumulate.

Attempts are being made to determine how Governmentium can be controlled to prevent irreversible damage, but results to date are not promising.

Adapted from a recent issue of the University of Calgary newsletter.

Gary Elovitz

September 1991

HOW DO YOU SPELL RELIEF

"I need your help in a hurry! I've got a tenant who won't pay his rent because he says the heat doesn't work. The whole problem is all his exhaust fans that put the building under a negative pressure. Go up there and see what the story is so I can get this guy paying rent again.

At the site, the building was under a negative pressure alright. The superintendent took me up to the roof - Nearly a dozen exhaust fans, about half of them running, including the "emergency" exhaust that the tenant said would run only to clear the lab after an accidental spill.

"Sure looks like excessive exhaust," I posited. "Especially with that 5000 CFM emergency exhaust on all the time. Just to be sure, though, let's total up the exhausts that are on right now and compare that to the minimum outside air on the rooftop units."

Standing in the shelter of one of the rooftop units, we added up the exhaust and outside air CFMs. "Looks like a couple thousand CFM of outside air available to pressurize the building even with the emergency exhaust on," I shouted. "What in the world is that roar behind us?"

"Oh, that's just a fan behind this damper. It runs all the time. That's its normal noise."

"If that fan runs all the time, then that explains the excess exhaust! Even with the damper closed, you're going to get some leakage. The pressure behind that damper is pretty high when the fan is dead headed.

These units come from the factory with the fan set to start when the outside air damper opens to a certain point. In a building like this that needs extra outside air, the start point has to be field adjusted so the fan doesn't come on too soon. All you have to do is adjust an end switch, and you'll solve your negative pressure problem. You'll save on fan energy at the same time, too."

"Easy for you to say," the superintendent snorted as he opened an access door. "Take a look in here and you'll see that you'd practically have to disassemble the whole economizer section to get to that switch. I don't have time for that shenanigan."

"I see what you mean, and I can't argue with you. But you don't have to give up, either. Since there are two rooftop units, you could disable the fan in one of them. With all the process exhaust in this building, one relief fan should have more than enough capacity. You'll not only solve the negative pressure problem and save some fan energy, but next time I come up here we won't have to yell over the roar of dead headed fans."

Ken Elovitz

October 1991

APOLOGIES TO GILBERT & SULLIVAN

When I was a lad I served a term
As Plant Engineer in an airplane firm
I cleaned the windows and I swept the floor
And I polished up the handle on the big front door

> He polished up the handle
> on the big front door

I polished up the handle so carefully
That now I am consulting on H-V-A-C.
As time went by, I joined 3-C
To solve space problems in Facilities.
I built new buildings, leased large and small,
And made the cooling systems OK over all.

> He made the cooling systems OK
> over all.

I made the cooling systems run pleasantly
So now I am consulting on H-V-A-C.

When Honey-well bought out 3-C,
It soon became not the place for me.
I worked for BALCO as sales engineer.
And wrote up lots of orders in a hand so clear.

> He wrote up all the orders
> in a hand so clear.

I wrote so many orders so handily
That now' am consulting on H-V-A-C.

I sold so many systems I was commandeered
To help get out the drawings as Chief Engineer
As Chief Engineer, I inspired such esprit,
I soon took on construction and became VP

> He soon took on construction
> And became VP

Design and construction, I was VP
And now I am consulting on H-V-A-C
At MASCO on the project for their big new plant
I was selected as the commandant.
I figured out the cycles, so efficiently,
We could serve all the members with economy

> He could serve all the members
> With economy

I served the MASCO members so efficiently
That now I am consulting on H-V-A-C.

When I left MASCO for EEl,
The pattern of my workday did transmogrify.
O! clients call me early,
and clients call me late
But almost no one calls me not in frantic haste

> No, no one ever calls him
> Not in frantic haste!

There's hardly time for lunches, can't watch TV
'Cause now I am consulting on H-V-A-C

Dave Elovitz

November 1991

GREMLINOLOGY

"A month ago," the building manager explained, "the bank installed new backup emergency generators, transfer switches, and big new UPS systems for its data center. Ever since, we've been chasing a gremlin that trips a main breaker when we switch from one primary feeder to the other. Not switching to emergency generators; Not switching to the other end of the double ended substation; Only when the utility's transfer switch swaps over."

"We've checked everything," explained the chief engineer. "Swapped the circuit breaker's 'brains' with another breaker, and even swapped the whole 600 pound breaker frame with one from another spot. We have to shut down the whole building to run these tests. Even though we only shut down 10 PM Saturday to 8 AM Sunday, the tenants are complaining about the disruption. We can't keep shutting down looking for this thing."

A few Sundays later I was standing in an electrical room at 3 AM with a cast of thousands, studying a recorder trace. "See those harmonics when the UPS units are first energized?" the breaker manufacturer's engineer pointed. "Even though the current is well below the breaker rating, all the harmonics must look like thousands of amps to the breaker's brains. I'd like to talk with the design engineers at the factory, but I am sure that changing to our new digital programmer will cure the problem."

"That sounds plausible," I agreed, "but how can we assure the bank and the building that this is 'for sure' and they are not going to face more shutdowns even with that change? As long as we're here, let's disconnect the sequencing controls and test again. That way the bank's engineers can at least be sure that it is definitely something internal to the circuit breaker." "OK, I guess we can do that. C'mon, let's get it over with so we can go home and go to bed.

After twelve tests without tripping the breaker, I was sure that the gremlin had to be in the sequencing controls. "Hook that control circuit back up and run the test again." Boing! The breaker tripped on the first try. Definitely the sequencing controls, but where? "Try that again," one of the breaker service engineers said. "I think I saw something funny on the X1 relay." Sure enough, one contact on a multipole relay was hanging up and opening just a fraction of a second after it should. When it was closed, that contact told the breaker to open.

After disassembling, cleaning and lubricating the X1 relay, no more tripping, and our exotic harmonic inrush high tech gremlin wasn't a breaker trip at all: The breaker had opened because a sticky $40 relay told it to!

Dave Elovitz

December 1991

AND TO THINK THAT I SAW IT ON MULBERRY STREET

"I've terrible problems, the client did say.
My home heating system is far from OK.

"We get lots of odors that smell like burnt gas.
The guy who installed it must be a real ass.

We think that the odors are harmful to health,
And the cost of repair could exhaust all our .wealth."

So down to the basement we went, him and me
And two separate systems is what I did see.

But as I examined and made my inspections
I saw they had common return air connections.

Now common return air is not always trouble
As long as the ductwork is sized at least double.

So I ran some tests to find out what went on
First one fan then two, and I knew what was wrong.

When Fan One was running while Fan Two was off
I saw what was making those furnaces cough.

You might not believe this, but there was no doubt
That the air would go in where it should have come out!

Now Fan Two was off when I espied that fan
And found out the answer, as backwards it ran!

To prove out my point I removed a few screws,
Took off a panel and proclaimed the news.

With this panel open, the fan can draw air
From close by right here or from way over there.

But when this here panel is mounted in place,
The air cannot flow down your too narrow chase.

Instead it must make a long, tortuous run
Through ducts from Fan Two and on into Fan One.

So here is the message of what I did learn
That supply is important, but so is return.

It's often the case when a fan will not work
That the problem is not just a miserable quirk

But rather it's likely a duct that's too small.
Check out the supply, but don't think that's all

For return is important and sometimes can be
The root of the problem like when this guy called me.

Ken Elovitz

January 1992

FAREWELL TO '91

Looking back on Ninety-One
We find the year was not all fun.
Times were hard on many friends
For some fine firms, it was the end.

All around, things weren't too bright,
Feds seized banks in dark of night.
Construction slowed down to a crawl.
For many friends, no work at all.

But many prospered, thrived and grew.
We hope that group included you.
That you're OK, you made the grade;
From lemons you made lemonade.

Elsewhere, things were getting hot.
Send in troops? Or just boycott?
Desert Shield, then Desert Storm.
Could UN make Saddam conform?

Crowd the TV, watch our missile
Track, then down the chimney whistle.
Patriot destroying Scud,
Like TV games: no mess, no blood.

Stormin' Norman, Kelley. Powell
Brief reporters, cheek by jowl.
Iraq's force crushed, Kuwait set free.
As for Saddam, we've yet to see.

Amazing year, you must agree,
Cold war's o'er; East Europe free.
Democracy where all was Red.
'Tho ethnic strife has raised its head.

And Nicaragua, Salvador
No longer seem immersed in war.
Will all swords now be turned to plows?
Doves ev'rywhere, with olive boughs?

A visitor on Aug nineteen
Knocked power out, made quite a scene.
Bob brought winds and Bob brought rains.
That's what you get in hurricanes.

Red Sox started out in first,
By 4th July, they played their worst.
In Fall they did come hustling back
To fold again - they lost the knack!

So now we turn to Ninety Two;
We pray it will bring much to you:
Fun, and pride, and joy, and health,
Maybe e'en a little wealth.

We bounced around in Ninety-One.
But now a new year we've begun.
From Ken and Gary, Dave and Fran,
A Great New Year to your whole clan.

Dave, Ken, and Gary Elovitz

Page 170

February 1992

WHEN I WAS A BOY ...

About this time every winter, you are sure to hear someone say "Whatever happened to real winters, like we used to have?" Not only do we hear talk about how much more wintry it used to be, but we also read about "global warming" and speculation about the polar ice caps melting. So it was kind of fun when I came across some long-term average Boston weather records published in 1925 and compared them to recent ones. I thought you might find them interesting, too:

Monthly Mean Temperature

Month	1925 20 yr avg	1990 10 yr avg
January	27.9	29.6
February	28.8	30.7
March	35.6	38.4
April	46.4	48.7
May	57.1	58.5
June	66.5	68.0
July	71.7	73.5
August	69.9	71.9
September	63.2	64.6
October	53.6	54.8
November	42.0	45.2
December	32.5	33.7
Annual	49.6	51.5

Normal Daily Maximum Temp

June	75.7	76.6
July	80.0	81.8
August	77.8	79.8

Normal Daily Minimum Temp

December	25.1	27.1
January	20	22.8
February	20.8	23.7

So over the last 65 years, things have indeed warmed up! About a couple of degrees year round. And, as for snowfall, I also seem to recall racing down the Lookout hill on my new birthday toboggan even before Thanksgiving, but I have to confess memory can be a very tricky thing:

Monthly Mean Snowfall

	1925 20 yr avg	1990 10 yr avg
November	1.7"	1.4"
December	8.4"	7.5"
January	11.2"	12.2"
February	12.5"	11.4"
March	7.9"	7.4"
April	2.5"	0.9"
Annual	44.2"	40.8"

OK, so the snow may not have been all the way up to my waist when I walked to grammar school every day, but winters really were snowier back then. Besides, my waist was not only smaller in those days, it was closer to the ground.

Dave Elovitz

March 1992

KEEP ON PUMPIN'

"We want to protect our building from fire
But one little problem has us in a mire.

We're putting in sprinklers before it's the law,
And a problem arose when the fire chief saw

That our fire pump now would no longer provide
Enough water flow to the opposite side.

A new fire pump's an expense we can bear,
But emergency power to run it is where

We run into trouble and can't seem to find
A method to help us get out of this bind.

A new generator is one good solution
But noise and exhaust are unwanted pollution.

A big diesel engine to make the pump go
Presents the same problems, and our boss would say, "NO!"

On top of all that we just haven't the space
To fit the machines we would need for this place.

I looked at my codes and the answer was clear:
Electrical drive was the way we should steer.

And as for emergency power' knew
This building already had services two.

A different substation fed each of the lines
So that made our pow'r the reliable kind.

A switch is the only new thing we would need
To let either service each fire pump feed.

The switch would just transfer should power go dead
And feed from the opposite service instead.

In fact, I did say, this arrangement is best
For some generators will fail in a test.

But this way you'll have a dependable source
To keep the pumps turning with twenty-five horse

And the cost, you will find, of this creative scheme
Is less than depending on diesel or steam.

Ken Elovitz

April 1992

GUIDE TO SAFE FAX

I am indebted to Judy Lovins in Phoenix for sending me these tips:

It generally helps to know the person you are faxing, especially if you want meaningful fax, but there are many people who safely fax complete strangers every day.

You will not go blind if you fax something to yourself, but you should not allow faxing yourself to interfere with faxing others.

Even if your friends say you fax too much, there is nothing abnormal about faxing many times per day. Just be careful to allow enough time for other activities such as eating or sleeping. Your friends are probably just jealous.

There is nothing wrong with going to a place where you pay to fax when you are away. Many people find themselves in situations with no other outlet for their fax drives and must pay a professional when their need to fax becomes too great.

Unless you are really sure of the person you are faxing, a cover should always be used to ensure safe fax.

Transmissions can get mixed up if you have a personal and a business fax. Being bi-faxual can be confusing, but as long as you use a cover with each one, you won't transmit anything you're not supposed to.

When faxing a co-worker, remember that, while faxing is okay between consenting adults, you must be certain you are not engaging in faxual harassment or else you could lose your job, end up in jail, or even become a Supreme Court judge.

There is no minimum age for faxing, as long as the person knows which switches to flip, which buttons to push, and what is inserted where.

ON A VERY DIFFERENT NOTE

If you are at all involved with kitchen exhaust systems, you should read ASHRAE paper AN92-16-2, "Design Considerations for Master Kitchen Exhaust Systems," a symposium paper Gary presented at the ASHRAE winter meeting in Anaheim last January. Give us a call or drop us a note, and we'll be proud to send you a copy.

Dave Elovitz

May 1992

TO THE LIMIT

"I have a real challenge for you this time," the GC's project manager said in a hushed voice. He was trying not to get excited or let the world know about what he thought might look like a big goof. "The HVAC contractor doing the tenant fitout claims he can't balance the system because the base building has pressure dependent VAV boxes. I checked the specs, and I don't know how it happened, but that's what we bought. Now I have a tenant ready to move in and a system my sub says can't be balanced.

Pressure dependent boxes don't regulate airflow the same way pressure independent boxes do. However, making balancing impossible is not one of their problems. So off to a meeting I went, armed with a standard specification on balancing pressure dependent systems.

The HVAC sub described the problem in painful, but characteristic, detail. Some zones were starved for air. Others had two or three times the desired airflow and were noisy to boot. "You're describing a typical unbalanced system," I started to explain, thinking this was just a case of a person facing something new. "The first thing you need to know is the airflow in each zone at peak demand. Some zones will be at maximum, so you set their thermostats for full cooling. Others will need less air, so you set their stats to an intermediate position. Then you set the volume dampers just like balancing a constant volume system. It's kind of trial and error, but there's no mystery to it."

"You might think so, Sherlock," the HVAC man retorted, "but did you know there are no volume dampers? And I'm not putting volume dampers on 54 base building boxes. My job was just to install runouts and diffusers!" He was probably right that the system could not be balanced properly without volume dampers. Adding dampers on 54 boxes would be costly. Even that would not be ideal, because manual volume dampers do not compensate for system airflow and pressure changes the way a flow controller does. Some box inlet ducts were too short to add dampers, anyway.

"Without dampers I agree there's a problem," I conceded, "but I have an idea that will not only be easier to implement but will actually work better than adding dampers. With just a sensor and a pressure switch on each box, you can add velocity limit control. Velocity limit prevents a box from opening further once it reaches design airflow. Since we'll sense actual duct velocity, the limit control will respond to changes in system pressure and airflow, regardless of thermostat setpoint or box damper position. Besides working better, I think you can add velocity limit controls for less money than cutting in dampers."

Ken Elovitz

June 1992

THIN AIR

"Our glycol ... It's gone'" The building engineer was mystified. "We started last spring with a 40% solution, but yesterday it measured 0%. I've checked all over for leaks, but everything seems tight. All that glycol couldn't just disappear into thin air!"

The nursing home's new heat pump system had a piping loop to circulate a water/glycol solution to all of the heat pumps, to a boiler, and through a closed circuit fluid cooler on the roof. The fluid in the loop flows through a closed tube bundle in the fluid cooler (the glycol protects the tube bundle from freezing in winter). A separate spray system sprays regular water over the outside of the tube bundle for the evaporative cooling effect.

The building engineer and I started tracing out the system piping in detail. "Here are the main loop supply and return pipes," I pointed out, "and there's your expansion tank. This valve is for the make-up water to the cooler spray system on the roof, and here's the valve for the loop's automatic make-up water line.

Automatic make-up is not a good idea on glycol systems. Any solution that leaks from the loop is automatically replaced with plain water, continuously diluting the glycol concentration. Even with a small leak, all of the glycol could be gone after a few months, and no-one would notice until a pipe froze. You should keep that valve closed from now on; if there's a leak, you'll find out right away because the system pressure will drop.

"Here's your backflow preventer and pressure regulating valve. Looks like the contractor saved a few dollars by using the same backflow preventer for both the loop make-up and the spray system. They're both nonpotable water, so I guess that's OK."

"Wait a minute!" I said. "One backflow preventer may be OK, but both make-ups come off the low pressure side of the PRV. There's nothing to prevent fluid in the loop from flowing backwards through its make-up water line and out to the spray system. That spray system uses a lot of water; every time it calls, a little solution from the loop bleeds into the spray system. Once in the spray system, your glycol is being sprayed into the air. Reconnecting the spray system make-up line to the high pressure side of the PRV should solve that problem once and for all."

Gary Elovitz

July 1992

SHAKE, RATTLE, & ROLL

We set out to check out the exhaust system in a new gourmet coffee emporium. Everything looked good in the shop: the exhaust grille over the coffee roaster would keep the airflow patterns under control, drawing air toward the roaster to contain smoke and odors generated by the roasting process.

"How's the air balance?" I asked. "Well, we're about 8% below your recommended airflow," the contractor admitted, "but it works okay, doesn't it? Besides, that fan is maxed out. We had to put in a lot more duct twists and turns than we originally expected. I don't think we could speed that fan up any more even if we wanted to."

We climbed a couple stories up to the roof to check out the fan. It sounded like that fan was a bit beyond "maxed out". We could hear that little fan roaring from over 100 feet away, and once we got within 15 feet of it, we could feel the roof shake beneath our feet. I was almost afraid the fan wheel would fly apart any minute.

Back in the office, I tried to sort out why the system should be so close to the edge. The contractor had reported a very high fan static pressure. Even with the extra twists and turns in the ductwork, my calculations said the static pressure should have been much lower. I rechecked the fan's performance data in the catalog. According to the data, the fan was nowhere near its maximum speed. Why should it vibrate so violently?

Then I remembered how the fan had been installed: The fan itself was a small squirrel cage blower, sitting on 4"x4" "sleepers" on the roof. The inlet duct came up through the roof about five feet from the fan and turned horizontal, but this put it about a foot too high to make a straight shot into the fan. Rather than lifting the fan up so the inlet duct would be straight, the contractor had added two elbows to offset the duct down to the fan inlet.

Those sharp turns right at the fan inlet didn't give the air enough time to straighten out before entering the fan. The resulting turbulence meant that the fan wheel was loaded unevenly. This led to vibration not unlike that of an airplane going through a patch of "rough air". All that noise and vibration used up so much fan energy, there was not much left to move the air.

Lifting the fan up on blocks so the inlet duct would be a straight shot not only eliminated the noise and vibration, it also decreased the static pressure. That freed up enough fan power to bring the airflow up to the specified CFM.

Gary Elovitz

August 1992

apologies to
HIAWATHA

High above the cooling towers,
Snow white plumes that last for hours
Billow upward, white pristine-ness
But in basin, Ugh! a big mess.

Slimy, stringy, sticky green goop
Water thick and looks like pea soup.
Fouls the basin, fouls the system,
Condenser tubes no longer glisten.

Upward! Upward climbs head
    pressure,
Straining ever more compressor.
'Til at last it hits the limit,
No more cooling 'til we trim it.

Pull off heads; don't tear the gasket.
Don't forget the strainer basket.
Rod the tubes with wire brush,
Dislodge the ooze then rinse and
    flush.

Sweaty, stuffy, hot machine room,
Grunting, straining, cursing this
    gloom.
Squatting there on burlap sack
Forcing brushes forth and back.

Up to tower -- boots and shovel,
Clean out slime, and then above all,
Scrub the basin. Flush the piping.
Finish up the task by wiping.

Fill the system with fresh water,
Clear and sparkling like it oughta.
Start up pump and check the tower,
Flow as clear as spring time shower.

Would that chiller starts up easy.
So much trouble makes me queasy.
Boss is here, his brow he mops.
Phone is ringing, never stops.

Add the chlorine, kill all microbes.
Clean the water and the test probes.
Adjust pH to stop corrosion;
Balance flows to halt erosion.

Do the tests and do them daily.
I will gladly, even gaily,
Add the potions right, and then
Tubes will never block again.

Dave Elovitz

September 1992

SALESMAN

Three real estate developers went down Maine to do some bird hunting but hadn't brought a dog. They stopped at a farmhouse and asked the farmer if he had a dog they could rent. "Ayuh. Young pup we call 'Salesman' -- Cost ya $5 fer the day. Just say 'Find, Salesman!' and he'll flush game fer ya." They handed over the $5 and off they went.

Dog was fantastic! They'd stroll along, say, 'Find, Salesman!' and time and again, the dog would look around, sort of dance in a little circle, then freeze and point, flushing game on just the right path for a perfect shot. Each of them had bagged the limit in less than three hours, and they brought the dog back to the farmer delighted.

Next year the same three stopped at the same farm: "You still have that dog we rented last year?" "Ayuh, $25 the day." $25 seemed like a lot of money for a dog, but this was no ordinary dog, and they had such a great hunt the year before. So they paid willingly and set off into the woods.

Needless to say, they had a super day again. Just say "Find, Salesman!" the dog would look around, sort of dance in a little circle, then freeze and point, and the birds would burst out of the brush. Got their limit in about two hours, but kept working the dog, flushing birds just for fun, not even shooting at 'em there were so many.

Returned the dog to the farmer, happy as could be.

As you might expect, they returned again next year to find the dog cost $50, but they paid and had a terrific day. Even the year after, when the price had gone to an unheard of $100 a day, they were glad to pay it, the dog was so great.

The fifth year they headed down to Maine again, and stopped at the same farm. "You still got that dog for rent?" "Ayuh," the farmer replied glumly. "$2 fer the day, eff'n you want 'im." "TWO DOLLARS!" the three exclaimed together. "How come? Last year he was $100. And why wouldn't we want him?"

"Well," replied the farmer shaking his head sadly, "Few weeks ago, some of them city folks came up here from New York City, all dressed up in fancy outfits, with brand new guns in fancy cases. Rented the dog and took him out, but one of them dang fools called him 'Sales Manager'. All the dog's done ever since has been sit on his rear end and bark!"

Dave Elovitz

October 1992

FAX FACTS

The other day, I read a fascinating magazine article about how a photograph was transmitted from Boston to San Francisco over telephone wires in just seven minutes. The article went on to say that since that experimental transmission, this new process had been applied to more and more uses: News pictures, criminal's fingerprints, advertisements, mechanical drawings, and much other material was all being sent over telephone lines.

The article explained the transmission process. It involved an intense beam of light that would shine through a negative of the image onto a photocell. At the other end, an image was formed by exposing film to another beam of light. The two machines were caused to rotate at exactly the same speed, so the light pulses at the sender's machine transmitted the original image onto the unexposed film in the receiver's machine.

The article waxed enthusiastic about the prospects for this new and unusual use for electricity developed by AT&T and Bell labs. They called it telephotographs. Sitting by your modern fax machine in 1992, you might wonder what the big deal is. The big deal is that I was reading the February 1927 edition of *Edison Life*, the Boston Edison employee magazine.

A WORD OF CAUTION

The Massachusetts legislature recently passed a new law that makes any person who obtains a building permit responsible for work place safety. Along with the building permit now comes liability to any worker or other person injured on the construction site as a result of violations of the building code or other codes. The law was intended to counteract a recent court decision that limited the liability of homeowners who act informally as job supervisors on their own projects.

Normally, a general contractor obtains the building permit. However, there may be special situations where an owner or a consultant obtains the permit. The broad language of the new statute could impose unexpected liability on anyone who obtains a building permit.

Ken Elovitz

November 1992

EVERYONE WINS
(except the lawyers)

The electrical engineer laid out the system with building standard light fixtures: Three lamps, but two ballasts -- one for the center lamp, the other for the two outboard lamps. By providing two circuits to each fixture, the engineer reduced the number of contactors required to control the inside and outside lamps separately.

The owner's stock of fixtures was not enough for the entire job. The supplier, thinking he was doing a favor, furnished the additional fixtures with 3-lamp switchable ballasts. These ballasts consist of a 1-lamp ballast and a 2-lamp ballast in the same enclosure.

After the tenant moved in, the electrician noticed that several lighting circuit conductors were hot to the touch. He measured currents and found loads as high as 28 amps on circuits that should not have been over 16. The owner feared his new building would burn down, panicked, and started thinking about who he should sue.

The first thing to do was tabulate the loads on each circuit. Of 90 circuits, 15 had overloads. Four were overloaded because the engineer had miscalculated. Eleven were overloaded because of different characteristics between the "building standard" ballasts and the 3-lamp switchable ballasts the supplier had selected. It looked like a real mess.

The contractor submitted a proposal to solve the problem for $30,000 by rewiring the system completely. That didn't seem necessary, and some calculations plus a little juggling proved it could be avoided.

"There's nothing wrong with the ballasts themselves," I suggested. "It's just that they aren't right for this application. You have several more floors to build out. so why not replace these fixtures with new ones, and reuse these on a floor that does not require 1-2-3 switching? Changing the fixtures will still leave some circuits overloaded, but others nearby have capacity to spare. We can balance the circuits and eliminate the overloads by just shifting a few fixtures over to adjacent circuits."

The contractor only charged $7500 to replace the fixtures and make the circuiting changes. The fixture supplier was happy because he kept his customer and even sold some more fixtures. The engineer gratefully completed the simple redesign. The owner and tenant were happy, because the problem was solved with a minimum of disruption. Looks like the only ones to lose were the lawyers.

Ken Elovitz

December 1992

THE TURKEY ZONE

This month Gary shared a cartoon called "The Turkey Zone" by Sandra Boynton. The cartoon depicts a man and a woman at their breakfast table with dialog like this:

Woman: You know, winter's coming, so I think we should get some of those insulated drapes to cut down draft.

Man: I hardly think the low R-value of the drapes and their up front cost would compare favorably with bricking up the windows.

In the next pane, the man, now shown as a turkey, says, "Why are you staring at me?"

Gary Elovitz

January 1993

to the tune of
AULD LANG SYNE

Should past fuel wasting be forgot
And never brought to mind?
Should past abundance be forgot.
'Fore fuel hurts bottom lines.

> CHORUS:
> We've got to find a way, my friend,
> Lest fuel hurt bottom line.
> We'll do a bit of saving yet,
> To aid your bottom line.

We've found a lot of ways to scrimp,
And made things more refined;
But we need lots of other ways
To aid your bottom line.

> CHORUS

Together we have tuned controls,
Gotten systems to run fine.
Sometimes we stayed there half the night,
To aid your bottom line.

> CHORUS

Sometimes we work out clever schemes
To shift loads from peak times;
Sometimes we work out other ways
To aid your bottom line.

> CHORUS

We check, we look up E-E-R's
Our calcs we do refine
To pick the one that does the most
To aid your bottom line.

> CHORUS

It's always our ambition to
Create a scheme divine
For energy efficiency
To aid your bottom line.

> CHORUS

But always we appreciate
Your trust in me and mine.
We always try to think of ways
To aid your bottom line.

> CHORUS

So we join hands, dear friends, and then
We send this wish in rhyme.
For each and everyone of you
A year exceeding fine!

> A year so filled with health and joy,
> And projects where you'll shine!
> From Gary, Ken, and Dave and Fran
> For auld lang syne!

Dave, Ken, and Gary Elovitz

February 1993

ONCE UPON A TIME ...

.... there was an old plant engineer who worked in a plant that took its cooling water from the sea. From time to time, the flow would drop off so low that the process didn't work. Then the old plant engineer would have to go down to the screen house and clean the debris off the intake screens. One day, when he went to clean off the screens, he saw that it was not debris, but a huge fish clogging the intake. "Aha!" he said to the fish, "I have you trapped in the screens, and I'll just take you over to the cafeteria and get them to cook you up for my dinner."

But the fish cried out, "Wait! Wait! I am an enchanted fish. If you set me free instead of cooking me, I can grant you a wish." Well, the old plant engineer was so startled by the talking fish that he lost his grip on the screens. The fish tumbled back into the intake sluice and swam back out to sea. Now you don't survive to be an old plant engineer by being dumb. Even though the fish got free accidentally, the engineer figured the fish couldn't know that, so he decided to try and collect his wish. The engineer stepped to the edge of the sluice and called out:

> "Flounder, flounder in the sea
> Prithee hearken unto me.
> When you asked I set you free,
> Now I beg a boon of thee."

Quick as a wink, the fish appeared, responding, "OK, a promise is a promise. What can I do for you?"

"The top brass are driving me crazy," the plant engineer lamented. "Human Resources wants me to do something about the complaints of headaches and scratchy throats and burning eyes. And Legal won't let me use our new computer system because of the potential liability question."

"I know just what you mean," replied the flounder, who (fortunately) had been an ASHRAE member before a wicked comptroller turned him into a fish for overspending his utility budget. So, with a snap of the flounder's flippers (after all, he was an *enchanted* flounder, why shouldn't he be able to snap his flippers), the old plant engineer found himself at the ASHRAE Winter Meeting in Chicago. There he heard Dave's paper on control of minimum ventilation air in VAV systems and Ken's paper on liability in the use of computers and lots of other papers on things a good plant engineer needs to know. After that, the old plant engineer returned to his plant and his sluiceway, where he lived happily - or if not always happily, at least more often happily than not - ever after.

Dave Elovitz

March 1993

THE RETURN OF SALESMAN

Remember Salesman, the great hunting dog you met last September? He's the one who lost interest when some fool called him Sales Manager. Salesman saw the folly of his ways and soon went back to work. He even took some courses and learned to count. The farmer who owned him started renting him out again, and he was even better than before.

When Salesman went duck hunting, he'd run up ahead and count the number of ducks sitting on the river. Then he'd run back and bark once for each duck. Five ducks, five barks. He'd even make a softer bark for the younger ducks.

One day another big city slicker came to the farm house and rented Salesman for some duck hunting. The man was so impressed he absolutely insisted on buying the dog. He offered $200. "Sir," the farmer said, stroking his chin, "I've had that dog for years, and he's stood by me through thick and thin. He's like family. I couldn't bear to part with him."

Not one to give up, the city slicker offered two thousand dollars. "Sold," the farmer sighed reluctantly.

A few weeks later, the city slicker telephoned the farmer. He was truly upset. "That dog of yours was terrible," he complained. "I had to shoot him on our first trip!"

"Shoot him? Why on earth would you shoot him?" the farmer asked incredulously.

"I took him out with my friends," the city slicker explained, "and wanted to impress them with how that dog could count. So we went out into the marsh. When the dog came back, he picked up a stick and started hitting me with it on the ankles. I wanted to show him off and told him to go back out and count the ducks. But he just picked up a bigger stick and started hitting me even harder. I took that stick, threw it into the woods, and ordered the dog to go back out and count the ducks. This time he came back with a tree limb and started smashing me with it in the legs. So I took my gun and shot him."

"You fool!" the farmer said. "The dog was trying to tell you there were more ducks than you could shake a stick at!"

Ken Elovitz

April 1993

STILL MORE FROM MURPHY

We haven't shared any Murphy's Laws in a couple of years, so here are some new ones along with a few old favorites. You remember the basis of all of Murphy's Laws, of course: Anything that can go wrong, will (and at the worst possible moment.)

LUCIANO'S LAMENT: Almost anything is easier to get into than get out of.

CARLI'S COROLLARY: When things can't possibly get any worse, they will.

TURNER'S TAUTOLOGY: An ounce of image is worth a pound of performance.

DEAN'S DILEMMA: The one who does the least work gets the most credit.

FRANK'S FALLACY: The first myth of management is that it exists.

DAMESEK'S DICHOTOMY: The truth of any proposition has nothing to do with its credibility, and vice versa.

ONUFRAK'S ONTOLOGY: The more innocuous the change appears to be, the more the plans will have to be redrawn.

HANDREN'S HARANGUE: Friends may come and friends may go, but enemies accumulate.

KAUFFMAN'S KISMET: In any bureaucracy, the higher the level, the greater the confusion.

BARBER'S BACKLASH: Trivial matters are handled promptly. Important matters are never solved.

WILLIAMS' WARINESS: It is difficult to win an argument when your opponent is not encumbered by facts.

CHERYL ON CHARLATANS: Following the path of least resistance is what makes men and rivers crooked.

KIEL'S COMPLAINT: Man has less tenacity than crabgrass.

RENKAS' THEORY OF RELATIVES: The number of a person's relatives is directly proportional to his fame.

DARRISH'S DICHOTOMY: Anything that begins well ends badly. Anything that begins badly ends worse.

MAURA'S MAXIM: An expert is a person who avoids the small errors while sweeping on to the grand fallacy.

FOULKES' FOLLY: Variables won't. Constants aren't.

BERNIE'S BELIEF: If builders built buildings the way programmers write programs, the first woodpecker that came along would destroy civilization.

Dave Elovitz

May 1993

COOKING ODOR RAP

Come gather all around
And permit me to tell
You the story how I chased
An elusive smell.

All the condo owners,
They just did not want to know
What was cooking from the menu
In the restaurant below,

"The brand new kitchen hood
Is surely up to snuff,
Can you please tell me why
I am getting all this guff?

"The building staff,
They have all thrown up their hands;
For the residents
And their damn olfact'ry glands

"Are making all our lives
A pernicious living hell,
Will you please come
And tell us how to stop the smell?"

So I went and I looked
And I measured all the flow
And I said to the chef,
"What could make the odors go?

"There is a basic rule,
And it's one I'm sure I know:
The odor will not waft
Where the air it cannot flow,

"When exhaust from the kitchen,
It exceeds the make up air
All the odors from the cooking
Should be captured inside there,

"Then the pressure that I measure
In the kitchen should be lower
Than the pressure that I measure
In adjacent corridors,"

But as I gently opened up
The kitchen exit door
I could feel the air rush out;
I could see the pressure's more,

For that simple kitchen exit door
Into the stairwell leads
And the stairwell is the magic route
Up which the odor speeds,

The exhaust from the condos
Is removing air galore
But there's very little make up
To the common corridor,

That makes pressure in the stairwell
Less than pressure in the kitchen
A relationship that sets
All the residents to bitchin'

"Now here's the fix,
If you listen to my plan:
Just reduce the flow of air
Through the kitchen makeup fan,

"Then the kitchen hood exhaust
Running at its total power
Will confine the ugly smells
Like from cooking cauliflower,"

At long last no more complaints,
For the problem it was eased;
Both the chef and building staff
Could relax and were quite pleased,

Dave Elovitz

June 1993

GOVERNMENT

Once upon a time, a king had many wise advisers. One day he was preparing for a picnic and asked his wise men what the weather would be. The wise men huddled together, but none could predict the weather. Not wanting to disappoint the king, they told him it would be a beautiful day and not to worry about rain.

As the king was leaving the castle, he saw a farmer walking along the road with his donkey. "Good day, Farmer," said the king. "Good day, your Majesty," replied the farmer, "what brings you out today?" "Why it's a fine day, and I'm off to a picnic," the king replied. "I hope you have a tent," the farmer warned, "it's going to rain." "Nonsense. My advisers said it would be a beautiful day."

The king went off into the woods, and the rain began to pour as soon as the food was out.

The next week the king asked his advisers again what the weather would be. They couldn't predict the weather any better than before, but after their last embarrassment, they weren't taking any chances. So they told the king to expect rain. Accordingly, the king loaded up his boots, umbrellas, and tents and set off down the road.

Soon the king saw the farmer and his donkey again. "Where to today, your Majesty?" the farmer inquired. "Another picnic," the king replied, "but this time I'm prepared for the rain. I have plenty of tents and umbrellas." "Won't need 'em," the farmer told the king.

The king didn't listen and set up his tents and umbrellas at the picnic. But the farmer was right, and it turned out to be a beautiful day. On the way back to the castle, the king saw the farmer with his donkey. "Farmer," said the king, "you are much wiser than all my wise men. Come with me and be my adviser."

"It's not me," the farmer replied humbly. "It's my donkey. When the weather is going to be good, his ears stand up straight. When rain is on the way, they flatten out."

The king thought about the farmer's reply and decided to buy the donkey. He took the donkey with him back to the castle, and ever since then, the government has been run by jackasses.

Ken Elovitz

July 1993

PUFFED UP

"What's going on here!?" The shopping center's manager was perplexed. "The dress shop complains that odors from the hair salon next door are ruining their business. But whenever I go there, I can't smell a thing!"

I met the manager at the dress shop, and sure enough, no odor. The sales clerks couldn't identify any pattern: not time of day, not weather. No clue at all.

What caused the odor in the dress shop? Odors need three conditions to migrate: a source, a driving force, and a path. The source was simple: the odors clearly came from the hair salon next door.

A pressure difference was the most likely driving force. But I measured no pressure difference at all between the two shops, regardless of whether the HVAC systems were on or off.

The most likely path would be through the partition that separates the two shops. However, the partitions looked pretty well sealed; even the space between the top of the drywall and the corrugated roof deck had been stuffed with compound. Sure there were a few gaps, but it would take a pretty big pressure difference to push smelly air through them. I was stumped.

So I went up to the roof to check the HVAC units on the chance that odors might cross from one to the other. That seemed unlikely because the units were pretty far apart.

Then I saw the "economizer" hood on the hair salon's rooftop unit and knew exactly what was going on. If the hair salon needs cooling on a mild day, its "economizer" damper opens, sending cool outside air to the space. The unit also has a relief damper to let that outside air escape from the space. But the relief damper is really too small to work effectively. Since the relief can't keep up with the economizer, pressure builds up in the space, forcing excess air out through doors and gaps in the partition.

I was there on a warm day, so the economizer was closed. The hair salon was calling for cooling, but there was no odor in the dress shop because the economizer damper was closed. By adjusting the economizer controls, I forced the damper open and created both a pressure difference and a smell in the dress shop. At the same time, I understood why the odor was so elusive: the weather had to be cool, the hair salon had to call for cooling, and they had to be using smelly "perm" chemicals. A new exhaust fan in the hair salon, set to run with the economizer, would keep the hair salon from getting too puffed up.

Gary Elovitz

August 1993

INADVERTENT SAUNA

"We need the Red Adair of HVAC," the caller from the prestigious theater declared. "Our new chilled water system seems to be working fine, but we can't cool the upper level seats. By the end of the show, temperatures are in the low 80's. With the big prices people pay for tickets, they aren't happy being hot."

The drawings showed 150,000 CFM of supply air and 130,000 CFM of return. That meant there should be 10,000 CFM left to pressurize the lobby after making up all toilet exhausts. Yet at opening time, a torrent of steamy outside air accompanied the steady stream of people flowing in through the open doors. That made the lobby hot. The hot air found its way into the theater itself and rose to the upper portion. Coupled with body heat from almost 4000 people, it was easy to see how the upper part of the room could overheat.

10,000 CFM should have been more than enough to combat infiltration. Even with all the entry doors open, air should have been pouring out of the building. Was the 10,000 CFM actually there?

The recent renovations included adding a cooling coil to a fan system that previously had none. The design didn't call for speeding up the fan, so the added resistance due to the new coil would reduce the supply air flow from that system. The balancing report not only confirmed that suspicion, it showed that the other supply fans were moving less than their design airflows as well.

The balancer may have just met the minimum on the supply fans, but he was much more successful with the return fans! They were all at least 10% over design. That meant 10,000 CFM more taken out of the theater than supplied to it. Instead of a surplus of 10,000 CFM relieving through the lobby, the resulting negative pressure sucked hot outside air right in through the open doors. Then it went across the lobby and into the theater where it made the balcony patrons feel like they were in a sauna!

"The explanation is very interesting," the theater president responded, "but what can I do for tonight's show?" "Just shut off the lobby return fan while people are coming in. Then you won't suck hot outside air in along with them. For a permanent cure, we'll just have the balancing contractor slow the return fans and adjust the dampers to bring in 10,000 CFM more outside air than the toilets exhaust. That should end your inadvertent sauna."

Ken Elovitz

September 1993

TECHNOLOGY TRANSFER

A young architect had lost his job and couldn't find another. He contacted all the other architectural firms in the area, but they, too, had been caught up in the real estate recession and were cutting back, not hiring. As an architect, the young man thought he knew something about construction. So, when he couldn't find design work, the young architect started looking for jobs with contractors.

The young architect had been a designer on a project for a former client who took a liking to him. The client was one of the few developers who was still building. When the client heard that the architect was unable to find work, he put the young man in touch with the general contractor on one of his projects. The client convinced the GC to hire the architect as a carpenter.

The young architect reported to the job site ready to work, with his hammer hung from a loop on his carpenter's apron. He looked very professional. Looks didn't impress the job super, who eyed him skeptically and sent him out to attach siding on the outside of the north wall.

A few days later, the general contractor called the site to ask how the new recruit was doing. The super responded, "That damn fool! I warned you it was no good taking on one of those college smart alecks. I finally had to let him go."

"Why? How bad could he be?"

"Well," the super said, shaking his head, "the first job I gave him was to attach siding to the outside of the north wall. A few hours later I walked by his work area and saw him nailing up the siding OK. But when I watched for a minute or two, I saw what he did with the nails. As he took each one out of his apron, he looked it over, and about every other one he threw over his shoulder!

"'What in tarnation are you doing?' I called to him. And he replied, 'Well, almost half of these nails are no good. They have the heads on the wrong end.' "That convinced me he didn't know what he was doing, and I let him go right there on the spot. Anyone would know those nails were for the other side of the wall!"

Dave Elovitz

October 1993

CLEARING THE SMOKE

Smoke control is one of the most contentious concepts in our industry. Plant engineers, fire chiefs, and building inspectors sometimes expect HVAC systems to keep the fire zone clear of smoke during a fire. The system doesn't do that, and they are disappointed. Understanding the physical principals behind smoke control helps explain what those systems are supposed to do.

A smoke control system is not intended to blow smoke out of the space during a fire. The fire department sometimes uses smoke purge systems to clear smoke after the fire is out. However, blowing air into a space with a burning fire can actually make the fire worse. The supply of new air feeds oxygen to the fire.

The idea behind a modern engineered smoke control system is to confine smoke to the area of origin. By keeping the means of egress clear and reducing the spread of smoke to other parts of the building, a smoke control system helps people evacuate safely.

A smoke control system does its job by creating a pressure difference between the fire zone and the rest of the building. Smoke, like air, will not flow against a pressure difference.

Creating the pressure difference relies on both exhausting from the fire zone and supplying clear air to the rest of the building.

The exhaust sucks air and smoke out of the fire zone. It also counters the tendency of expanding hot air in the fire zone to pressurize the space. The exhaust removes a great deal of smoke, too, but it will not clear the room of smoke.

Supply is introduced to exit stairs and other parts of the building surrounding the fire zone. The supply pressurizes those areas against the flow of smoke. Without supply, smoke might enter the stairwell when people leaving the fire zone open the doors to the stair. The fire zone does not need any direct make up. Introducing make up directly to the fire zone reduces the pressure difference so is counterproductive.

If you are interested in this topic, you might like to read the article we wrote for the ASHRAE **Journal** on smoke control systems. Just call or drop us a note, and we'll be happy to send you a reprint.

Ken Elovitz

November 1993

SELECTION

Last summer, I chaired an architect selection committee for a remodeling project for my congregation. It was an interesting experience. Even though I am no stranger to the designer interview process - either as a candidate or as a member of a selection committee - I learned some things worth sharing.

Getting the committee members to list all the things the project should address was pretty easy. Fortunately, they all acknowledged from the outset there likely would not be enough money to do everything that would be nice to have. So, setting priorities was difficult but went quickly. Preparing a background paper to brief the prospective architects was not hard. Even scheduling interviews when all five busy committee members could attend and arranging pre-interview site visits for the architects was not a real problem. Nor was keeping the interviews flowing and on schedule.

But then came time to decide on a recommendation. How could I help this disparate group of bright people with little or no construction project experience focus on what made one candidate better than another for this project?

I finally settled on making a big flip chart sheet divided into four squares for each architect. Each square had a topic heading:

Creativity/Ingenuity - Will the architect come up with thoughtful, insightful solutions for this project or just apply cookbook solutions that worked elsewhere in the past?

Discipline/Organization - Will the candidate keep the project on track, on budget, and on schedule, pushing the building committee to make hard decisions when they are needed, or will the building committee have to push him?

Responsiveness - "Whose project is this any how?" Will he see his role as a tool to solve what the congregation sees as its problems, or as an opportunity to display his architectural virtuosity?

Experience - Has he accomplished similar assignments smoothly and consistently? Were they happy experiences for both architect and client?

As we discussed each characteristic for each candidate, I recorded the comments. Then we reviewed the completed sheets and reached a decision surprisingly quickly and with little dissension. As I thought about it, those four categories are a pretty good measure for selecting any kind of consultant. I hope you will feel free to use them, should the occasion arise.

Dave Elovitz

December 1993

THE ONE THAT GOT AWAY

"You won't believe this, but sometimes the negative pressure in this building is so bad people can't open the front doors!" the building manager told me over the phone.

"Wow!" I said. "How long has this been going on?"

"Since the building was built in the '70's," he replied. "But it's serious now that the executive offices have moved here."

"Well," I said, "why don't you send some drawings over, and I'll take a look."

The drawings arrived with a very complicated "Request for Proposal" and bid form. Apparently the job was going out to several firms for bids. The building manager had said that the design called for much more exhaust than supply, so I checked that first. The design was actually pretty much in balance, though I could see how the schedules might be misleading. So I traced out the main systems: big variable volume air handlers, with separate return fans. I had a hunch and called the building manager.

"How are your supply and return air fans controlled?" I asked.

"The supply is on duct static pressure, and the same signal goes to the return fans," he said. This means the two fans always run at the same speed relative to each other.

"And your outside air intake louvers?" I probed.

"Funny thing about them," he said. "Sometimes they're outside air intake, and sometimes they're exhaust."

Bingo! But should I tell him? After all, there are other firms bidding for this work, I thought. Oh, what the heck.

"That's exactly your problem. You see, using the same signal for both supply and return fans usually doesn't work well. The return system is too different from the supply. Reducing both fans to half speed doesn't reduce both airflows the same. The supply might drop to 1/3 (depending on lots of conditions), but the return would still be up around 1/2. At that point, the return fan moves more air than the supply fan can use, and the extra air blows out the outside air intake louver. The return fan is depressurizing the building, which makes it hard to open the doors. Changing the controls should solve this problem in no time."

Funny thing, though. I never heard from that guy again. Maybe he figured it was just too good to be true.

Gary Elovitz

January 1994

THE BALLAD OF EEI

There's songs about men
>who go down to the sea;
Of the heroes with cannon and swords;
And stories of valor and dead chivalry,
And the bravery of old knights and
>lords.

Don't sigh for romance
of knighthood long past
'Cause there's just no ideals anymore;
Don't say that this world is just
>rolling too fast
To develop that "esprit de corps."

We don't wear bright armor or
>ride on a horse;
We don't brandish weapons of war.
But when you have trouble,
>we'll be there in force,
We'll rush o'er like heroes of yore.

We'll climb on your chiller,
>we'll figure your loads.
We'll measure your amperes and volts.
We'll study your drawings,
>decode control modes.
Get right down to nuts and to bolts.

We study your system with painstaking
>care
For energy dollars to save.
We search out the gremlins
>that lurk everywhere,
Bring comfort the occupants crave.

Our weapons aren't fearsome:
>Equations and skill
As onto the problems we zoom.
Our battles aren't fought to capture
>some hill
But to straighten out your machine
>room.

We may not wave banners
>embroidered in gold
In Latin nobody can read.
Our calendars tell you of deeds oh so
>bold,
And give you good tips you should
>heed.

As now one year wanes, another starts
>new
We wish you the best life can bring.
"Good Health! and Good Cheer!
Good Fortune to you!"
The four of us sincerely sing.

From Fran and from Dave,
>from Gary and Ken,
To calendar mates the world o'er:
May we be together as clients and
>friends
For many long happy years more.

Dave, Ken, and Gary Elovitz

February 1994

THE "EXPERT" LEARNS A LESSON

When I decided to install central air conditioning in "My Old House", I figured I might as well get rid of my old steam radiators at the same time. I selected a gas furnace and cooling coil, then laid out the ductwork in my attic. I'd seen enough condominiums where heating with warm air from overhead was a disaster, but I was sure I could avoid all those pitfalls.

When I described my plans to my friend (and oil dealer) Roger Litman, he had two comments: First, why not convert the existing oil boiler from steam to hot water and run hot water up to a heating coil in the attic? That way he can keep the account. Second, make sure the return air inlet is at floor level.

The hot water coil idea was brilliant. It was a lot cheaper, and it allows for better control. But I knew enough engineering to know that the location of the return isn't very important -the return can't suck the supply air down from the ceiling. As long as the ceiling supply diffusers do a good job of mixing, the return can be anywhere. So I bought a hot water heating coil, installed it in my air conditioning fan-coil unit, and installed the ductwork and diffusers in the ceiling. I used one main return at ceiling level in the center of the house. Converting the steam boiler to hot water went without a hitch. All summer, the air conditioning was great, and we were thrilled to be rid of those old radiators. Then, on the first few days of cold weather I got a shock! The heat just didn't seem to work. It took forever to get the house comfortable, and the temperature at the ceiling was about 10 degrees higher than at the floor. My ceiling diffusers just couldn't push the buoyant, warm supply air all the way down to the floor. A good deal of the supply air hovered in the top portion of the room. With the return at the ceiling, the system simply recirculated air in the top half of the house, and the bottom didn't get enough.

I immediately realized why Roger told me to put the return at the floor: It's not practical to select diffusers that can push warm supply air all the way to the floor in my high ceiling house. Diffusers with that throw would be noisy, and drafty to boot. But with the return near the floor, air from the top half of the space drifts down to replace the cooler air drawn into the return. That way the warm supply air can't hang near the ceiling.

Fortunately, I found a way to drop a return air duct down to the floor without much disruption. Since then everything has worked fine. Of course when I told Roger, he wasn't too impressed. He already knew he was right.

Gary Elovitz

A REAL BULL STORY

Once two brothers owned a dairy farm. When they wanted to increase their herd they had always purchased calves from other farmers. Being sophisticated businessmen they performed a "make or buy" analysis and determined they should try breeding some of their cows. They agreed that one of them would go to Texas to shop for a bull, and after the deal was struck, the other would drive down and help bring the new bull home.

So one of the brothers traveled to Texas and went from ranch to ranch, shopping for just the right bull. When he found the one he wanted, he shook hands on the deal and asked the rancher if he could use the phone to call his brother to come and help take the bull home. The ranch was in a very rural part of Texas and did not have a phone. The rancher told the man there was no phone but he could go into town and send a wire.

In town at the Western Union office, the man told the clerk his story and asked about sending a wire. The clerk said that would be no problem but the cost would be $10 per word. At $10 a word, the man knew he would have to get right to the point. After thinking for a minute, he gave the clerk his brother's address, took out a $10 bill, and told the clerk to send a one word wire that said: "COMFORTABLE".

"Comfortable?" asked the clerk, not understanding how that one word could get the message across.

"Comfortable," the man confirmed. "You see, my brother reads very slowly. When he gets the wire, he'll read it and know I need him to 'come for ta bull'."

OPINION

You might appreciate the following quote from an opinion by the Massachusetts Appeals Court. A building inspector had brought criminal charges against a contractor who erected a silo to store cement in violation of the zoning bylaw. The defendant, found guilty, appealed on the ground that Massachusetts Law requires a person aggrieved in a zoning matter to go to the Board of Appeals before seeking court action. The Appeals Court upheld the building inspector's right to seek criminal charges directly, saying, "The building inspector has not been aggrieved by any public action which he might appeal to the Board of Appeal. It is not in the nature of his work which, one might say, is to give grief, not suffer it." 'Nough said.

Ken Elovitz

April 1994

A CURRENT TOPIC

The two office buildings sit side by side and are similar in many respects. Both have water loop heat pumps, and both operate on the same schedule. Yet one building consistently used 20% more electricity, even though it is 10% smaller.

Neither the design drawings nor a tour of the building showed any unusual loads or special equipment to explain the difference. The heat pump system and its night setback control worked properly. Driving by at night proved that after hours lighting was not the problem.

To see where the electricity was going, I measured the actual current on every feeder in the building and used this data to project average electricity usage. Surprisingly, my projection fell right in line with data from the other site but was about 30% short of this building's metered usage!

Could the meter be reading high? The electric company had already changed the meter once, so a faulty meter was unlikely. I checked the meter wiring, the current transformer (C/T) ratios, and the calculation of the meter "multiplier", all of which looked OK. That left the C/Ts themselves, but C/Ts have no moving parts to go out of calibration.

I could test the C/Ts by measuring both the "primary" current in the feeders that pass through each C/T and the signal that the C/T sends to the meter. The ratio of these currents should match the C/T's nameplate rating. Measuring the signal was a snap, but the primary was a chore: I had to "amp" each of 15 big wires all crowded together in the switchgear cabinet and sort them by phase. Two phases came out pretty close, but the third phase was way off!

When I told the electric company, they agreed to test the C/Ts in their lab, and they found that one C/T actually had a ratio of 600:5, exactly half its rated ratio! That proved the meter was reading high. The customer was entitled to a refund of the overcharges.

'With one of three C/Ts at half the ratio, that means the error is 1 part in 6", the utility representative and I initially agreed. $(1 + 1 + ½)/3 = 5/6$. This was good news, but I was looking for a bigger error than 17%. 'Wait a minute!', I said, "A C/T with half the ratio means the secondary current is *doubled*, not halved! The equation should be $(1 + 1 + 2)/3 = 4/3$ That makes the error 1 part in 4, or 25%!" The electric company replaced the C/Ts, so the meter readings should be accurate from now on. And the owner got a refund of 25% of his electric bill from the day the building opened.

Gary Elovitz

May 1994

HOT WATER

The client was frantic on the phone. "The tenants have been complaining for months, and now they've called in their lawyers. We've got to solve this hot water problem."

The complex had recently replaced an aging, inefficient central boiler plant with boilers in each building. As soon as the new systems went on line, the tenants started complaining about insufficient hot water. Actually, keeping the new boilers on line was part of the problem, and the first order of business was getting the boilers to run reliably. After all, we explained at a meeting with the hostile tenants, you can't make hot water unless the boilers work. But even after ace mechanic Bernie Beauchemin eliminated the control failures, the tenants continued to complain.

Calculating hot water requirements according to ASHRAE guides showed the system should have been adequate. Field testing verified that the water heaters performed close to the their design rating. Still, some evidence indicated these tenants expected and used much more hot water than design guides anticipate.

Telling the tenants to conserve water didn't seem like it would fly. Replacing the entire system just didn't seem right and increasing storage might even have been a mistake. Doubling storage would mean twice as much hot water when everything works, but it would also mean twice as long to recover if there's a problem.

How could we be short of capacity when I watched the boilers cycling on and off every couple of minutes? A cycling boiler means the boiler generates more heat than the system can use. If I could just suck more of that heat out of the boilers, I could increase the amount of hot water available.

Several pages of log mean temperature difference, U-value, BTU, and flow calculations later, I had it: "The manufacturer's rating for these particular heat exchangers uses a much lower heat transfer factor than other water heaters. But there's an easy way to get more heat out of these heat exchangers. Just use a bigger pump. By increasing boiler water flow, we can increase turbulence in the tubes, which increases heat transfer. At the same time, we'll raise the average temperature in the heat exchanger, and that increases heat transfer even more."

Installing the new pumps was a simple but rewarding job. "We haven't had a hot water complaint since the new pumps went in," the client announced proudly. "Things have run so smoothly, the night mechanic even thought his beeper was broken!"

Ken Elovitz

June 1994

THE CARIBOU FACTOR

An architect, an engineer, and a contractor had become a tremendously successful project team over many years of working together. They attributed their success in part to finely honed negotiating skills and in part to the personal friendships they had developed.

Each summer the trio vacationed together on a hunting trip to northern Canada. They flew to Toronto, where they met the small, twin engine charter plane that would take them to the north woods for a week of stalking caribou. Before the pilot dropped them off, she reminded them that the plane's weight restrictions meant they could only take home one caribou.

At the end of the week, the pilot returned and was surprised to find each hunter had his own caribou. When she asked if they had forgotten about the plane's weight restrictions and the limit of one caribou, they told her what happened: "We negotiated for six days over how to allocate the one caribou limit. When we realized we were about to run out of time before we even started the hunt, we decided to let market forces prevail. We're offering you an extra $50 if you let us take all three caribou home."

"I'm sorry," the pilot said, shaking her head, "but one caribou is the limit."

"But last year you let us take the extra caribou," they protested.

"That was last year, and this is this year," she replied.

"OK, how about $500?" the hunters tried. At $500 the pilot couldn't resist, so the hunters crammed their gear into the plane and strapped the two extra caribou to the wings. To compensate for part of the weight of the extra caribou, they left some nonessential luggage behind.

As the pilot pushed the throttle forward, the engines roared, and the plane started rolling down the runway, creaking and shuddering as it built up speed. After a few bumps and thumps, the plane lifted off but then unexpectedly crashed to the ground.

Though no one was hurt severely, the pilot and passengers were all knocked out. When they came to a few hours later, the architect asked, "Where are we?"

"I don't know," the engineer answered, "but it looks like the same place we crashed last year."

Ken Elovitz

July 1994

MEDIALYSIS

As court dockets become more crowded and the time and expense of litigation increase, the construction industry is giving alternative means of resolving - or even avoiding - disputes more and more attention. Arbitration was one of the earliest alternatives to litigation. It has now been joined by mediation, mini-trials, partnering, and just about any other" alternative dispute resolution" process the parties can think of.

A recent DART newsletter described a fairly new dispute resolution technique called MEDIALYSIS. DART is the Dispute Avoidance and Resolution Task Force, a non-profit organization dedicated to making the construction process work more smoothly. In medialysis, the parties jointly hire an expert or team of experts to analyze the dispute and present a neutral, unbiased analysis and opinion. The opinion and any recommendations in it are not binding but can be the basis for a negotiated settlement. Like other alternative dispute resolution techniques, medialysis is private. The big difference between medialysis and conventional mediation is that the parties jointly select and retain the medialysis expert or experts on a standby basis before any dispute arises.

When and if a dispute arises, it is referred at once to the medialysis experts before differences in opinion become festering antagonisms. That way, the parties maintain positive working relationships. Medialysis is hardly a new concept to us at Energy Economics. We just never knew it had a special name. When we have been called on to perform that sort of service - either on a standby basis at the beginning of a project or when the difference first arises - we have found that "medialysis" is largely a process of having a mutually respected third party educate the disputing parties about the facts and spell out the implications of those facts. Since the disputing parties then all have the same information, they work out a mutually acceptable resolution on their own, rather than simply accepting a pronouncement of how the baby will be hacked in two.

One limitation of medialysis is that medialysis experts are primarily educators - seekers after truth, if you will- and not conciliators. Their role is to assist the parties in arriving at an equitable outcome, not to cajole them or motivate them to keep talking. For that reason, medialysis really works only when the parties want to resolve the matter. If either digs in his heels or drags his feet, the dispute may require a conventional mediator to keep the process going.

Dave Elovitz

August 1994

NEW INVENTION

It's a little known fact that automotive air conditioning was invented by the three Rosenberg brothers in Milwaukee, back in 1934. The Rosenbergs were in the appliance service business and spent most of their time repairing ovens, washing machines, and refrigerators. One very hot summer day, while driving out to a service call in a sweltering car, one of the brothers hit upon a brilliant idea:

"I bet we could rig up the workings of a refrigerator to cool our service truck!" he exclaimed. So the brothers set to work and soon came up with the basic workings of the automotive air conditioning system. The new system used an extra fan belt, rather than an electric motor, to spin the compressor.

Once they got their "Rosenberg Cooler" working, visions of riches began to dance in their heads, so they piled into the car and drove off in air conditioned comfort to the Ford Motor Company headquarters in Dearborn, Michigan.

With a bit of luck and some persistence, they managed to get Henry Ford himself to take a ride in their air conditioned car. Mr. Ford immediately recognized the potential of the idea, and negotiations began.

"I'll give you two million dollars for all rights to your invention." Mr. Ford offered. The Rosenberg brothers were pretty impressed, but their interests went beyond mere money. "We'll accept your two million," they responded, "on condition that a nameplate with the name of our invention, 'Rosenberg Cooler', is attached to every car with our invention in it."

"Three million, with no name!" Ford replied. Negotiations continued, with the three brothers pushing for some public recognition of their contribution to technology, and Mr. Ford fighting to keep their names out, or at least relegated to some inconspicuous place. The two sides finally agreed on a compromise: 2½ million dollars, with the Rosenberg brothers' first names only listed on the control panel on the dashboard.

And that is why, to this day, every Ford automobile with air conditioning has the three Rosenberg brothers' first names - Max, Norm, and Hy - prominently displayed on the air conditioner control panel.

Gary Elovitz

September 1994

TEN TO ONE

"I'm sending you three proposals I want you to review for new boilers," the caller said. "But Richard, you just converted those boilers to gas. Why replace them?"

"I know, but now the gas company says they never should have been converted. They say the flue gas temperature is over 700 degrees, robbing us of efficiency. On top of that, the flame burned through the face of the boiler and jumps out into the room. We're afraid the boilers are going to blow up."

"It sounds like you need some diagnosis and a competent service company more than you need new boilers. Before you spend thousands of dollars on new boilers, let me go down there and take some measurements."

First I checked the chimney. A blocked flue could build up back pressure and push flame out even tiny openings. After the maintenance man shoveled a 5-gallon bucket's worth of soot out of the clean out, I looked up the chimney and found a clear path. No reason for back pressure. The draft measured OK, too, and the barometric dampers were controlling properly. No draft problems.

Then I measured boiler efficiency. It was slightly low, which I determined was due to too much excess air. A good mechanic could solve that problem with a simple adjustment.

Clocking the gas meter showed the firing rate was right on the money. Yet the flame still poked out through the observation door.

"This boiler is supposed to operate under negative pressure," I explained to the maintenance man. "And we have plenty of draft, so air should be flowing into that port, not flames sneaking out. Let's shut off the gas and run the burner fan alone. That way I can check around the burner setting without the risk of getting burned by the flame."

Shining my flashlight into the observation port, I saw the problem right away. "Look inside here, and you'll see exactly what happened!" I announced proudly. "The end of the burner blast tube burned right off. With the end burned off, part of the flame escapes to the space between the refractory combustion chamber and the boiler shell. And I'll bet the problem started with improper installation. The distance the blast tube slides into the combustion chamber is a critical dimension. A skilled burner technician should be able to fix this problem in less than a day. While he's at it, the mechanic can adjust the air settings and the shape of the flame to boost efficiency and improve heat distribution. A few hundred dollars in smart repairs and you'll be just as well off as a few thousand for a new boiler!"

Ken Elovitz

October 1994

DILEMMA

The calendar text
At the printer is due;
    Ken, Dave, and Gary
    Are feeling so blue.

They need a case hist'ry,
Surprising but true,
    A humorous story,
    To bring smiles to you.

They've borrowed from Murphy.
His laws they have told.
    And told you of buildings
    Too hot or too cold.

They've plagiarized stories,
Your sides you would hold.
    And energy wasters
    They've hastened to scold.

Well, what's left to tell?
What more can we say?
    The printer's expecting
    Our copy today!

But not something boring,
No worn out cliche.
    If I had an idea
    I could write straightaway.

I could tell it in prose;
I could tell it in rhyme;
    I could parody music
    In three-quarter time;

I've got it!
A story that will be a hit.
    With even a moral
    So you'll think a bit.

Two woodsmen were chopping
And cutting up trees.

By lunchtime Bob's pile
Had reached just to his knees.

But Bill had a pile
That was tall as the trees.
    Bob just didn't get it
    Bill had taken his ease

On sev'ral occasions
While Bob chopped with verve.
    It bothered poor Bob
    'Til he got up the nerve

To ask how friend Bill
Could poss'bly deserve
    To have done so much more
    And with resting preserved.

"Your question just tells me
You don't have the facts.
    "It's clear you weren't watching
    You just thought I was lax.

"I didn't just pause
During your fierce attacks.
    "I stopped," Bill smiled kindly,
    "To sharpen my axe."

Dave Elovitz

November 1994

OSHA IAQ PROPOSAL

If you have not followed the press reports on proposals by OSHA to regulate indoor air quality, you probably need to take a long hard look. OSHA, the Occupational Safety and Health Administration, is part of the Federal Department of Labor. Their proposed new regulations are OSHA's first foray into *nonindustrial* work environments. Secretary of Labor Robert Reich has touted the program as "part of the most ambitious standard setting agenda in OSHA history. Here are some of the highlights from the Federal Register:

For multi-tenant buildings, the regulation seems to require that individual tenants, not just building owners and managers, establish written indoor air quality compliance programs. Each tenant must designate a specific person to take necessary steps to assure the program is implemented. The "designated person" must be knowledgeable in the regulations and have technical expertise in HVAC system functioning. Apparently a tenant could "designate" the building manager's" designated person" to act for the tenant with regard to the massive amount of documentation required on the building HVAC system. However, tenants would remain responsible for substantial reporting requirements on activities within their own space.

The required reports include written records of employee complaints of signs or symptoms that may be associated with building related illness. There is no guidance to define what complaints fall into this category. Records must include the nature of the illness reported, the number of employees affected, the date of the complaint, and remedial action, if any, taken to correct the source of the problem.

Other reports required for each space include a description of the building, its function, and known air contaminants released into the space. Reportable air contaminant sources include new furniture or wall coverings, painting or remodeling, pest extermination, cleaning materials, solvents, and" other airborne substances which together may cause material impairment." OSHA acknowledges that the definition of airborne contaminants may be broader than that applied to industrial environments.

Where smoking is permitted, compliance essentially entails establishing a separate enclosure, exhausted directly to the outside, and maintained under a negative pressure.

If these regulations are promulgated without major revisions, they will substantially change the way building occupants and managers do their jobs. Better get familiar with them, or write your Congressman!

Dave Elovitz

December 1994

GREEN LIGHTS

Many business people consider government agencies the enemy. And with good reason, considering that sometimes it seems like we spend more time and effort navigating the regulatory morass than running the business of our businesses. But one agency actually has a cooperative program that benefits businesses. Surprisingly enough, that agency is the US EPA, and the program is called Green Lights.

The EPA launched the Green Lights program in January 1991. Its purpose is to reduce pollution by providing technical assistance and encouraging lighting energy conservation. Green Lights has three types of participants:

> PARTNERS are businesses that use lighting energy and sign up for the program. The benefits include access to technical assistance in lighting energy conservation and permission to use the Green Lights logo that identifies the business as "green". The only obligation appears to be an agreement to implement lighting upgrades determined to be profitable and report project completions to the program administrator so EPA can keep track of how well the program is doing. Hundreds of businesses of all sizes are partners.

ALLIES are electric utilities, lighting manufacturers, distributors, and energy service companies that help customers implement lighting energy conservation measures. Boston Edison and New England Electric are Green Lights Allies.

ENDORSERS are associations and societies that promote the goals of the program.

The EPA's 1993 Green Lights annual report says almost 500 million square feet of building space have been surveyed under the program, and 100 million square feet have been upgraded with more efficient lighting systems. The EPA boasts pollution reductions of 200,000 tons of carbon dioxide, 1500 tons of sulfur dioxide, and 750 tons of nitrogen oxides per year as a result of Green Lights upgrades. These pollution reductions come from 371 million KWH per year of electricity savings, which save customers almost $30 million on their electric bills.

Green Lights looks like a win-win program. For once a government initiative benefits the people who pay the bills. The Green Lights hot line is 202/775-6650.

Ken Elovitz

January 1995

to the tune of
RUDOLPH THE RED NOSED REINDEER

Energy Economics
Charming, witty engineers
Who hope that you, our clients
Have had happy, joyous years.

When all the office tenants
Ring the phone and call you names.
We'd like to solve your problems;
Change complaints into acclaim.

When we go out to a site
We look all around.
Then come home and calculate
'Til the problem we have found.

Then will the tenants love you,
And we hope you'll shout with glee,
"Energy Economics,
You have solved my mystery!"

Energy Economics
Wants to help you build it right.
What we find on your drawings
We'll explain in black and white.

HVAC, Electric,
Energy, or Smoke Control.
We'll look it over for you
Watching out for your bank roll.

When we spot a hidden flaw,
We'll step up to say,
"Here's what we think you should do:
Saves some cash and works well,
too."

Then how the clients thank us
As they shout out loud with glee,
"Energy Economics,
You're the engineers for me!"

Energy Economics:
Three guys with the same last name.
Energy Economics:
Guys that make your problems tame.

Each year we send our greetings
And our wish that all of you
Meet with success and prosper
All throughout the year that's new!

David, Franny, Gary, Ken
Simply want to say,
"Thanks for bringing us to mind
When you find you're in a bind."

Now is the time for sending
As we do this way each year
Our deep and heartfelt wish for
Health and joy in the new year!

Dave, Ken, and Gary Elovitz

February 1995

CAN'T FIGHT CITY HALL
(or can you?)

Architects and engineers are supposed to design systems that comply with the design aspects of building codes. Contractors are supposed to install systems that comply with the workmanship aspects of building codes. And inspectors are supposed to check the whole thing for code compliance.

Unfortunately, with today's fast changing technologies and the consensus and review process that most code writing bodies follow, codes don't address every situation directly. From time to time, when faced with new or unfamiliar situations, an inspector rejects an installation based more on lack of understanding than actual code violations. The rejection letter puts everyone on the horns of a dilemma - accede to the inspector's demand (if it's even possible to comply), perhaps increasing costs or compromising system function and design concept, or appeal and risk delay.

That type of situation occurred on a large project when an electrical inspector rejected an installation because some of the built in furniture contained electrical devices but was not itself a UL listed assembly. The owner and designer decided to appeal. Reading the code carefully showed that the furniture at issue was not "equipment" as defined in the code and therefore was outside the code's jurisdiction. Simply providing physical support and a mounting surface for electrical devices did not make the furniture electrical equipment. Even if the furniture could be considered "equipment", the code does not require all equipment to have a UL listing. Listed equipment can always be used, if applied properly, but UL listing is not the exclusive test for suitability. In fact, the code provides an eight step check list to determine whether "equipment" may be connected to an electrical system.

After a lengthy hearing where both sides presented their cases, the appeals board took the matter under advisement. A few weeks later, they reported that the inspector had agreed to withdraw his rejection.

The appeal process was relatively painless, took much less time than many had feared, and solved a big and potentially very costly problem for this particular project. More importantly, perhaps, it established a vital precedent that could head off similar conflicts on future projects.

So, not only can you fight City Hall, but if your facts are straight and you are well prepared, you can even win!

Ken Elovitz

March 1995

NOT WITHOUT RISK

I read an article in a recent issue of *The Firesafety Designer* discussing a new miracle fire fighting agent. *The Firesafety Designer* is a quarterly newsletter of the Firepro Institute, an organization dedicated to improving the delivery of fire safety information. The article quoted an announcement from a major chemical company, "We hereby announce the discovery of a new fire fighting agent. It augments, rather than replaces, existing agents such as dry powder and carbon dioxide, which have been in use from time immemorial ...."

"Though required in large quantities, it is fairly cheap to produce and it is intended that quantities of about a million gallons should be stored in urban areas and near other installations of high risk, ready for immediate use."

According to the article, the new product is not without controversy: A noted professor pointed out that if anyone immersed their head in a bucket of the new agent, it would prove fatal in as little as three minutes. A Fire Service expert pointed out that the material is a constituent of beer. Might fire fighters become intoxicated by the fumes?

Preliminary lab tests by the Institute for Chemical Safety show the substance is a near universal solvent. How long will fire hoses last, and what will we hold it in?

The Friends of the World reported that they had tested a sample and found that it caused clothes to shrink. If that is what it did to cotton, what would it do to people? Other investigators reported severe deterioration of steel components exposed to the new material.

As a result, one political candidate announced that he would strongly oppose storage of the new extinguishing agent in his district unless the most stringent precautions were followed. Open reservoirs would definitely not be acceptable, he assured his constituents: What would prevent people from falling into them? He called on the Legislature for a full investigation and appointment of a blue ribbon Major Hazards Group to study the problem and prepare a report.

Have you already guessed the name of this

> **W**onderful
> **A**nd
> **T**otal
> **E**xtinguishing
> **R**esource?

Dave Elovitz

April 1995

TROMPE L'OEIL

"We had a broken water main recently, and I think it's fixed, but something strange is going on," puzzled Barry, the building manager. "I think you should take a look."

First I got some history: there are two water mains from the street - domestic water and fire protection. About a week ago, water started gushing from the ground. Barry's crew "pressure tested" both lines and determined the leak had to be in the domestic water main. They dug down to find the leaking pipe, but the backhoe couldn't dig deep enough to uncover all the piping; both mains had "u" shaped expansion loops that dipped well below the building. When they didn't find a leak, they concluded the leaks were deeper underground and simply bypassed the expansion loop.

Two problems remained: First. although the leak was much reduced, there was still a trickle that didn't seem to come from the domestic water main. Second, they "pressure tested" the domestic main after the repair, and its pressure dropped significantly overnight. With the crew ready to fill the holes and patch the parking lot that day, Barry needed to be sure the leaks were really fixed.

I asked Barry to describe the pressure tests. "The pressure in the street is 100 psi," he explained, "so we let street pressure build up in the pipe, shut the valves, and watch the gauges. The first time, the domestic main went to zero in no time, but the fire main held 100 psi all day. We tested the domestic main again after making our repairs: We filled it to 100 psi last night, but it was down to 30 psi this morning. Could we have a pinhole leak?"

"I doubt it." I answered. "You see, when you pressure test with water, you have to account for expansion and contraction with changes in temperature. Last night was cold; if the temperature in the pipe fell just a few degrees, the water would contract enough to drop the pressure significantly. You're lucky last night was cool. If it had warmed up, the expanding water might have burst the pipe."

Not convinced the fire main was tight, I turned my attention to it and asked to repeat the pressure test. Barry closed all the same valves, and we watched the gauge carefully. It stayed rock steady at 100 psi. Then I noticed the small valve ("cock") at the base of the gauge. I turned the knob slightly, and the gauge quickly dropped to 0 psi.

" Here's your culprit," I said. "This gauge cock was closed the whole time, so you never saw the real pressure in the fire line. The fire line is leaking." More digging uncovered a crack in the fire main just below the floor slab. Once fixed, they were high and dry again.

Gary Elovitz

May 1995

PLAIN TALK

Relief at last!! Bill Cavanaugh of Cavanaugh Tocci just sent us a clipping about how Congress is working to make government talk to us **in** plain English. The proposed" Job Creation and Wage Enhancement Act" says that agency heads can't issue a new regulation unless the OMB certifies that it is written in reasonably simple and understandable language and is easily readable - short sentences, no double negatives, confusing cross references or convoluted phrases. Sounds great!

Apparently Congress was concerned that the Director of the OMB might not recognize objectionable language, because later on the law spells out "Prohibited Regulatory Practices" and I quote:

(a) DEFINED. - For purposes of *this subtitle, "prohibited regulatory practice" means any action described in subsection (b)(i), (ii), or (iii)* of *this section.*

(b) PROHIBITION. - (1) No employee of *an agency who has authority-*
(A) to take or direct other employees to take,
(B) to recommend, or,
(C) to approve,

any regulatory action shall -

(i) take or fail to take, or threaten to take or fail to take,
(ii) recommend or direct that others take or fail to take, or threaten to recommend or direct, or

(iii) approve taking or failing to take, or threaten to so approve

such regulatory action because of *any disclosure by a person subject to the action, or by any other person.* of *information that the person believed indicative of ... [mismanagement coercion, etc.]*

(2) An action shall be deemed to have been taken, not taken, approved, or recommended because of *the disclosure* of *information within the meaning* of *paragraph (1) if the disclosure* of *information was a contributing factor to the decision to take, not to take, to approve, or to recommend.*

Aren't you glad that once this law is passed you won't have to cope with confusing government language any more?

Dave Elovitz

June 1995

LOOP THE LOOP

"We installed some new boilers, and they haven't worked right since day one," the manager complained. "We've even had to jump out the flow switches to keep the boilers on line. And we've had so many shut downs for high temperature that we replaced the manual reset controls with automatic resets."

"Before we head out to the site, let's look at the drawings to see how the system was supposed to work," I suggested. The drawings made the problem obvious - each boiler served a space heating loop plus an indirect hot water heater, each with its own pump. The boiler manufacturer recommends a certain minimum flow rate for adequate circulation through the boiler. Otherwise, portions of the boiler will get too hot, and the safety controls will shut the boiler down. The 20 GPM heating pump provided just enough flow to meet that minimum. But the domestic hot water pump, sized at only 8 GPM to match the heat exchangers, moved less than half the recommended minimum flow. If the domestic hot water pump came on when the space heating pump was off, the boiler would overheat and shut down.

Clearly, the system needed increased flow in the domestic hot water loop. But a bigger pump alone would not solve the problem. Pushing 20 GPM through the domestic hot water loop would require almost 40 feet of head. The heating loop pump, on the other hand, developed only 17 feet of head. With two pumps piped in parallel, the head across them must always be the same, even if they serve separate loops. After all, the pressure drop from point A to point B has to be the same, regardless of which route the water takes.

We could create a primary/secondary loop, but that would involve an extra pump and quite a bit of piping. There had to be a better way. "Wait a minute! I only need 20 GPM of flow through the boilers, not through the heat exchangers. I can just install a bypass line with a ball valve for balancing to bypass the extra 12 GPM around the heat exchangers." With the bypass line, the new hot water pump could even be the same model as the heating pumps. Besides solving the low boiler flow problem, the new pump and bypass line increased hot water system reliability. Since the heating and hot water pumps would now be identical, the heating standby pump could also be a standby for the hot water system.

Ken Elovitz

July 1995

TWO OUT OF THREE

The other day a friend of mine told me he had met a management consultant who handed him a business card:

I. M. SMART
Management Consultant

GOOD, FAST, CHEAP
Choose any two

I chuckled heartily at the time, but the message on that card keeps popping up in my mind. I'm sure that I.M. Smart hadn't intended a serious philosophical message about the construction industry, but it strikes me that" Good, Fast, Cheap - Choose any two" sums up in a six words the way our industry works.

Our clients span the full range of approaches, so we see all kinds of thinking. Some owners believe the best value comes from a fast-track project with a negotiated price from a professional construction manager: That the financial benefits of getting the project on-line earlier, and the quality benefits of cooperation throughout the design and construction team, more than make up for the costs of more design and construction management services.

Other owners rely on what they hope is a complete and tight set of construction documents to get what they want from the lowest bidder. Some even build in long design and planning lead times before the start of construction and plenty of construction time to analyze, check and recheck, and plan out on paper so that everything can be done just right once the work starts in the field.

We even have one multi-project client that says the best approach is not to expect things to be done right anyway, but accept the lowest bid, get the contractor off the job as quickly as possible without long periods devoted to pursuing punch list corrections, and include enough money in the project budget to go back and correct what he believes will be the inevitable goofs on the owner's account.

Each believes the path they follow is the best, and the chosen path seems to work for each. The disappointments and disputes seem to arise when someone unrealistically expects to get all three: Good, Fast, and Cheap.

Dave Elovitz

August 1995

LANGUAGE BARRIER

A client told me about his recent trip to Montreal. He pulled a matchbook from a well known Montreal night club out of his jacket pocket. Holding it for me to see, he grinned and said: "You know, this language thing really gets me. I went to this night club and asked a girl to dance. She didn't speak English, but I asked her to join me, and she came over to my table. I drew a cocktail glass on my napkin, and she smiled. So I called a waiter, and she ordered a drink. With signs, we were able to establish that her name was Cherie, and mine was Joe. The she drew some stick figures. It was a couple dancing, so we got up and danced a bit. With just these two little symbols, we had a most enjoyable evening."

Montreal is supposedly more inviting to US tourists than it was a decade ago, but some residents still resent anything that isn't French. So I asked my client how he happened to wind up in a part of the city where no one speaks English.

"That's the main point of my story," he continued. "She didn't speak English, and I don't speak French. But we had a most enjoyable time. Then - and get this - after we danced and had another drink, she turned the napkin over like this, and drew a bed with two stick figures in it. I'm still trying to figure out how she knew I'm a hotel engineer!"

SURVIVING

An experienced outdoors man went hiking in California and got lost. He used up all his food and grew very hungry. At that high altitude he found some berries but not enough to sustain himself. So he got desperate. Using shoe laces, he made a snare and captured a wild bird. He lit a fire, cooked the bird, and was just finishing the meal when the environmental police swooped in. "You killed a wild condor, an endangered species!" the Ecop cried. The outdoorsman explained that he did not mean any harm but just wanted to survive. The Ecop arrested him anyway.

The outdoorsman had his day in court. He explained the situation to the judge, who pronounced him guilty. The judge imposed the mandatory 5-year jail term but suspended it due to the extenuating circumstances, warning the man not to go into the wild without adequate food supplies.

The outdoorsman was the last to leave the courtroom, and as he got into the elevator, he found himself alone with the judge. The judge complimented him on his resourcefulness before saying, "I'm curious. What does wild condor taste like?" "Well," the outdoorsman replied as the elevator door opened, "somewhere between bald eagle and spotted owl."

Ken Elovitz

September 1995

BREAKER BREAKTHROUGH

"Our little hot water pumps are overloaded and trip the circuit breaker. We switch to the standby pump and it runs OK for a while, then POOF! the breaker trips again."

The breaker was tripped when I got to the boiler room, so I reset it, put my ammeter on record, and left it to monitor the breaker while I checked the pumps and traced out the piping. After a half hour, the meter showed a maximum of 5.7 amps. I didn't see how 5.7 amps could trip a 20 amp breaker, so I moved my meter over to the pump motor.

The motor was slightly overloaded: 3.5 amps measured compared to 3.4 amps nameplate. But with its 1.4 service factor, the motor could safely run with up to 4.7 amps. Still, the motor seemed awfully hot. It was! The casing measured 156F. The boiler room was almost 100F when the boilers came on, so maybe room temperature was the problem.

Suddenly, BOOM, followed by silence as the pump stopped. The breaker had tripped! Naturally, I was off looking at the boiler room ventilation, and my ammeter wasn't on the breaker at the time. I thought the motor thermal protector might be the problem, so I checked the motor winding resistance. It measured 3.6 ohms - it would show an open circuit if the protector had tripped.

With no wiring diagram I resigned myself to tracing out the rat's nest of power and control wiring. As I pulled the cover off the boiler control, I noticed a little carbon stain on it. "Must have arced," I thought to myself. Sure enough, one wire had a piece of tape covering a nick in the insulation, and there was a small burn mark on the insulation of the wire next to it. But the tape looked OK. "That explains the carbon stain," I thought. As I jiggled the wires around in the box to trace them, a small piece of tape fell off the end of one blue wire. And the end of that blue wire was right next to the carbon stain on the box cover!

It took a little more detective work to find out that the blue wire came from the normally closed contact of a relay. That meant the blue wire was live when the boiler reached its temperature setpoint and the burner shut down. The loose tape was no problem when the burner was on, but when the boiler shut down, a live wire brushed against the control box cover. When a spark jumped, the short circuit would trip the breaker.

I wonder how many wire nuts the electrician could have bought for the cost of all the trouble that loose piece of tape had caused?

Ken Elovitz

October 1995

SECOND CHANCE

It looked like a perfect candidate for a "Design/Build" job: The existing HVAC systems in the 20,000 sf office building were outdated and unreliable. What's more, with very tall glass curtainwall on all sides, both cooling and heating the perimeter offices would be a challenge. The old systems were clearly not up to the task.

After an initial walk through, we proposed a design/build approach. We would prepare a Scope of Work that defines the overall design concept with measurable performance criteria, including the required cooling and heating capacity in each room. The Scope would allow the contractor flexibility to fill in the design details and build the project efficiently. To minimize cost and disruption to the tenant's ongoing operations, the Scope would permit reusing as much of the existing ductwork as possible. We would keep an eye on the project for the Owner to see that the contractor followed the Scope.

But we never got a response to our proposal. About a year later a new client asked us to look at HVAC problems in a small office building they had just purchased. To my surprise, it was the same building! The building came with a full set of plans and specifications to gut the existing systems and replace them with heat pumps. There were even four bids - all in the order of $330,000.

The prices seemed high, so we suggested the same design/build approach to the new owner and got the go-ahead. The Scope called for new rooftop units with enough capacity to meet the peak loads. It included converting the existing distribution to VAV for improved comfort and economy plus new perimeter heating systems to offset drafts from the tall glass curtainwall. The successful bid came in at $235,000.

During construction, we kept an eye on the job and sorted out the few inevitable complications that always come up. There were almost no "extras" because the contractor was responsible for the outcome, not for specific hardware in a specific arrangement. The new systems are now up and running. For the first time in years, the building is comfortable.

The other day I had lunch with the former owner. Out of curiosity I asked why he chose the "plan-and-spec" approach. "We thought design/build would be too expensive," he replied. "It has to cost more because the contractor's price includes engineering plus enough slack to cover all the unknowns." Should I have told him?

Gary Elovitz

November 1995

DIGIT FIDGET

"I need your help on our electric bill. My cost per KWH went from $0.10 last year to $0.13 this year, and I can't get anywhere with the utility. They either ignore me or give me the run around. You talk their language, so see what you can do."

I looked through the past two years' bills and noticed that something had happened in August 1994. The rates had not gone up, so I used the meter reading data on the bills to calculate each component of the bill and see where the money was going. Usage had increased slightly over last year, but the August demand stood out - it was almost twice anything in the recent past. That demand spike cost $2600 by itself, but the "ratchet" clause in the rate meant the customer would continue to pay for that demand spike for the next 11 months.

It seemed odd that demand would nearly double in August and then return to normal in September. What kind of fluke could create that big a spike? I looked at the total connected electrical load and wondered if turning everything in the building on at once could create a demand as high as the August bill indicated. And if demand did increase, why was there no corresponding increase in usage? The problem was beginning to look like a billing error, but what type of error and how could I convince the utility to refund something like an $18,000 over charge?

I studied the demand readings. As I typed the data into my spreadsheet program, I suddenly realized what must have happened! Like me, the utility's billing clerk probably used the number pad on the computer. With dozens of bills to key in, the clerk might not have noticed the slight slip when the customer's actual demand of 252 kW got entered as 552 kW - an easy mistake to make with the "5" key right above the "2" and no other digits in the demand reading.

That explanation was obvious to me, but how could I convince the utility? Looking at the billing history, I saw that demand had never even been close to 552 kW. In fact, it never even reached 300. My report showed that an August demand of 252 KW fit the pattern with July and September and was consistent with the previous summer's data. The client presented my report to the utility and offered to meet for what we expected to be a wrestling match over the real August demand, the actual over payment, and whether we could get any of it back.

To their credit, and the client's delight, the utility simply reviewed our report, acknowledged the error, and issued the appropriate refund.

Ken Elovitz

December 1995

QUOTH THE MAVEN

Once upon a morning dreary,
While I pondered, weak and weary,
While I struggled, nearly weeping,
Suddenly there came a beeping,
Then the pages quickly heaping,
Heaping on my office fax.
'Tis some client sending data,
Something that can wait 'til later.
    This it is and nothing more.

On and on the sound went longer,
Hesitating, then e' en stronger,
Slowly from my desk I turned;
Curiosity in me burned.
Lifting pages to the light,
I could scarce believe my sight!
Anxious, urgent, was the pleading.
Can I be at Tuesday's meeting?
    "Without you, our chance is poor."

Patiently, I dialed the numbers,
Tuesday's slate to disencumber.
One by one, rescheduled clients,
Most of them were quite compliant.
Then call the sender of the plea:
"Explain just what you need of me."
Insurer of a boiler vendor,
Being sued and needs defender.
    Quoth the lawyer, "Save his store!

Sold steam boiler, header, piping.
To a tann'ry that is griping.
Says steam's wetter than it oughta,
Pasting tanks are full of water."
"What's a pasting tank?" I scream.
Flat steel tank all full of steam.
Wet new hides are "pasted" on
There to dry 'til moisture's gone.
    "What's the problem? Tell me more."

Claims their expert, we did wrong:
Steam brings big wet drops along.
The way he knows the steam's too wet:
With biggest steam trap they can get
They can't drain all water from
Says that's proof the steam's no good
Boiler don't work like it should.
    That's the cause and nothing more.

My friend, their expert is a jerk!
This is how steam systems work:
Boiler heat makes water steam,
But that is only half the scheme
Steam flows where for heat there's need,
Then for the process to proceed
Steam gives up heat just where it oughta
All the steam turns back to water!
    That is what you use steam for!

To do its job the steam condenses,
That is how its heat dispenses.
Steam that's used its heating force
Turns to water, that's the source.
Not from steam that came in wet
But from heat the tank did get
The B-T-U's into the hide
Made water from the steam inside!
    Quoth the maven, "That's the score!"

Dave Elovitz

January 1996

CYBERTALE

'Twas the end of the year, and it seemed a safe bet
We'd find our good friends out cruising the 'Net
So we booted up Windows, clicked once, clicked again
Entered our username, password, and then
There arose on the screen to our shock such a clatter!
We leaped into cyberspace: "What is the matter?"

The high resolution red, blue, and green
Gave the luster of mid-day to words on the screen
When what to our wondering eyes should we see
But an E-mail from Plant Engineer Dot P-E.
His tale it was sorry, his mood cast a pall,
His troubles were many, we downloaded it all:
"I got automated, slashed maintenance cost

Put the whole thing on CADD; no data got lost.
I replaced my mechanics with Pentium chips,
Downsized my whole workshop, did away with work slips;
Instead of that cheery gal saying hello
I installed a new voice mail that says, 'Ho, Ho, Ho!
'You've reached Engineering, for service just hold,

Press one for too hot, press two for too cold.
Press three for strange noises, press four to back up ... '
And so on and so on 'til people crack up.
I stopped going on calls, fixing things face to face,
Just installed a new fax and stayed here in one place;

I entered it all on the latest spreadsheet,
Then my hand slipped a key and I hit the delete.
Now I'm sitting alone tapping out this epistle
My plans have all flown like the down of a thistle;
The news is not good from the basement tonight,
My systems are down, and there's no help in sight.

And to you who are reading, 'ere I log off the 'Net,
As my battery fades, I have one message yet:
Don't count on high tech to fulfill all your dreams,
There's a place for us old fashioned humans it seems."
Then we read on the screen 'ere he slipped out of view,
"Happy New Year to all, may your dreams still come true." :-)

Dave, Ken, and Gary Elovitz

February 1996

PIPE DREAMS

The problem was in one of those elegant Commonwealth Avenue mansions that had been divided into condominiums. The ground floor unit didn't have any cooling even though all the units directly above it had plenty. The building manager had already figured out that the problem was no flow but didn't know why or what to do.

The new cooling system used "high rise" type fancoils. These units have built in risers so they stack one above the other. The risers on one unit" plug in" to the risers on the unit below, so chilled water distributes through the building vertically. With flow through the upper level units, why didn't the ground floor unit have flow?

The ground level apartment wasn't part of the original design. During construction, a separate fancoil was added for it at the bottom of the stack. To check for flow, we measured pressures. We had 17 psi difference between supply and return in the mains, which confirmed flow to the upper levels. But there was no pressure difference and therefore no flow at the ground floor unit.

The chilled water mains snaked their way through the small space above the ground floor ceiling. We couldn't really see into the small gap above the unit. We could see that the return on the ground floor unit was connected to the return on the upper floor units. It looked and felt like the supply main elbowed up into the riser for the floors above, while the return main elbowed down to the supply of the ground floor unit. That really seemed odd: Piped that way, the ground floor unit was in series with the upper floor units. Units piped in series have to have the same flow, so how could the upper floors have flow when the ground floor did not?

"Bobby, our only choice is to tear open this wall so we can see how this unit is actually piped." Once the gypsum board was off, we found the ground floor unit sitting on a 3" high wooden box. "Why would they do that? What's under there?" A neat copper loop connecting the supply and return together!

The installer who came back to add the ground floor unit must not have taken the time to analyze the system. He must have assumed the system was the type that requires a U-bend at the end. But these units had coils and valves, so the U-bend did not belong. It let water bypass the ground floor unit instead of flowing through its coil.

Sometimes calculations and test results alone can't solve the problem. When you don't have drawings or other records of what was done, you might have to take things apart to see what's there.

Dave Elovitz

March 1996

CLUELESS

It's funny how many clues you can ignore or explain away when you start out with the wrong hypothesis. The other day my wife told me that the left front turn signal on our car had burned out.

"The turn signal is clicking fast (which usually means a dead bulb)," she told me, "and the plastic lens is cracked too. Whatever broke the lens must have killed the bulb, too."

I looked up my car model in the book at the auto supply shop and bought the appropriate replacement lamp (even though it looked like it would be too big to fit into that small flasher housing). Sure enough, when I removed the old bulb, it was obviously wrong. Funny that the book would list the wrong bulb, I thought.

On my second trip to the auto supply I brought along the old bulb and got a dead match. (Funny, though, the old bulb didn't look dead, but you can't always tell just by looking.) I plugged in the new bulb, taped up the plastic housing neatly, put everything back together, and tested it by turning on the emergency flashers (it was easier than turning on the ignition to test the turn signal). I could see from the driver's seat that the damned bulb still didn't light! But even worse, the front right signal was dead, too! I knew the emergency flashers worked, though, because the rear lights flashed just fine.

So I disassembled the right front flasher, too, and pulled out both front flasher bulbs. Both bulbs looked OK and even tested OK when I checked them with my ohmmeter. It must be a wiring problem! I didn't relish the prospect of tracing out all that wiring, but I had already sunk half the afternoon, and I wasn't about to give up then.

With about half of the car's wiring disassembled (or so it seemed), a thought crossed my mind: if the right flasher was dead, too, why didn't the right turn signal flash quickly like the left one had? I turned on the ignition and tested the right flasher again. Sure enough, it clicked nice and even.

Boy, was that frustrating! I didn't even bother shutting off the turn signal or the engine when I got out of the car. And a good thing, too, because once out of the car, I could see that the right front flasher was flashing just fine! But the flasher is the little light below the bumper (which you can't see from the driver's seat)! All this time I had been monkeying with the parking lights! A third trip to the auto supply to buy back the "wrong" bulb I had just returned, and I was back in business.

Gary Elovitz

April 1996

IT ALL DEPENDS HOW YOU LOOK AT IT

Definition of *a Cow*

In 1966, "American Red Angus" magazine described a cow this way: A cow is a completely automatic milk manufacturing machine, It is encased in untanned leather and mounted on four vertical, movable supports, one on each corner.

The front end contains the cutting and grinding mechanism, as well as the headlights, air inlet and exhaust, bumper, and foghorn. The rear contains the dispensing apparatus and an automatic fly swatter.

The central portion houses a hydrochemical conversion plant consisting of four fermentation and storage tanks connected in series by an intricate network of flexible plumbing. This section also contains the heating plant complete with automatic temperature control, pumping station, and main ventilating system. The waste disposal apparatus is located at the rear of this central section.

In brief, the externally visible features are: two lookers, two hookers, four stand-uppers, four hanger-downers, and a swishy-wishy.

There is a similar machine known as a bull, which should not be confused with a cow. It produces no milk but has other interesting uses.

Spell Checker

I halve a spelling checker,
It came with my pea sea.
It plainly marks four my revue,
Mistakes I dew knot sea.

I've sent this message threw it,
And I'm shore your pleased too no.
Its letter perfect in its weigh;
My checker tolled me sew.

ASHRAE

We thought you would like to know that Dave Elovitz has been elected a Fellow by the American Society of Heating, Refrigerating, and Air Conditioning Engineers,

Dave Elovitz

May 1996

WHAT A GAS

The new nursing home had opened in February. Each resident room had a combination gas heat/electric cool throughwall unit. The kitchen, the corridor heat, and the domestic hot water boilers used gas, too, for a total connected load of more than 6 million BTU/hour. Most gas companies would love a new load that size, but there was no natural gas in the area. This building used propane from four 1000 gallon underground tanks the propane supplier had furnished and installed.

The problem occurred in cold weather when the water heaters fired. They drew so much fuel they sucked down the gas pressure in the rest of the building. When the gas pressure dropped too low, the through wall units shut down, and Maintenance had to go around to every one and reset it. As a temporary measure to get through the cold weather, the propane supplier parked a tank truck on the site and used it to supplement the feed from the underground tanks.

"You're way short of vaporizing capacity," I concluded. "Propane is stored as a liquid but used as a vapor. The heat to vaporize the liquid and push it out of the tanks comes from the surroundings. If you try to pull too much vapor from an underground tank, the surrounding soil gets too cold, the moisture in it freezes, and heat transfer to the tank slows way down. You can increase vaporizing capacity by adding more storage tanks, but to be on the safe side, I'd install a mechanical vaporizer."

"No way we can be short of vaporizing capacity," the propane supplier snorted. "We have dozens of underground tanks and have never needed a vaporizer. The problem last winter was a kink in the 1/2" underground line that connects the four tanks. We replaced it with a 3/4" line above ground, and it's guaranteed to work OK this winter. The only thing we have left to do is go out to the site next week and dig up that 1/2" line. I'll bring the kinked piece back and give it to you."

A month or so later the owner called again wanting to know what I thought about the propane supplier's proposal to solve the problem by installing a mechanical vaporizer. "I'm definitely in favor of it," I told him. "That was my original suggestion. By the way, what did they find when they dug up the 1/2" line?"

"You were right. Vaporizing propane made the tank so cold, the ground froze so solid they couldn't even dig the first shovel full, even in July." Sometimes those fancy calculations really do predict what goes on in the real world!

Ken Elovitz

June 1996

ENGINEER STORIES

Dan Holohan, steam heating guru par excellence, told me this one: An engineer is walking down the road when he comes across a frog sitting right in the middle of the road. "Kiss me!" says the frog. "I am a beautiful princess put under a spell by a wicked witch jealous of my beauty. Kiss me and I will once again become a beautiful princess.

The engineer bends over, picks up the frog carefully, studies it, and puts it in his pocket. The engineer starts walking along again, and the frog keeps shouting from inside the pocket. "Hey, I'm a beautiful princess. Kiss me!" come the muffled cries, over and over. Finally in exasperation, the frog scrambled up far enough to get her head out of the engineer's pocket and croaked, "What's the matter with you? I'm telling you I'm a beautiful princess. All you have to do is kiss me and I'm yours."

"Maybe so," observed the engineer, "but I am just a simple engineer. What would I do with a beautiful princess? But a talking frog! That's something I can use."

* * *

A mechanical contractor, an HVAC engineer, and a factory rep are driving down a mountain road when the brakes give way. The car screams down the mountain, faster and faster, but they finally manage - more by luck than anything else - to stop it just inches from a 1000 foot drop to jagged rocks. They get out of the car shakily, and the contractor looks under the hood and says, "I think I can patch it up." "No," says the engineer. "We should take it into town and have a specialist look at it." "But first," says the factory rep," I think we should get back in and see if it does it again."

* * *

Ken Gill of HDR Architects explained to me how you can tell an extrovert engineer from an introvert engineer. It's simple: An extrovert engineer looks at your shoes when he talks to you.

* * *

And finally, for those of you who haven't saved your copies of all the calendars we have been sending you, we have compiled another anthology of the most popular messages. Just give us a call or drop us a note and a copy of Still More Cracked Ice is yours.

Dave Elovitz

July 1996

TOWER POWER

"We think we need a new cooling tower. All summer we barely got 85 degree water. We had to keep bleeding and adding city water to keep our units on line. Let's get a new tower with more capacity so we won't have this problem again. The utility even said they'd chip in part of the cost as long as the new tower fan motor is smaller than the existing one."

The building was all electric with water cooled VAV package units. The cooling tower was rated to cool 1500 GPM from 95F to 85F at 78F wet bulb. It had two cells with 10 HP fan motors. The total connected load was only 1343 GPM, leaving a 10% cushion for safety factor and supplementary units. Something else had to be wrong.

A site visit revealed nothing unusual about the installation. There were three pumps, one for each tower cell plus one standby. A differential pressure bypass valve let water rat race back over the tower if too many of the package units did not call for cooling. The operator told me he had no idea how the bypass valve worked and didn't know whether to keep it open or closed.

"The rat race flow is supposed to maintain flow through the pumps and prevent excess pressure build up," I explained. "Too much flow through the valve could starve the air conditioning units. Maybe the controller needs to be adjusted so the valve maintains constant flow over the tower, opening up only when the units downstairs do not need the cooling water."

I thought back to the pumps. Were those balancing cocks wide open? Maybe the tower had more than design water flow. With but the leaving water temperature will not be as low. That might make the operator think the tower was not performing up to design.

Those ideas all made sense, but they didn't seem to tell the whole story. "Can you measure the fan motor amps for me?" I asked the building engineer. "11 amps on #1 and 8 amps on #2." The 10 HP motors are rated for 13 to 14 amps at full load. At 8 amps, the one tower fan was down to about 2/3 of its design amps or something like 85% of airflow. That means tower capacity about 15% less than design. The tower couldn't make temperature, even if the flow and bypass valve were OK!

Replacing worn sheaves, tightening fan belts, and adjusting some controls was a lot quicker and less expensive than replacing a cooling tower on top of an 11-story building, and it got the system working just as well.

Ken Elovitz

August 1996

WHAT A RELIEF!

"This job is a disaster! We can't keep the offices comfortable, and half the time the front doors blow open by themselves! The tenant is withholding rent and threatening to move out!"

The specs called for a Variable Air Volume (VAV) system. To cut costs, the contractor had installed a "bypass VAV" system, in which the supply fan always runs at full capacity, and a main "bypass damper" dumps excess supply air into the return duct when that capacity is not needed.

Unfortunately, the contractor also cut some corners: The flimsy bypass dampers he installed tended to bind and hang up. But worse than that, the dampers were in the wrong place! They were in branch ducts, so they did not relieve excess capacity evenly. Instead, they "robbed" capacity only from the branches that had a damper. And to top it all off, the contractor used gravity relief hoods on the rooftop units instead of the specified exhaust fans. When the system took in 100% outside air in its economizer mode, the gravity relief couldn't handle all that air. The extra air pushed its way out the front doors.

Fixing the problem would not be easy: Better bypass dampers could be installed in the right places, but it would be a squeeze, and they would be a bear to service. The gravity relief hood on the rooftop units could be replaced with exhaust fans, but clearance was tight, and the wiring to the rooftop units was too small to add the relief fans.

I had an idea - retrofit the supply fans with variable speed drives. "Variable speed drives slow down the fans as load decreases," I explained, "so we can get rid of the bypass dampers entirely. The existing controls won't work if we just disconnect the dampers, but we can 'trick' the controls into thinking the bypass dampers are still there.

"The drives will also help the relief problem, because the building won't be at peak load when the system is on economizer. Even if the load is as high as 80% of peak on economizer, slowing the supply fan would reduce the need for relief by 36%, which should be enough to keep those doors closed. The drives will even save energy, because fans draw less power as they slow down. Of course, the tenant will reap the savings from your investment, but you can probably use some good news for the tenant right now."

The client remained skeptical, so he decided to price the job both ways. The drive retrofit came in only a bit higher than replacing the bypass dampers and adding exhaust fans. But, considering the other advantages, the drives were well worth the extra money.

Gary Elovitz

September 1996

MORE ENGINEER STORIES

Our June calendar inspired my friend and retired mentor Eric Hammond to send me this story, which he attributes to Garrison Keillor:

Four engineers were speculating on the nature of God in the creation of Adam. The structural engineer spoke of the marvels of the skeletal-muscular system, its incredible strength to weight ratio, use of lever arms, and elasticity. "There can be no question but that God is a structural engineer," he proclaimed.

"Not so," cried the electrical engineer. "Only an electrical engineer could devise such a highly efficient transmission/communications network serving the whole body."

"Nonsense," said the HVAC engineer. "Look at the elegant pumping system and elaborate ventilation and temperature control systems. Only an HVAC engineer could think through such things to come up with such a result."

The fourth engineer had been sitting silently in a corner. The others finally turned to him to settle their dispute. "My friends," he said, "I am a civil engineer, and I can prove beyond a doubt that God thinks like me and my fellow civil engineers. Who but a civil engineer would run a waste pipeline right through the middle of a prime recreational area?"

* * *

Friends David and Randy Klatzker sent us this related story they snagged off the Internet:

Abraham (as in Genesis) decides he wants to upgrade his PC to Windows '95. Isaac is incredulous. "Pop," he says, "you can't run Windows '95 on your slow, old 386! Everybody knows you need at least a fast 486 with a minimum of 16 megs of memory to multi-task effectively with Windows '95."

But Abraham, ever the man of faith, gazed calmly at his son and replied, "God will provide the RAM, my son."

* * *

We have reprints of some fascinating articles from the ASHRAE Journal. The titles are "Understanding Smoke Management and Control," "Design for Commissioning," and our most recent addition, "Evaluating Smoke Control and Smoke Management Systems." If you would like copies of any of these articles for yourself, just give us a call or drop us a note, and we'll be happy to send them out.

Dave Elovitz

October 1996

HOW TO BE A SMART CLIENT

Over the years, we have worked with many different clients. Some seem to have gotten more for their money than others. We identified six characteristics that those "smart" clients seem to share:

1. GET HELP EARLY

It may seem contrary to common sense, but engaging a consultant sooner rather than later can save money both in the short run and in the long run. Recognizing the need for outside help and getting that help started before other events foreclose options often makes the final solution simpler and less expensive.

2. TELL THE WHOLE STORY

It is surprising how often clients withhold information from consultants. Maybe they want to test the consultant's abilities. Maybe they're afraid they'll influence the consultant unduly. But consultants are on the client's side. Clients who aren't comfortable sharing information and communicating openly need a different consultant.

3. BE OBJECTIVE

Problems and projects are rarely black and white. The most successful solutions arise when both the client and the consultant remain objective, putting personal preferences and prejudices aside. Answers that you don't want to hear or wish were not the case are not necessarily wrong.

4. BE DEFINITIVE

Consultants need a well defined scope. That scope might be specific questions to answer or specific problems to solve. Other times it might be identifying the problem or defining the scope for further study. Either way, clients and consultants need to agree on goals and objectives at the start of the relationship. Otherwise, neither will have a way to measure success and achievement.

5. PAY ATTENTION

We always appreciate the trust that clients put in us. At the same time, blind trust can be as bad as distrust. Clients need to pay attention and understand their consultants' recommendations. Consultants advise. Only clients can decide.

6. BE REALISTIC

Consultants are professionals, which means they use knowledge and experience in connection with professional judgment to reach conclusions and formulate recommendations. Consultants are not magicians or miracle workers. We cannot change the facts. We can help you see what they mean.

Ken Elovitz

November 1996

RE-ENGINEERING

To: Orchestra Trustees
Re: Recent Concert of Schubert's *Unfinished Symphony*

After attending a recent concert of Schubert's Unfinished Symphony, I developed the following recommendations for our corporate re-engineering program:

(A) We do not need the conductor at public performances. The orchestra has obviously practiced and received prior authorization from the conductor to play the symphony at a predetermined level of quality. Considerable money could be saved by merely having the conductor critique the orchestra's performance during a quality circle meeting.

(B) For considerable periods, four oboe players just sat there. Their numbers should be reduced and their work spread over the whole orchestra, thus eliminating peaks and valleys of activity.

(C) All 12 violins were playing identical notes with identical motions. This is unnecessary duplication: the staff of this section could be cut drastically with consequent savings. If a larger volume of sound is required, this could be obtained through electronic amplification, which has reached very high levels of reproductive quality.

(D) Much effort was expended playing 16th notes, or semi-quavers. This seems an excessive refinement, as most listeners are unable to distinguish such rapid playing. All notes should be rounded up to the nearest eighth, which would allow the use of trainees and lower grade operators with no loss of quality.

(E) No useful purpose would appear to be served when the horns repeat the same passage that has already been handled by the strings. If all such redundant passages were eliminated, as determined by a QUAT (Quality Action Team), the concert could have been reduced from two hours to 20 minutes, with greater savings in salaries and overhead. In fact, if Schubert had attended to these matters on a cost containment basis, he probably would have been able to finish his symphony.

Dogbert (Dave) Elovitz

December 1996

OUGHTA BE A LAW

How many times have you seen something and said, "There oughta be a law!"

Well, you might be surprised to know about some of the laws that are on the books in Massachusetts. It's against the law to:

> Use tomatoes when making clam chowder.
>
> Direct profane, obscene or impure language at a participant or official in a sporting event ($50 fine).
>
> Wrongfully remove a dog's collar.
>
> Play the "Star Spangled Banner" as dance music or to play only part of it.
>
> Pullout or dig up a mayflower (the state flower). The $50 fine is doubled if the offense is done in disguise or secretly in the night time.
>
> Transport a colony of bees without a permit.

Foolish laws are not limited to criminal matters. The Legislature enacted a law making the corn muffin the official muffin of Massachusetts.

Of course, not all ridiculous laws remain on the books forever. The 1659 act that outlawed Christmas has been repealed.

And Massachusetts is not alone in harboring these peculiar statutes. In Atlanta, Georgia, it is illegal for a smelly person to board a streetcar. Along the same lines, in Gary, Indiana, it's against the law to take a streetcar or go to a theater within four hours of eating garlic. Pickle lovers will be relieved to know that Connecticut law makes it illegal to sell pickles that will fall apart if dropped from 12 inches.

Government bureaucrats can even find a way to make rational laws absurd. The November 1920 issue of *Compressed Air Magazine* reported that an American firm had been denied an export permit to ship some threaded pipe to Australia. The denial was not because we were at war (we were not), or because pipe was in short supply at home (it was not). Rather, the export permit was denied because the application was not complete. It did not state whether the thread was to be linen, silk, or cotton.

Ken Elovitz

January 1997

A LOOK BACK

'Twas back in nineteen-eighty
    We got in the cal'ndar game
With stories, puns, and plagiaries
    Without a trace of shame.
Case histories: Such tales we told
    Of things that we had done!
And sometimes off'ring sage advice
    Of use to everyone.

It's seventeen years later,
    And my! How time has flown!
Computers fast as lightning,
    With fax, not just a phone.
Back then we had one office,
    But now we've grown to three.
Just one guy then, but now two more,
    Each one with a P.E.

We hope that you enjoy our words,
    And laugh or share a grin.
If we have eased your day at work,
    Our pleasure it has been.
To share ideas, to make a point,
    To bring, we hope, that smile
And just remind you that we're here,
    And say to you with style:

We thank you for your confidence
    When offering a chance
To help you with your problems,
    Your systems to enhance.
We'll analyze and calculate
    Before we recommend
And then make sure you understand,
    Before your bucks you spend.

Gary's moved to Newton-town,
    Ken's office is enlarged.
Fran and Dave have traveled wide,
    Their batt'ries to recharge.
We're busy and we're happy,
    'Cause helping you is fun.
For which, on *eve* of brand new year,
    We thank you ev'ry one.

We value you as clients,
    But most of all as friends,
And so we send this message
    Just as the old year ends.
With wishes for the coming year,
    For joy, for health, for cheer,
Prosperity and happiness,
    For all that you hold dear.

We send to you these stanzas,
    In meter and in rhyme
To bring our fervent wishes,
    This is our paradigm:
May nineteen ninety *seven*
    Bring you all the best.
A year when you'll do all great things
    But know the joys of rest!

Dave, Ken, and Gary Elovitz

February 1997

ANOTHER GREAT AIR HUNT

The last adventure on every job seems to be a scramble to find out why the air flows are so much less than design. The new office uilding was no exception, but besides the usual efforts to show how poorly the system performs, the balancer tried his hand at a little ex post facto engineering. "The velocities are way too high," he cried. "Over 2000 feet per minute. There's no way the box can work with that high velocity. I have over 2" of pressure in the main duct and still can't get design airflow. These boxes are under sized!"

The design did indeed call for those high velocities, but I had calculated pressure drops and knew the high velocity was not a problem. I also knew that this system would operate most of the time at less than half the design air flow, so larger boxes would have been a control problem. The boxes were not too small.

At the site I measured pressures while the balancer measured the airflow at the outlets. I compared his flows to the box airflow sensor. About 25 to 50 CFM leaked out the box fan inlet, and using the flow hood properly accounted for another 100 of the missing CFM. By controlling the air handler fan manually, we proved that the box could deliver the design air flow.

The rooms could all get design airflow, but the pressure drop between the main duct and the outlets did seem high. I wanted to check the pressure right at the box inlet, but the flexible duct was not perfectly straight, so my readings were all over the place. The flex duct was no longer than 2 feet, so I doubted it was the problem. Must be the box itself.

We removed the flex duct to examine the box innards but were disappointed to find nothing unusual. For the sake of workmanship, I had the contractor eliminate the dip and offset when he reinstalled the flex duct. The new airflow measurements were real eye openers - where we previously needed more than 2" to get 890 CFM of airflow, we now got 890 CFM with only 1" of static in the main duct. And boosting the pressure in the main to 2" pushed more than 1300 CFM through the box -- 30% over design.

Examining the situation more closely revealed that the dip and offset in the 8" flex duct left it with an effective area equal to a 6" duct. That restriction accounted for the excess pressure drop and reduced airflow. I've often seen long runs of kinked and twisted flex duct restrict airflow, but now I know that even short connectors have to be pulled tight, especially at high velocities. At least one contractor agrees with me and knows why, but how can I convince the rest?

Ken Elovitz

March 1997

ODE TO GIL CARLSON*

We have put in all your check valves
And I swear they do not hold.
When we open up the system,
These here pipes, they should stay cold.

They get so hot I cannot bear
To hold them with my hand.
It's backward flow through these new valves
I'm sure you understand.

So I traveled to the far off site
And traced out all the pipes.
They first were cold but got so hot
All I could say was "Yipes!"

The pipes had all been drained bone dry
While boiler plants were plumbed.
They started burning oil again
Before the vents were thumbed.

When I opened up the hand valve
I heard gurgles and a whoosh.
The water, it was flowing,
But what gave it such a push?

And then the thing that Gil would say
Came into my mind's range.
Each system can have only one
Point of no pressure change.

As the boilers made the water hot
It wanted to expand.
Its pressure can build up quite high
'Less expansion space is planned.

The pipes all full of air
Were like a big expansion tank.
The water in the smaller tank
Worked like a pressure bank.

The diaphragm expansion tank
Became a reservoir.
It pushed water to the air-filled pipes
Whose volume was much more.

The problem went away
As soon as all the air was bled.
For then there was but one
Location where it can be said

That pressure remains constant
When the system is on line.
And that is what old Gil had said
Makes systems work just fine.

Ken Elovitz

*Gil Carlson was a prominent engineer who wrote a series of authoritative articles for the ASHRAE *Journal* in 1968 and 1969 that explained what makes hydronic systems work.

April 1997

AS THE HINGE CREAKS

"My furnace is sooting up my house," the caller complained." We made an appointment to evaluate the system, and I asked to have the oil burner technician there. He was knowledgeable and helpful, and we checked the boiler and burner thoroughly. I saw traces of soot around the boiler but nowhere near enough to be a problem upstairs. There was no soot in the fire box, and a combustion test revealed a good 83'% efficiency with no smoke. Nothing wrong there.

Then I checked the air conditioner. Even though the A/C and boiler don't run at the same time, the filter and coil would pick up residual soot in the air. None found.

The only unusual conditions were short run time and high oil pump pressure. Short run time increases the number of starts, which is a time when soot can form. High oil pressure increases the fuel input rate, possibly overloading the nozzle for that burner and boiler. Over firing can also overload the flue pipe, contributing to soot.

Just to be sure, I had the technician adjust the oil pump pressure and the burner air damper. A new combustion efficiency test showed 85% efficiency, even better than before.

I headed upstairs to look for soot deposits. Heat attracts dirt, so I checked the top of the baseboard heat. There were some carpet fibers but no soot. The client showed me a few tiny dirt spots on the flat surfaces of some panel doors, and there was some black dust on the floor under a door. She absolutely resisted the idea that the boiler might not be the source of the soot. "I know soot," she declared, "and that's soot!" I used a napkin to wipe up a sample to examine under a microscope in my office.

That afternoon I started my report. Nothing seemed wrong with the boiler. I described what I saw under the microscope - though mostly black, the "soot" had some shiny, metallic looking particles, making me question whether they came from the boiler. I decided to finish in the morning, and closed up the office wondering if this would be another one of those jobs when clients don't want to pay because they don't like the answer.

I woke up at 4:45 AM, and it hit me: The "soot" was only near hinges, and those shiny particles not only looked metallic, they were. The dirt wasn't soot from the boiler but wear particles from hinge surfaces rubbing!

In the morning, I finished up my report and included my new discovery. I still wondered whether I'd get paid for this one, but at least I think I solved the puzzle. Now, if I could only figure out that vane axial fan problem .....

Ken Elovitz

May 1997

DR. SEUSS TECHNICAL WRITER

Here's an easy game to play.
Here's an easy thing to say.

If a packet hits a pocket on a socket
 on a port,
And the bus is interrupted as a very
 last resort,
And the address of the mem'ry makes
 your floppy disk abort,
Then the socket packet pocket has an
 error to report.

If your cursor finds a menu item
 followed by a dash,
And the double-clicking icon puts your
window in the trash,
And your data is corrupted 'cause the
 index doesn't hash,
Then your situation's hopeless, and
 your system's gonna crash!

 You can't say this?
 What a shame, sir!
 We'll find you
 Another game, sir!

If the label on the cable on the table
 at your house
Says the network is connected to the
button on your mouse,
But your packets want to tunnel on
 another protocol,
That's repeatedly rejected by the
 printer down the hall,

And your screen is all distorted by the
 side effects of gauss
So your icons in the window are as
 wavy as a souse,
Then you may as well reboot and go
 out with a bang,

'Cause sure as I'm a poet, the
 sucker's gonna hang!

When the copy of your floppy's
 getting sloppy on the disk,
And the microcode instructions cause
unnecessary risc,
Then you have to flash your memory
 and
you'll want to RAM your ROM.
Quickly turn off the computer and be
 sure to tell your mom!

Dave Elovitz

June 1997

HORSE TRADING

One night a peddler went to see his mystic. "Master," he cried, "I am going to kill myself!"

"Perish the thought!" cried the mystic. "What could give you such a horrible idea?"

"Is it better I should starve to death? Today my horse died, and without a horse I cannot earn my living!"

"Look," the mystic replied. "The spirits will provide for you. Tonight, at midnight, meet me at the stable of the Count." The peddler had no idea what the mystic could mean, but obediently he arrived at the Count's stables at 12 midnight sharp. The mystic took him to one of the stalls and told him to take the beautiful white stallion standing there.

"No way!" cried the peddler. "I can't do this. The Count will have me hanged!"

"Not to worry," the mystic assured him. "Take the horse and go in peace." Since in those days one did not disobey a mystic, the peddler did as he was told.

When the peddler had gone, the mystic lay down in the stall and went to sleep. The next morning the Count arrived with his groom, and seeing the man asleep on the floor, kicked him and cried: "Hey you, who are you, what are you doing here, and where is my horse?" The mystic sat up and rubbed his eyes. "Praise be! The spirits have forgiven me!"

"What's this?" cried the Count. "What is going on, who are you, and where is my horse?" "Don't you understand?" asked the mystic. "I was your horse! I used to be a famous scholar. But one night I succumbed to the Evil Impulse and went to a prostitute. In punishment the spirits turned me into your horse. But in my misery I repented and prayed for forgiveness. Finally my prayers were heard and I have been changed back into a human being. Thanks be to the spirits."

Now the Count was a superstitious man and a respecter of miracles, so he also cried, "Thanks be to the spirits!" and let the mystic go.

Several weeks later the Count was riding through the town. Suddenly he spied the peddler leading his beautiful white stallion. The count leaped from his carriage and ran to the beast, struck him brutally on the rear end with his riding crop and shrieked, "Scoundrel! Ingrate! Going to prostitutes again?"

Dave Elovitz

July 1997

QUOTABLE QUOTES

Once at a social gathering, Gladstone said to Disraeli, "I predict, Sir, that you will die either by hanging or of some vile disease". Disraeli replied, "That all depends, sir, upon whether I embrace your principles or your mistress.

There are two classes of people in the world: Those who believe the world may be divided into two classes and those who do not.
    -Robert Charles Benchley

Television is called a medium because it is neither rare nor well done.
    -Ernie Kovacs

In the long run, we are all dead.
    -John Maynard Keynes

Civilization is the process of setting man free from men.
    -Ayn Rand

No matter how far you have gone on a wrong road, turn back.
    -Turkish proverb

I know not with what weapons World War III will be fought, but World War IV will be fought with sticks and stones.
    -Albert Einstein

Procrastination is the thief of time.
    -Miss Hubbard
    Jones Junior High School

If they don't stop letting foreigners into this country right now I'm gonna move to a country where they don't allow no immigrants.
    -A Bunker

In Germany they came first for the Communists, and I didn't speak up because I wasn't a Communist.

Then they came for the Jews, and I didn't speak up because I wasn't a Jew.

Then they came for the trade unionists, and I didn't speak up because I wasn't a trade unionist.

Then they came for the Catholics, and I didn't speak up because I was a Protestant.

Then they came for me, and by that time no one was left to speak up.
    -Martin Niemoeller
    Lutheran Pastor, former Nazi sympathizer, and eventual victim of Hitler's Nazis

    Ken Elovitz

August 1997

THAT BIG SUCTION SOUND

"Could you check out the heat pumps in our condominium building?" the property manager asked. "Sure," I answered, "What's the problem?" "We just had two compressor failures. The contractor says the systems are at the end of their useful life and all need to be replaced. He wants to use similar units, but we want to make sure his proposal is okay."

"I could check that for you," I replied, "but it sounds pretty extreme to replace all of the systems because of a few service calls. I've seen systems a lot older than these, and most still work fine. Why not replace units as they fail rather replacing perfectly good units now?"

Two reasons: First, the heat pump outdoor units were on the roof of the tony Back Bay brownstone; replacing them means hiring a police detail and closing off Beacon Street to bring in a crane. That's big bucks and costs the same whether you replace one unit or all six. Second, the condominium felt that the units were old and inefficient and hoped that more efficient units would have a good payback.

On the roof I saw a little rust on the unit casings, but they were far from falling apart. I opened up one of the units that had had compressor problems. The original compressor was still there! The suction accumulator was new, though. The situation was the same in the other unit with the "compressor failure." At another unit, sure enough, the original accumulator was on the verge of rusting through. (The accumulator collects cold refrigerant before it returns to the compressor. Because it is so cold, the accumulator tends to sweat. Apparently, the original accumulators were not very corrosion resistant, so they rusted through.)

"Look," I said, "You don't need to replace these units. Other than the suction accumulators, they are in pretty good condition. If you replace the rest of the suction accumulators (which should be relatively short money), you might get another 5 to 10 years out of these units. Of course, you might lose a compressor or two over the next 10 years, but you can replace a failed compressor if you need to. (That might cost almost as much per unit as replacing the whole unit, but you save the cost of the crane and the police detail.) As for efficiency, more efficient equipment would cost a little less to run, but it would take 10 - 15 years of energy savings to recover the cost."

You don't get to own a Back Bay pieds a terre by throwing money away, so you can guess which option they chose.

Gary Elovitz

September 1997

TRIVIA

Heron of Alexandria invented the first steam turbine as a toy in the 1st century. He heated water in a globe, and water jetted out of two openings, making the globe rotate.

Denis Papin of France designed the first boiler with a safety valve in 1679. He also designed the first steam engine in 1687.

Thomas Newcomen invented the first piston steam engine in England in 1712. James Watt improved the efficiency by adding a separate condenser to the engine.

The first electric chair was used in the Auburn Prison in New York in 1890.

A furnace heats air, but a boiler heats water.

Dr. Willis Carrier invented the world's first air conditioning system in 1902.

Air conditioning was first used in the early 1900s by the textile industry.

The Reaumur thermometer uses alcohol instead of mercury. It was invented around 1730.

In 1834 American inventor Jacob Perkins received a patent in England for mechanical refrigeration.

Clarence Birdseye, the co-founder of General Foods, is known as the father of frozen food.

A common mercury thermometer would freeze at -38°F.

The metal vanadium was discovered in 1801 by Mexican chemist Andres Manuel del Rio, who at first mistook it for chromium. Vanadium is an undesirable impurity in some heavy fuel oils but is useful as an alloying element to add strength, toughness, and heat resistance to steel.

10 average cows belch enough hydrocarbons to heat a small house for a year.

The Celsius temperature scale, which is named for Anders Celsius (1701-1744), is based on 0° for freezing water and 100° for boiling water. It was first suggested by a Greek physician, Galen, in the second century.

Ken Elovitz

October 1997

LINGUISTICS

Ken, Fran, and I had an interesting dinner with our friend Victor Neuman the other night when he was in Boston on business. Besides being one of the most accomplished laboratory ventilation experts around, Victor is a third generation professional engineer who can knowledgeably discuss a wide range of technical and non-technical topics, so it is always a delight to spend an evening with him.

This particular evening the conversation turned to the word engineer. Because the term has a different root in French than in English, Victor explained, engineers are accorded higher social and professional status in France than in the US: The French call people who do what we do ingenieurs from the French verb ingenier, "to exercise one's wits." In other words, to be ingenious. My big Webster's New International Dictionary tells us that the English word engineer is an alteration of an earlier Middle English word, engineer, a constructor or operator of engines, especially military engines. Linguistically, then, we engineers here in the US are basically glorified mechanics or machine operators, while our learned European colleagues are brilliant, clever fellows. We engineers are held in relative esteem by our respective societies accordingly.

According to The Encyclopedia Americana, engineering did not originate as a recognized or well defined human activity at any single stage of the world's history. From Archimedes (2250 years ago) through the Roman legionnaires who built battering rams, catapults, boiling oil throwers, and similar civilizing devices for enforcing the Pax Romana, and on to Leonardo DaVinci and Galileo Galilei (500 to 300 years ago)), every civilization contributed to the development of engineering. Achievements before the middle of the 17th century, however, were the results of ingenious men with natural ability and first hand practical experience rather than the outgrowth of a store of workable scientific engineering facts in the modern sense. It was only after about 1650 that schools of applied science began to take their places alongside schools of pure science and training schools for military engineers, and our modern idea of the field of engineering came to be. Not until 1818 would the term" civil engineering" be officially defined to differentiate civilian engineering work from military.

Just thought you'd like to know.

Dave Elovitz

November 1997

SURGERY

The President of one of our hospital clients tells of being awoken at 3 AM by the sound of running water. Descending to his living room, he was dismayed to find water pouring out of one wall. He searched the Yellow Pages for plumbers claiming 24 hour service, calling one after another without success as the water spread all over. Finally, one fellow promised to come right over.

Twenty minutes later, he stopped mopping long enough to answer the door to a smiling man in sparkling white coveralls who wiped his feet carefully before entering, handed him a business card, surveyed the situation, and disappeared into the basement to turn off the water.

Carefully spreading out a drop cloth, the visitor started neatly cutting a hole in the wall, vacuuming up every bit of plaster dust as he sliced away. Placing a fire extinguisher next to the opening, he unsoldered a cracked fitting and soldered in a new one, ever careful to leave no scorch marks or solder drips on the surrounding wall. He tested the line, turned on the water, and proceeded to patch the plaster opening so neatly, the difference in color was the only clue the wall had ever been opened. Then he took his vacuum back out, meticulously cleaned the whole area, carefully folded up his drop cloth, and put away his tools.

"All set!" He smiled reassuringly.

Needless to say our client was both amazed and extremely grateful. "I have never seen anyone work so neatly and efficiently! And I can't tell you how much I appreciate your rescuing us from this problem so quickly in the middle of the night. hat do I owe you?" he asked.

"$1500" came the reply.

"I don't want to seem unappreciative," our client said, somewhat taken aback, "But you have been here less than two hours. That's $750 per hour. Our brain surgeons in our hospital only get $500 per hour!"

"I know," the plumber nodded. "That's all I made when I was a brain surgeon."

Dave Elovitz

December 1997

THE YOUNG ENGINEER

My five year old son, Jesse, spent part of last summer at "gymnastics camp," where he learned tumbling, balance beam, and swimming and generally had a great time. One afternoon, when I picked him up at the end of the camp day, one of the counselors related the following story:

One of the kids' favorite activities is playing in what they call the foam pit. It is shaped like a small swimming pool, filled with blocks of spongy foam. The bigger kids climb down into the pit, and the counselors lift the smaller ones down. They have a great time piling the blocks of foam up into the most fantastic structures imaginable. This particular day, the counselor noticed that Jesse had climbed out of the pit and gone over to another part of the gym where a pile of big blocks of foam had been stored. She saw him busily dragging a big block of foam toward the foam pit. That didn't appear to be dangerous, so she decided not to intervene and just watch what developed.

Jesse dragged the big block, which was shaped like a set of steps, to the edge of the foam pit. He jumped down into the pit and carefully cleared some small foam blocks away out of one area 'and pulled the big step-shaped block down into the pit. Once he had the steps firmly based in that cleared away are on the floor of the pit, he showed the smaller children how to climb out using the new steps.

"You know," the counselor said, "we've had this foam pit here for years. The kids love it, but we have always had this problem with having to lift the smaller children in and out. It has bothered us for years but it never occurred to anyone on the staff to put steps in! But Jesse just saw the problem and set out to solve it."

When I got home with Jesse and told the story to Dena, she just said, "Well, there's no doubt about who his father is."

So if there are problems in your facility that have been bothering you for years, give us a call - we'll send Jesse right over!

Gary Elovitz

January 1998

THE WALRUS and THE CARPENTER REVISITED

"The time has come," the father said,
    "To write of many things:
Of cales - and loads - and pressure
    drop - Of CFM's - and swings-
And how to make a boiler hot - and
    when to change O-rings."

"No, wait a bit," the sons replied,
    "About that we must chat;
For there is much to think about,
    before we write of that!"
"Like clients," said the younger one,
    "for whom we don cravat."

"Much gratitude," the older said, "we
    should now send with speed.
To clients, friends, and vendors
    who've been very kind indeed.
So, if you're ready, Father dear, let us
    begin the screed."

"But, everyone," the old man cried,
    saying what was true;
"Expects our rhyming at this time to
    greet the good year new!".
"You're right, you're right," the two
    boys said, "So what else can we
    do,

"But go on rhyming, we and thee,
    rhyming as best we might
And we will do our best to make the
    verses smooth and sprite-
We'll write and write, e'en if it takes
    'til halfway through the night.

"We'll send our greetings happily, to
    each and everyone
Who called on us to help them out,
    and get their projects done
And told their friends, 'These are the
    guys:

They're good and they are fun!'

"We thank you all for being friends,
    for trusting us to try,
To straighten out your problems, and
    to do it on the fly;
We pledge to do our very best - your
    woes to rectify."

As it is said this time of year, we too
    say, "Ho, ho hoi
To you and yours for happiness and
    health our wishes flow.
For peace and joy, - fulfillment - and
    a year that's cheerio!"

Very best wishes to you and yours
    from us and ours,

Dave, Ken, and Gary Elovitz

February 1998

RATS ON THE RUN

Absenteeism. Discrimination and sexual harassment suits. OSHA inspections. These and other problems plague American business, but electrical contractors may finally have a way to avoid them. It's Judy's rat.

Judy is Judy Reavis, a technology consultant. Judy's rat is a real rat. She is a laboratory rat with white fur and pink eyes. When she was left over from a research experiment, Judy took her home as a pet.

Judy's consulting firm had volunteered some time to help wire schools to connect kids to the Internet. Wiring existing schools is a tough job that involves snaking cable through small spaces.

One day back in 1995, a colleague had told Judy he tried unsuccessfully to train a rat to pull cable through some tough spaces. Judy had trained rats in research labs, and she thought it could be done. Since she had an unemployed rat, she decided to try to train her rat to pull cable through tight spots.

Judy started by rolling some screen into a tube. She put some of the rat's favorite food at one end and encouraged the rat to crawl through. Little by little, Judy made the tube longer. She added bends and other obstacles. Judy would tap near the end of the tube so the rat would know where to head. The rat found the exit every time.

After a few months, Judy tied a string to the end of a Category 5 cable and put the string in the rat's mouth. The rat ran through the tube, pulling the string all the way to the end where its favorite food was waiting. Judy took the string and pulled the communication cable through the tube. Success!

Since then, Judy's rat has helped wire several schools. It jumps into a hole in one room and finds the shortest path to a hole in the next room where the food is placed. Through this experience, Judy's rat has learned a lot about wiring schools and the Internet and is willing to answer questions. You can learn more about Judy and her rat and even ask your own questions at www.judyrat.com.

Ken Elovitz

March 1998

NEW AIA DOCUMENTS

New 1997 editions of the American Institute of Architects (AIA) forms B141, *Standard Form of Agreement Between Owner and Architect* and A201, *General Conditions of the Contract for Construction,* were published last fall. Last revised in 1987, the new documents include some striking changes.

New sections 4.3.10 in A201 and 1.3.6 in B141 contain waivers of consequential damages. Under these clauses, the owner, contractor, and designer cannot sue each other for damages such as lost business opportunity arid extended home office overhead. These claims are typically complex, making them difficult and expensive to resolve. Everyone except the lawyers should be glad to see them go.

The Associated General Contractors of America (AGC) approved the new A201, but only after expressing concerns about section 3.12.10. That section clarifies the architect's ability to delegate responsibility for design of manufactured components or subsystems like curtain walls. The architect usually does not design the details of manufactured components. Rather, the manufacturer designs and furnishes them according to the architect's performance specification. The new A201 requires that those components or subsystems be designed by a licensed design professional.

The new documents also include a provision requiring mediation as the initial step in resolving disputes. Arbitration is still available, although consolidation is now prohibited. Therefore, under the new documents, an owner who wants to bring claims against both the architect and the contractor will have to proceed separately against each.

One change to B141 not adopted was a controversial proposal to return the words "general supervision" to the description of the architect's services. The term was eliminated from contracts years ago because courts had held design firms liable for workplace injuries on the basis of their responsibility to "supervise". The change would have allowed architects to expand their scope of services by recapturing an activity that architects traditionally had performed but lost to non-architects who offered "observation" services.

The new documents contain many changes, and, like any contract, should be reviewed carefully and tailored to the specifics of each project.

Ken Elovitz

April 1998

TALK TO THE BOSS

Every year we have to give our insurance company a detailed statistical analysis of the prior year's projects broken down into various categories. A rather tedious and time consuming task, but not really difficult, except the completed renewal application has to be submitted by January 15 with data for the year that just ended.

Until a couple of years ago, our insurance agency was a small company where we talked to the boss, and he read our mail himself. January 15 was no problem because he always saw that we had the renewal application forms in plenty of time. Then that agency - and our account-was acquired by one of the big guys. Now, the letter notifying us of the acquisition assured us, we would get better service than ever, with a large support staff to look after routine things so our agent would be free to concentrate on personal service. Are you surprised that now we have to call and call to get the applications in time?

Why should you care about our troubles with the big bureaucracy? Because the story illustrates one of the basic principles on which Energy Economics, Inc. is based: When you call, you talk to the boss. Not some clerk who neither knows nor cares about your problems. You talk to Ken or Gary or me. (Unless we are out or on the phone. But when you leave a message, one of us will almost always be back to you before the end of the next business day, even if it is to say "Dave is chasing orangutans in Borneo, can I help?") Your problems are important to us for a good reason: They are our living. We don't turn your project over to someone you never met and will never see or talk to. One of us crawls over your pipes and looks inside your ducts. And one of us studies your prints, does your calculations, writes your report, and answers your calls. Because that is how we know your project will be done in the professional, dedicated way we want it to be done.

I guess that's why we are just as responsive on small projects as on big ones: In 1997, we had 38 clients who each used less than $1,000 in services. They are just as important to us as the client who sends us $38,000.

And that's probably why 72% of the clients we served in 1997 were clients we had helped before.

Do you think the big guys will ever figure that out?

Dave Elovitz

May 1998

TRAPPED!

"We just lost another compressor and want to make this one our last. Can you take a look before we pull the corpse out?" The building had 3 identical rooftop units. Identical, that is, except that one had now destroyed 6 compressors in a year and a half.

The refrigeration piping seemed to break every rule. Each unit had two coils on one compressor. The coils had separate suction lines but shared a single expansion valve. With the expansion valve bulb on the suction line of one coil, the other coil might flow more liquid refrigerant than it could evaporate. On the other hand, the piping in the other units broke all the same rules, and they still had their original, 1985 vintage, compressors. The piping may have been questionable, but it did not cause the compressor failures.

As I traced through the piping and checked the auxiliary devices, I noticed that filter driers had been installed after the first compressor failure. The cores had been removed from the suction line filters, so they were not likely suspects.

The contractor started the new compressor, and we began our vigil. We knew the failures occurred at light load. We also knew from the compressor autopsy that the likely cause of failure was liquid refrigerant mixing with the oil in the compressor crankcase. Where did the liquid come from? We set up refrigeration gages and thermocouples so we could monitor pressures and temperatures.

As we waited and watched, I noticed one suction line was consistently several degrees warmer than the other. Suddenly, we saw liquid refrigerant bubbling in the crankcase. A few seconds later, the temperature of the warmer suction line plunged. We leaped to shut the system down before the new compressor could crash. Aha! I knew what was happening, and now I could see why.

The empty housings from the suction line filter driers that had been installed after the first compressor failure trapped oil, and oil attracts refrigerant. When the refrigerant pressure got low enough, as it could at light load, liquid refrigerant dissolved in the trapped oil boiled out, carrying refrigerant droplets into the crankcase. Refrigerant in the crankcase dilutes the oil. Diluted oil does not lubricate well, so the compressor smashes up.

After the mechanic repiped the suction lines so oil and refrigerant could not accumulate, more testing showed the problem was gone. By the way, the new compressor has now lasted almost as long as its 5 predecessors combined.

Ken Elovitz

June 1998

YOGI BERRA

Lawrence Peter "Yogi" Berra was born in 1925 in St. Louis. He got his nickname from a childhood friend who noted a resemblance between "Yogi" and an Indian snake charmer in a movie. After serving in World War II, Yogi joined the Yankees in 1946. He was a World Series MVP 3 times and was inducted into the Baseball Hall of Fame in 1973. Yogi became famous for his fractured English quotes like these, which made him one of the most quoted people in our times:

It ain't over 'til it's over.

It's deja vu all over again".

When you come to a fork in the road, take it.

You can observe a lot by watching.

The future ain't what it used to be.

Nobody goes there anymore, it's too crowded. (referring to Rugerio's restaurant in St. Louis)

You better cut the pizza in four pieces because I'm not hungry enough to eat six.

It was impossible to get a conversation going, everybody was talking too much.

Right-handers go over there, left-handers go over there, the rest of you, come with me.

If the people don't want to come out to the ballpark, nobody's going to stop them.

To NY mayor John Lindsay's wife after she said Yogi looked nice and cool, "You don't look so hot yourself."

I couldn't tell if the streaker was a man or a woman because it had a bag on it's head.

Baseball is 90% mental, the other half is physical.

When asked what time it was: "You mean now?"

Never answer an anonymous letter.

I knew I was going to take the wrong train, so I left early.

Interviewer - "Why, you're a fatalist!" Yogi Berra - "You mean I save postage stamps? Not me."

You got to be very careful if you don't know where you're going, because you might not get there.

Even Napoleon had his Watergate.

He is a big clog in their machine.

I'd give my right arm to be ambidextrous.

Ken Elovitz

July 1998

PUMPED UP

"Oh, and one more thing," the property manager said as we walked through the boiler room," I need to extend a drain line from the relief valve to that floor drain over there. Can I use 1" PVC?"

"Why do you want to pipe the relief valve to the drain?" I asked, "The relief valve isn't supposed to let go unless there's a problem."

"Maybe so," he answered, "but this relief valve pops at least half a dozen times every summer. The system can't take the high temperatures. "

"It sounds like you don't have a large enough expansion tank. As the water loop temperature increases, the water expands. Your expansion tank is there to absorb that expansion. If the expansion tank doesn't have enough room for the water to expand, the pressure builds up until the relief valve pops. I see you have glycol antifreeze in the heat pump loop. Maybe the designers forgot to allow for the extra expansion of glycol when they sized the expansion tank."

"Great," the property manager replied, "Now I'll have to find money in the budget for a new expansion tank, right?"

"Let me do some analysis and get back to you."

I calculated the volume of the system and the required expansion tank capacity: The existing tank would normally be big enough, even with the glycol! Puzzling over my notes, I suddenly realized that the expansion tank was connected to the discharge side of the main pump instead of the suction. With the expansion tank at the pump discharge, it has to operate at a higher pressure than it would on the suction side - in this case, 65 psi rather than 25 psi. The air in the tank starts out more compressed, which leaves less room for expanding water. A quick calculation at the higher pressure confirmed that the tank was of marginal capacity, even if it was precharged for the higher pressure. Since it probably wasn't, expansion capacity would have been reduced enough to account for the relief valve lifting. Just reconnecting the tank at the pump suction would "create" lots more capacity.

I faxed the property manager a sketch showing the piping change with a note: "By the way, better check the glycol concentration after you repipe: Every time the relief valve popped you lost a little glycol. If you haven't been adding glycol from time to time, you may have lost all your antifreeze over time."

Gary Elovitz

August 1998

SALESMAN

I was really tired when I got on the red-eye coming back from the West Coast the other day, but, as luck would have it, the passenger in the next seat was an insurance salesman who wanted to talk. He asked me if I wanted to playa fun game, but I just wanted to sleep, so I politely declined and leaned my pillow against the window, eyes closed.

"Come on," he persisted. "This game is easy and lots of fun. I ask you a question, and if you don't know the answer, you pay me $5. Then you ask me a question, and if I don't know the answer, I'll pay you $5."

"No thanks," I grunted and snuggled further into my pillow.

He thought for a few minutes, then offered, "OK, if you don't know the answer you pay me $5, but if I don't know the answer, I'll pay you $50." I could see that this torment wasn't going to end, and the odds sounded interesting, so I agreed to play.

The insurance salesman asked his first question: "How many people were killed in automobile crashes in New Hampshire in 1996?" I didn't say a word but just opened my billfold and handed him a five dollar bill. I thought that would end the game, but he was not satisfied. "Now it's your turn," he urged.

I thought for a minute, then asked, "What goes up a hill with three legs and comes down on four?"

The insurance salesman looked puzzled. He opened up his lap top computer and searched all his references. I dozed contentedly while he picked up the Airphone to connect his modem so he could search the Internet and the Library of Congress. In desperation, he sent e-mail to several coworkers, all to no avail.

After a couple of hours, he woke me up and handed me $50. I smiled, turn back to the window, and closed my-eyes once more. The insurance salesman, more than a little miffed, shook me and asked, "Well, so what's the answer?"

Without a word, I reached into my pocket, handed him a five dollar bill, turned away, and went back to sleep.

Shamelessly plagiarized by
Dave Elovitz

September 1998

MAGIC WORDS

The phone rings, and I answer: "Hello?" I am greeted by all-too-familiar silence. Eventually a cheery voice begins: "Hi, I'm Dennis, and I'm calling from AB&B communications with a tremendously great offer which I'm sure you'll want to hear all about. May I speak to the person who makes decisions about long distance telephone service?"

I'm tempted to make a wise-crack: "I'm sorry, but this isn't a good time for me. Why don't you give me your home phone number, and I'll call you back at about 3 in the morning."

I'm tempted to retaliate: "Gee, can I put you on hold for just a second?" But then I have to remember to check if he's still there half an hour later.

I'm tempted to ask: "Is this a sales call?" (Amazingly, it never is! "Oh, no! I'm not trying to sell anything! I just want to make you aware of a terrific opportunity ... ").

I'm tempted to lecture: "Do you realize how many times every day I receive unsolicited phone calls selling something I don't want? Do you care that you're disturbing me? I have a telephone for my convenience, and for the convenience of my friends and clients - not to provide anyone with something to sell a cheap way to advertise."

I'm tempted to blow a loud whistle into the phone.

But then I remember the magic words: "Please put us on your 'do not call' list." It works! Hucksters for phone systems, copiers, and replacement windows; well-meaning solicitors for every imaginable charity (except the" Foundation to Prevent Invasion of Privacy by Telephone") all seem to understand that the game is over. (Though I occasionally need to follow up with: "I never respond to telephone solicitations of any type.")

You might also want to know that writing to the following two organizations and requesting to be put on their "exclusion lists" should reduce the number of calls you will receive:

> American Telephone Fundraisers Ass'n
> USPS Kilbuck Branch
> 1001 California Ave
> P.O. Box 99311
> Pittsburg, PA 15233-9998

> Direct Marketing Association
> Telephone Preference Service
> P.O. Box 9008
> Farmingdale, NY 11735-9009

Let's work towards making boiler room operations with automatic phone dialers a thing of the past!

Gary Elovitz

October 1998

PUMP and PITCHER

When I was a young boy, my grandparents spent the summer at my Aunt Mae's cottage at Crystal Lake in Connecticut. Each Sunday, our family and all my Elovitz aunts and uncles and cousins made their way from various points in New England to that cottage. I remember stopping for gas at a Gulf station that gave away free comics, and on the 4th of July weekend, stopping to buy fireworks that would be shot off over the lake. We little kids were only allowed to have sparklers, but there were Roman candles and pinwheels and skyrockets to delight everyone.

But mostly I remember the great cast iron pump just outside the kitchen door, with its long handle that I could not reach if someone left it all the way up. I would put all my strength into cranking on that handle, raising it as high as I could reach, then pulling it down as far as it would go. The cold, clear water would come gushing out of the spout with each stroke to fill a bucket, or just to stick your head under for a drink. It didn't make any difference, though, how hard you pumped if the last cousin to pump had forgotten to leave a pitcher full of water you could pour down the hole in the top to prime the pump. Someone who came before you had to set some of his or her water aside and leave it for those who came after, just as the cousin before that person had pumped one extra bucketful to be there for the next pumper.

A client asked me the other day why I devote so much unpaid time to ASHRAE, off to national meetings twice a year, traveling all the way to Turkey to teach a course last spring, then to Toronto to teach it again last summer, writing technical papers and chapters for the Handbooks, and providing technical review for other papers. I just told him about that pump and the pitcher I always found there so I could prime the pump and use it.

I was just a little kid, and I don't know if I always remembered to fill the pitcher again once I had the pump going, so the next cousin could pump when he got there. But when I came into this industry many years ago, there had been engineers before me who had set aside some of their time and effort to make their knowledge and skills available for me. Now I am obligated, in turn, to see that the next generation of engineers who come after me have the benefit of what I learned from the wise men who went before.

Like using that old cast iron hand pump, each of us needs to set aside a little of our effort to help the guy that comes after us get started.

Dave Elovitz

November 1998

ODD SPECIFICATIONS

Thanks to Stuart Teger of Detroit construction law experts Honigman Miller Schwartz and Cohn for sending an interesting clipping about railroad track specifications. Did you ever wonder why all standard gage railroad tracks in the United States happen to be installed 56½ inches apart? What an odd figure! And it doesn't translate to a nice, neat metric number, either.

US railroads were built with tracks 56½ inches apart because that is how they were spaced in Britain, and the first US railroads were built by British expatriates. But why did the Brits use that odd dimension? Because the people who built the first British railroads had been building pre-railroad tramways, which used that spacing.

And why did the tramways use that spacing? Because tramway builders used the same tools and jigs for building tramways that they did for building wagons that were all built with the wheels spaced 56½ inches apart. Wagons were built with that spacing because wheels with any other spacing would break. Wheels with other spacings would break because they would not match the wheel ruts in the old British roadways.

Who built those old roads in Britain? The Imperial Roman Army built all the first long distance roads in Britain and in the rest of Europe. Where did the ruts come from? The ruts were made by the wheels of Roman war chariots, which were pulled by a pair of horses. Everyone else who wanted to use the roads had to match their wheel spacing to the chariot wheel spacing or risk breaking a wheel themselves.

So the specification for standard railroad tracks in the US to be installed 56½ inches apart is based on the original specification for a Roman war chariot's wheelbase. (I always suspected an awful lot of specifications just got copied over from one project to the next.)

So next time you pick up a spec and find something that makes you wonder what horse's ass came up with that idea, you may be right. You see, Roman war chariot wheels were specified to be 56½ inches apart to accommodate the rear ends of two horses!

Shamelessly plagiarized by
Dave Elovitz

December 1998

AS THE PUMP TURNS

"We just reopened this supermarket, and the cooling tower doesn't make temperature on hot days. The contractor said he doesn't know how the tower can work because there's no fan. He thinks we should convert to air cooled condensers."

I got the cooling tower model number from the contractor and dug an old catalog out of my files. The tower was a spray type unit that has no fan. It uses momentum from a high velocity water spray to induce airflow. My first thought was an undersized pump. I ran a few numbers and figured the pump should be about 15 HP. My second thought was plugged spray nozzles, or even worse, that someone had removed them in an effort to increase water flow.

"Man, these pumps are noisy!" I shouted as we walked into the mechanical room. "I know," the contractor answered. "I greased the bearings, but it didn't seem to help." The system had 2 pumps in what looked like a prepiped factory package. I checked the horsepower - 15. Theory #1 down the drain. Then I traced the piping. It was tricky because the two pumps were side by side but pumped in opposite directions. Eventually I figured out how the thing worked.

"We cleaned the nozzles last week," the contractor offered while I was on my hands and knees looking at check valves. Theory #2 down the drain. I would have to search for the key to this mystery.

Having determined the direction of flow, I was ready to measure pressures. Some gage cocks were plugged, and others were stuck. When I found a place I could measure, the pressure was only 12#. The tower needed 47# to work as designed. Then I checked the pump motor amps. A restricted pipe should give low amps. So should a pump running backwards, just like a fan. The amps were right in range.

I decided to verify pump rotation visually. The pumps had been there for years and presumably ran without problem. Still, we shut down the pump and watched the shaft as it slowed to a stop. Then I had the contractor bump the motor a couple of times to confirm: Despite the normal amperage readings, both pumps ran backwards!

The contractor reversed the leads, and we restarted the pumps. The awful noise disappeared, and the pressure was where it should be. Apparently the electrician reversed the phasing when he put in a new electric service. The client was thrilled. We not only identified the problem, we even fixed it on the first visit.

Ken Elovitz

January 1999

TO THE TUNE OF ...

Hark! Let us some mem'ries bring
Back to mind, so many things.
Some are sad and some bring smiles
Some may roll you in the aisles.
We're not trying to seem wise,
Just three thoughtful, helpful guys
Listing things you should recall
'Ere the Times Square sphere doth fall.
    Come look back on ninety eight.
    All in all, we found it great!

The El Nino brought us rain
Floods ensued beyond the plain.
Elsewhere they fought forest fires
Texas temps went higher and higher.
Winter passed without much snow,
Mercury did not plunge low.
Can't recall a year so strange
Is the next year gonna change?
    Weather all through ninety eight.
    Was not dull, at any rate.

Wall Street led a crazy chase
Zoomed way up, fell on its face.
Russia, Asia stiffed the banks,
Dunked the market in the tank.
McGwire and Sosa up at bat
Roger's record just went splat.
Here we sit with broken heart
'Cause the Red Sox fell apart.
    Looking back on ninety eight.
    What a fickle thing is fate!

Special legal counsel sought
Things that soon made Bill distraught.
Linda taped her friend Lewinsky.
Starr won out as chief buttinsky.
And if that were not enough,
Lieberman was talking tough,
Adding to the White House pain.
Sorrow Bill had to proclaim.
    When we look back on ninety
        eight.
    Maybe this was not so great.

'Nother year we shared with friends
Many problems we did mend.
Every month we wrote to you.
We had fun; hope you did, too.
Gary, Dena, Ken, Dave, Fran
EEI is all one clan.
At this time of every year
We send you in voices clear:
    Peace and joy and all things fine,
    Our wish to you for ninety nine.

Dave, Ken, and Gary Elovitz

February 1999

A DATE WITH TIME

It seems every magazine and newsletter has had an article about computer problems when the calendar changes to the year 2000. Some tell us not to worry - the programmers who created the problem by figuring out that a 2 digit shorthand for the year would save computer space will be smart enough to solve the problem before it ruins us. Others said start worrying now because the programmers who failed to anticipate the problem are certain to overlook some other problem in their fix. A recent article in IEEE *Spectrum* identifies several dates in the near future that no one mentions but could be serious trouble for computers:

The 2-digit shorthand for the year is the most talked about "Y2K" problem, but the rules for leap year introduce a little wrinkle. The earth orbits the sun in just under 365-1/4 days. We handle that situation by adding an extra day to February every 4 years (leap year). But the revolution is just under 365-1/4 days, so adding a day to every year divisible by 4 builds up a surplus over a century or two. We solve that problem by skipping leap year if the year is divisible by 100. However, a little known third rule for leap years makes the year 2000 a leap year because it is divisible by 400. The last time that situation occurred was 1600. We didn't have the same calendar we have now, and we didn't have computers, so we didn't have the problem. If our computers survive January 1, 2000, are they ready for February 29th?

Long before computers face those Y2K problems, they will face other hurdles. The Global Positioning System (GPS) used for navigation and other purposes includes a precise clock. It keeps track of dates by counting the number of weeks from January 5, 1980. The counter is programmed to reset to 0000 after week 1024, which will occur on August 21, 1999. This rollover feature was clearly documented when GPS was introduced, but did programmers take heed? It's not too likely that ships navigating by GPS will become lost at sea or that airplanes will disappear into cyberspace when the counter resets to 0000 on August 21st. However, some financial systems use the GPS clock to synchronize international funds transfers and calculate interest to the second. The banks have told us all about their plans for Year 2000, but are they ready for Week 1025?

A little less than 3 weeks later, we'll reach September 9, 1999, which computers will designate 9/9/99. Some computer programs use a string of consecutive 9s to identify the end of a file. Those programs could derail on the 2nd Thursday of next September.

Ready or not, here we go!

Ken Elovitz

March 1999

FLUFFERNUTTER

"The folks who built our dehumidifier say it's too small now that we've added more beaters. They want us to spend $70,000 for a whole new system. I hate to spend that kind of money unless I'm sure it's necessary." The caller was the maker of Marshmallow Fluff, and I was about to discover what precise conditions are required to produce that delicious comestible.

Marshmallow was originally made from the root of a pink flowered herb called the "marsh mallow." Nowadays, the ingredients are largely corn syrup, sugar, water, and air. That sounds simple enough, and it probably is for making individual marshmallows. But one pound of Marshmallow Fluff has to fill a one pound jar exactly or customers will think they are being cheated, so density must be controlled exactly.

The critical ingredient in Marshmallow Fluff is air, which is added by "beating" warm syrup in what looks like a huge kitchen mixer. Getting the right amount of air to end up with just the right density depends on the temperature and humidity in the beater room. Which is complicated by the fact that while all that air is going into the mix, huge amounts of warm moisture are driven out! And nobody knew for sure how much moisture the process gave off. But they did know exactly what conditions they need in the beater room to get consistent product.

Which conditions they were not getting since they added the new beaters. The equipment salesman's solution was "More beaters, more dehumidifier." I recognized that might be true, but I had to understand the process. First we made a schematic of the dehumidification and air conditioning system for the beater room, identifying each point where air conditions changed. Then we measured air flow, dry bulb, and wet bulb temperatures while the beaters ran at maximum production rate. That data let me calculate how much moisture the process released, how much moisture the dehumidifier and cooling coil took out, and how much heat the dehumidifier put into the airstream.

When I plotted the data on a psychrometric chart it was obvious that the dehumidifier wasn't doing what it should. The unit should have had just enough capacity to do the job, but it was actually doing less than half that. Pressure drop measurements suggested the desiccant wheel might be clogged, and, sure enough, investigation confirmed a thin film of Marshmallow Fluff on the desiccant.

It cost about 1/5 the cost of a new system to change some ductwork, put in a new desiccant wheel, replace a failed regeneration fan, and adjust the system. Results: Not an hour's production lost to humidity the next summer, a happy client, and who knows how many thousands more Fluffernutter sandwiches.

Dave Elovitz

April 1999

DOPEY KIDS

Friend and steam heating master Dan Holohan tells a story of what he saw when a 'fellow troubleshooter took him into a boiler room a few years ago. The place looked familiar, but after you've seen dozens, if not hundreds, of systems, some are bound to look alike.

They noticed the steam traps were not piped correctly and made a sketch of how to fix the problem. Then they saw some lines that should have been trapped but were not. Those problems were just the beginning. There may have been more wrong with the job than was right, but somehow it had worked for years, and all the problems were fixable.

"Who installed this system?" Dan asked his buddy. "I did," the friend advised. "You did?" Dan replied incredulously, waiting for the other foot to drop. "Yeah, and guess who told me how to do it?"

Suddenly Dan knew why the boiler room looked familiar. Indeed, he had been there before and indeed he and his friend had combined their meager months of experience to put this job together. Now, older and more experienced, they were looking at what a couple of dopey kids had done years ago.

The world is full of dopey kids. Some of them work hard, read, pay attention, and learn from their mistakes and the mistakes of others. Those kids grow up to become respected authorities. As professionals, they look for ways to fix problems without finding fault. They focus on what's right rather than what's wrong and emphasize what can be saved rather than what has to be thrown away. After all, the people who made those mistakes might have been the dopey kids who used to wear their clothes.

Sometimes clients call us looking for help on new projects. We hope we come up with sensible, maybe even intelligent, solutions. Other times clients call us looking for help on existing systems that never worked right or maybe don't work at all. We hope we have the good grace to focus on how to make things better rather than what dopes the original designers or installers must have been. After all, we never know if any of those dopes were wearing our old clothes.

Ken Elovitz

May 1999

A BOOST FOR LESS BUCKS!

"Ouch!" the developer's Tenant Coordinator said, "This one is going to hurt! We ran a 1½" conduit for this tenant's electrical service, but now the tenant says he needs a 3" conduit! That's over 140 feet of new conduit, through finished space, all at union rates. And remember, this is New York City."

"Wait a minute," I responded, "The tenant couldn't possibly need a 3" conduit. It's a 500 sq. ft. coffee shop, for crying out loud!" But when I looked at the tenant's design drawings, I couldn't believe what I saw: First, the tenant had squeezed a mezzanine into his space, where he installed several large freezers and other equipment. Then, he had packed the main floor with more coffee making equipment than I had ever seen - enough to make two cups of coffee every second, 'round the clock! Too much load for a 1½" conduit.

"Listen," I told the T.C., "This guy has really packed equipment into his space. I can't make business decisions for a tenant, but he can't possibly need all this. The lease says the tenant has to pay for additional electrical capacity. Show him the bill for the new conduit, and half of this stuff will disappear."

"Yes," the T.C. replied, "but remember, this is New York. This deal is very political, and we're going to end up giving this guy whatever he wants."

"I see," and we hung up. A little while later, an idea occurred to me. After a bit of research and analysis, I called the T.C. back.

"Listen," I started, "About our coffee shop friend: we may be able to make the 1½" conduit work after all. The Landlord provides 120/208V power, requiring 4 wires to the tenant's space. But if we install "buck-boost" transformers in the Landlord's electrical room, we can boost the supply voltage to 240V, 3-phase, which needs only 3 wires. This allows us to run larger wire (because one less wire is needed); also, the wire can handle more capacity because of the higher voltage. Together, these factors allow the 1½" conduit to carry 53% more power, which should be just enough to handle this guy's load. In the tenant's space, he can provide a 240V, 3-phase panel to handle most of the big loads and a small transformer to make 120V for lighting and small equipment. That should be a lot cheaper than running a new conduit."

"Sounds great," replied the T.C., "I'll see if we can make this fly. But remember, this is New York City."

Gary Elovitz

June 1999

REDUCE, REUSE, RECYCLE

The old boiler in the 75 unit apartment building had sprung a leak. Yes, it could be repaired, but the boiler was already over 40 years old and because of tight quarters, the repair would be expensive. Of course, new boilers aren't cheap either! On top of that, operating costs were out of hand, as were tenant complaints about both overheating and underheating. Time for a comprehensive look at the heating system.

The apartment building had "one-pipe" steam heat (fed by the leaking boiler) on the top three floors, and hot water baseboard (fed by a second old boiler) on the ground floor. The second boiler also ran year-round to make domestic hot water.

Changing the way heat is delivered to the apartments clearly would not fly; the cost would be prohibitive, and savings would not justify that investment. The one-pipe radiators were there to stay. However, we could improve comfort and reduce overheating cost-effectively with better controls on the boiler and better venting at the radiators.

What to do about the leaking boiler? And what about the second boiler? It was much bigger than needed, and running that behemoth all year for domestic hot water was wasteful. In fact, the second boiler was almost as big as the leaking boiler - several times larger than needed for its load. Hold on! Why not swap boilers?

Could the second big boiler be converted from hot water to low pressure steam? Yes, it could - there was even a tee with a blind flange in the steam main right above the second boiler, which would make the switch a cinch! (Could the original pipefitters somehow have foreseen this need?)

The leaking boiler was demolished right after the heating season, and a new, smaller, more efficient boiler was installed in its place. The new boiler efficiently provides domestic hot water and heat for the ground floor. The old hot water boiler was then converted to steam and connected to the one-pipe mains - with new controls - in time for the next season. It was a "green" solution overall: more efficient, less old equipment to the landfill, and - most important - more "green" in the owner's pocket.

Gary Elovitz

July 1999

ANAGRAMS

An anagram is a word or phrase made by rearranging the letters of another word or phrase. No letter can be used more than once or left out. Here's a good one:

To be or not to be: that is the question, whether it's nobler in the mind to suffer the slings and arrows of outrageous fortune.

In one of the Bard's best-thought-of tragedies, our insistent hero, Hamlet, queries on two fronts about how life turns rotten.

That's one small step for a man, one giant leap for mankind.
A thin man ran; makes a large stride, left planet, pins flag on moon! On to Mars!

Now some for you to try (answers on the back):

Dormitory	_____
Evangelist	_____
Desperation	_____
The Morse Code	_____
Slot Machines	_____
Animosity	_____
Mother-in-law	_____
Snooze Alarms	_____
Alec Guinness	_____
Semolina	_____
A Decimal Point	_____
The Earthquakes	_____
Eleven Plus two	_____
Contradiction	_____

Answers:
- Dirty Room
- Evil's Agent
- A Rope Ends It
- Here Come Dots
- Cash Lost in 'em
- Is No Amity
- Woman Hitler
- Alas! No More Z's
- Genuine Class
- Is No Meal
- I'm a Dot in Place
- That Queer Shake
- Twelve plus one
- Accord not in it

And one final anagram:

President Clinton, of the USA
To copulate, he finds interns

Dave Elovitz

August 1999

NOAH IN THE NINETIES

The Lord spoke to Noah, "In six months I am going to make it rain until the whole earth is covered with water and all evil is destroyed. But I want to save a few good people and two of every kind of animal. So build an ark!" And He delivered the specifications in a flash of lightning. "OK," said Noah, fumbling with the blueprints. "Six months and the rain starts," thundered the Lord. "The ark better be done, or you better have learned how to swim for a long, long time." Six months passed, and when the rain began to fall, the Lord found Noah sitting in his front yard weeping. "Lord, please forgive me," sobbed Noah. "I did my best. First I had to get a building permit for the Ark construction project, but your plans didn't meet the local code. So I hired an engineer to redraw the plans, and he got in a big fight with the Fire Department over whether the Ark needed to be fully sprinklered. Then the neighbors claimed the zoning by-laws would not allow me to build the Ark in my yard, so I had to get a variance from the Planning Board.

"There was a problem getting enough wood for the Ark because there is a ban on cutting trees to save the Spotted Owl. I had to convince Fish and Wildlife that I needed the wood to save the owls. But they wouldn't let me catch any owls, so, no owls. Then the carpenters formed a union, and I had to negotiate a settlement with the NLRB. Now we have 16 carpenters but no owls. Then I got sued by an animal rights group because I was taking only two animals of each kind. Just when I got that suit dismissed, EPA said I couldn't complete the Ark until you filed an environmental impact statement on your proposed flood. They didn't take kindly to the suggestion that they had no jurisdiction over the conduct of a Supreme Being. When the Corps of Engineers wanted a map of the proposed flood plain, I sent them a globe. I still haven't resolved a complaint from the EEOC over how many Hittites I am supposed to hire, and the IRS has seized all my assets, claiming I am trying to avoid paying taxes by leaving the country. Not to mention the notice from the Commonwealth about some kind of use tax.

"Lord, I don't see how I can possibly finish your ark for at least five more years," Noah wailed. Suddenly the sky began to clear. The sun began to shine. A rainbow arched across the sky. Noah looked up and smiled. "You mean you're not going to destroy the earth, Lord?" "No," said the Lord sadly, "the government beat Me to it."

from *Power Engineering* Magazine, May 1998, by Charles Juster.

Dave Elovitz

September 1999

PIPE DREAMS

"Our plumber thinks a sewer line must be broken somewhere under the floor slab," the building manager told me, "but the building has no basement and he has no idea where to look. Can you help us out?" I'm not a plumbing expert, but since I had worked with this client on other projects (and after a little prodding by the client), I agreed to take a look.

The builder had used an interesting construction process for this building: The site abuts a wetland, which explains the lack of a basement. Before construction, the developer covered the site with a pile of dirt about 20 ft. high and left it several years to compact the soil. Then that burden was removed, pilings were driven for the columns, and utility lines for the first floor (waste piping and main electrical feeders) were laid out on the ground. Huge concrete "grade beams" were poured in place to support the first floor.

Then the site was backfilled, burying all the underground utilities and holding them in place. Then a 6" thick concrete slab was poured for the first floor. "Let's start by checking a cleanout," I suggested. But when we opened the floor cleanout in the Men's room, there was no pipe! Just a hole through the floor slab and an overpowering stench. I got out the plumbing design drawings, but as I traced out where the waste lines ran below the floor slab, it became clear that the piping did not match the drawings (... and you thought that only happened with HVAC systems!)

The only way to solve the mystery was to get under the slab. The trick was finding a good place to cut a big hole in the thick concrete slab: we had to deduce where the piping ran and where a hole would give us the best access to the piping, with minimal disturbance to an occupied building.

When the first chunk of slab was removed, we realized that the problem was even worse than expected: the fill under a huge section of the building had dropped - up to four feet in places! As the fill sank over the years, it pulled the underground piping down with it, snapping pipe and pipe hangers like twigs.

Needless to say, there wasn't a lot of enthusiasm for climbing down through the hole to deal with the problem. Several yards of fill had to be dug out by hand to reach the parts of the main that had not settled and to clear a path to allow the damaged piping to be replaced. This time all of the piping was securely supported from the slab above, but we fitted an access plate over the floor opening - just in case.

Gary Elovitz

October 1999

WHOOSH

"We continually need to bleed air from our chilled water system," was the complaint from the new owner of an old client's building. I remembered a good deal about the system and even thought I knew what the problem was - insufficient pressure at the bottom of the system that left the high points under a vacuum. A vacuum pulls air in between pipe threads and through tiny gaps that are way too small for water to leak out. I asked the client to have gages and fittings so we could measure pressures at the site.

I dug out my old file and collected the papers I thought would help. I found my schematic diagrams and looked up the elevation of each floor so I could easily calculate the effect of gravity on system pressure. A building that high needs at least 95 psi at the bottom to provide 5 psi at the top and keep air out.

We started in the chiller room. I traced the piping to confirm there had been no major changes since I was last there. Everything looked the same. Then I asked to shut the pump off to check the pressure at rest - 115 psi. That was good news for the client because it meant there should be ample pressure at the top to keep air out. It was bad news for me because the simplistic fix - increase system pressure - wasn't going to solve this problem.

We gathered our tools and headed upstairs. I wanted to verify positive pressure at the top floor. The mechanic climbed up and measured the pressure: 25 psi - much more than the minimum to keep from sucking air in. Then we measured pressures in 4 or 5 other locations. Everything checked out perfectly - no restriction or other explanation for all that air. I recommended installing automatic air removal devices at the top of the system and promised to follow up with a report. That fix should work, but it was treating the symptom without addressing the underlying problem, so I wasn't entirely satisfied with it.

As I wrote my report explaining why the gizmos I recommended would do the job, I realized they would indeed treat the underlying problem. The system has an air separator in the chiller room where the operating pressure is 120 psi. At 120 psi, water can hold 3 times as much air as it can at the 25 psi at the top. Unless the air separator can simulate the effect of a 95 psi pressure reduction (it cannot), it leaves some air in the water. A system as big as this one - 7000 gallons - could need years of manual bleeding to get rid of all the dissolved air that slowly comes out of solution of its own accord at the top. Air removal devices at the low pressure point at the top would strip that air from the water and remove it without manual bleeding.

Ken Elovitz

November 1999

OREO COOKIES

Matt was a cowboy, even though he lived in the heart of Minnesota. He owned a horse, wore a cowboy hat, and spoke in an odd drawl. His nephew was stationed at the North Lakes Naval Facility near Chicago. The nephew had pestered Uncle Matt to take a day off and ride the commuter train from Minneapolis to visit him in Chicago.

Matt was finally convinced. He found the train station and boarded. At the first stopover, passengers were allowed to disembark at the station for 20 minutes. Matt climbed off, taking his ticket with him. Wearing his cowboy hat, boots, and duster, he wandered inside. He stood in line at the snack bar and bought a box of milk and a six-pack of Oreo cookies.

Matt sat down at a table next to a businessman. The man wore a tie and was reading the *Wall Street Journal* through half glasses balanced on his nose. Matt thought the man looked like Ward Cleaver. Matt opened his milk, took a swig, and opened the cookies. Ward glanced at Matt, then took one of the cookies. Matt looked up at Ward, who'd gone back to reading, munching on a cookie. Matt ate a cookie himself. As he washed it down, Ward reached out and snared another cookie.

People in Minnesota are trained not to make others ill at ease, so Matt didn't say anything. But he wondered if this was now acceptable behavior inside the castle walls of Minneapolis/St. Paul. He did take a second cookie, though, and quickly stuffed a third in his mouth.

Like a striking snake, the businessman grabbed the last cookie, scarfed it, stood up in a huff, and stalked off. Matt had obviously offended him. Matt wondered if he would be turned in to the Social Services Police. "It ain't this way in the country," he thought. "Nobody'd take even one cookie without askin', and to top it off, the freeloader never said thanks!"

Matt got up to board the train, picked up the wrapper, and tossed it in the trash. He felt in his pocket for his ticket, and there were his Oreo cookies.

Ken Elovitz

From cowboy, poet, philosopher, former large animal veterinarian, and NPR commentator Baxter Black, who attributes this story to Douglas Adams in So Long and Thanks for All the Fish.

December 1999

PUMPING AWAY

A client called for help in a building where I had worked some years ago. I remembered the system had an unusual twist - when the chilled water pumps started, the discharge pressure stayed the same, and the suction dropped. Usually it's the other way around: pump suction stays constant because we are trained to pump away from the expansion tank, which is a point of constant pressure. I remember thinking, "By pumping toward the expansion tank, pump pressure doesn't add to the initial fill pressure, so the chillers don't need high pressure water boxes."

I reviewed my diagrams and tried to solve the problem on paper before going to the site. This time, my diagram raised a bigger question than it answered. The diagram clearly showed "pump away" piping. Return water passed through an air separator/expansion tank, then through the chillers, then out to the building. I was sure I had verified the piping at least once before. How could this be a "pump away" system if the pressure at the suction goes down when you start the pumps?

I walked back and forth through the chiller plant, sorting out the pipes overhead. My diagram was right, and the system was piped for "pump away". Maybe a big pocket of air was acting like an expansion tank. I'd seen that happen before, but it seemed unlikely, especially in a system that had not been drained in more than 10 years.

Out we went with our gages to make a pressure profile of the system. I corrected for changes in elevation, and everything looked normal. We traced all the way back to the chiller plant, where the pressure was 120 psi on the return. I had to find a 60 psi pressure change somewhere in that mechanical room. Something that big should not be so elusive.

Then I remembered that each chiller has a flow control valve. Those spring loaded valves automatically throttle to maintain the design flow through each chiller. The flow control valve was the only device between the chiller and the pump suction. It was also the only device in the system that could eat up 60 psi of pressure. Even at peak flow, the pumps developed more pressure than was needed, and the flow control valve dissipated the excess. That explained the drop in suction pressure when the pump started. It also told me the pumps were oversized, so the pump impellers could be trimmed. Trimming the impellers would let the pumps produce less pressure for the same flow, saving a lot of energy because the flow control valves would have to dissipate just a few psi at full flow.

Ken Elovitz

January 2000

to the tune of
GOOD KING WENCESLAS

Good King Wenceslas looked out
    on the winter season
While the snow lay round about
    visually pleasin'!
Brightly shone the sun that day,
    fighting frost so cruel;
"We've got all this solar gain,
    why're we burning fuel?"

Aye, the sun comes beaming in,
    but the wind is stronger.
So much cold air infiltrates
    burner runs much longer!
It's all emissivity,
    and which wave radiation.
Stefan-Bolzman, Planck and such
    exotic information.

Call the guys from Energy
    Economics, quickly.
They will help us find the leaks,
    and correct them, you'll see.
They'll explain that stuff to us
    so we'll understand it.
Their reports are nice and clear,
    not transcribed in Sanskrit.

No matter which one comes to help,
    Dave, or Ken, or Gary,
He'll analyze and calculate,
    whatever's necessary.
Then he'll tell us how to cure
    our puzzling situations.
They also send us calendars
    with wise elucidations.

Good King Wenceslas, he smiled,
    "Their cal'ndars are amusing!
"They may be poems, they may be
    prose
    "but always worth perusing!

"Those are the kind of guys I need
    to keep me out of trouble.
"To tell me what is what and why,
    When I start seeing double."

King Dub-ya is a monarch wise.
    Heed his indication:
Read these calendars each month,
    and when you face frustration.
Don't tear your hair and gnash your
    teeth.
    Call and let us help you
'Til then we send you ev'ry wish
    For joy throughout the Year
    New.

Best wishes for a wonderful new millennium from all of us and ours to all of you and yours.

Dave, Ken, and Gary Elovitz

February 2000

THERE'S NO AIR THERE

The new tenant (calling from Florida) was ballistic: "We built this beautiful new kitchen - and we're paying top dollar rent - but every time we try to cook, the kitchen fills with smoke! You told us we could get 2000 CFM from this hood, but look at the balancing report: Your lousy system is only giving us 467 CFM. How can we cook with that?"

The landlord (on site in D.C.) was in a quandary: "We already sped up the exhaust fan to its max. - it's starting to shake. We can't give them any more capacity. When you specified the new fan a few years ago, didn't you say the fan would have plenty of spare capacity?"

I (in Boston) was in the middle: "It sounds like there's a real problem, but we need some more information. First of all, what is the static pressure in the main exhaust duct at your point of connection? The system is supposed to provide 1" of static at the tenant's connection. If we have 1" in the main duct then the problem is in the tenant's ductwork."

A technician was dispatched immediately, and came back with the word: 0.8". "See," the tenant said, "I told you it was your lousy exhaust system."

"Not so fast," I countered. "True, the base building system is not quite up to snuff, but that's not the main problem. You see, airflow varies with the square root of the pressure. Let's say we raised the static to 1"; that's 1.25 times the current pressure. The airflow from your hoods would only go up by the square root of 1.25, which is 1.12 times. Your hood would go from 467 CFM to 522 CFM - hardly enough to notice. The problem must be in the tenant's ductwork. Are there accurate as-built drawings for this job?"

"You've got to be kidding," the landlord chuckled. "But I'll make some sketches." The sketches arrived by fax later that day. It was pretty easy to see what had gone wrong. "Look, the design air velocity in the tenant's branch duct is much higher than it should be. That means pressure losses will be high. And these sharp bends in the ductwork make things even worse. No wonder the tenant's airflow is so low - the tenant's duct is so restricted that it hardly lets any air through. Increase the duct size and provide smooth elbows, and that should solve most of your problems."

"But what about the pressure in the main duct?" the landlord asked.

"We'll get to that next month."

Gary Elovitz

March 2000

E-ENGINEERING

As you recall from last month's gripping cliffhanger, the new tenant's bitter complaints about the landlord's kitchen exhaust system quickly subsided once we proved that the problem was mostly due to the tenant's own ductwork. However, the tests did show a shortage of about 10% on the landlord's side.

When I reviewed the fan test report, I was puzzled - performance was not at all consistent with the fan I had picked a few years back. A note on the report indicated that the pulleys had been changed twice, so I asked the Landlord about it.

"Oh, yeah," he replied, "We sped the fan up as much as we could to get more air, but the fan started shaking like crazy. We had to slow it back down."

Something was seriously wrong. I needed to see the fan with my own eyes, but I couldn't justify traveling to D.C. just to look at a fan. Then I remembered the landlord's digital camera.

"Listen," I told the landlord, "Can you e-mail me some pictures of the fan? Take about 10 shots from every angle you can think of."

The photos arrived that afternoon, and the mystery started to clear up. The main exhaust duct - 5 ft. tall and 20" wide - runs across the floor of the mechanical room. The contractor had stabbed a 30" round into the side of the duct to connect to the fan. You might get away with that type of fitting on low velocity HVAC, but not on kitchen exhaust.

The photos also showed a row of plugged test holes in the rectangular duct a few feet before the fan connection. That was where the technician had measured the airflow. I asked him where he measured the static pressure.

"Right where we did the traverse", he replied, and things began to make sense.

"Do me a favor," I said, "Get me a static reading right at the fan inlet. Poke a tiny hole in the canvas connection if you have to."

The answer came back: 2" greater than at the pitot traverse point. Now the readings all fit together. The stabbed-in inlet fitting was "eating up" almost half of the fan capacity and creating unstable operation, to boot! I worked out the fix in less than a minute.

"Look," I said, "All you have to do is rotate the fan 90 degrees so the inlet lines up with the duct. Then install a smooth transition, and you'll get your capacity back. The fan won't shake, and you'll even save more than $800/year on power for the motor!"

Gary Elovitz

April 2000

FLIGHT TEST

According to *Feathers,* the publication of the California Poultry Industry Federation, this really happened:

One of the hazards of concern to aircraft safety is the possibility of a collision between an aircraft and a flying bird. The Federal Aviation Administration developed an unusual device for checking the strength of windshields on airliners. They have a sort of cannon that projects dead chickens at a plane's windshield at approximately the speed the plane flies. I guess the idea is that if the windshield doesn't crack from the impact of the carcass projectile, it won't crack if there is a collision with a real bird during flight.

Now you may be saying to yourself that such a test concept is interesting but probably not of much interest to anyone else. You are wrong, however. A few years ago, the British were hot and heavy into developing a new high speed locomotive. The designers were concerned about how the locomotive windshield would stand up to impact. They did some research on the topic and discovered the FAA chicken test. That seemed to be just the test they needed, so they arranged to borrow one of the cannons from the FAA.

When the cannon arrived, the Brits hurriedly set it up on the tracks at the prescribed target distance from the locomotive windshield, dropped in a chicken, and fired. Not only did the ballistic chicken shatter the windshield, it went right through the engineer's chair, broke an instrument panel, and embedded itself in the back wall of the locomotive cab.

Fortunately, no one was in the cab at the time, but the Brits were stunned! They carefully re-checked the distance and the amount and type of propellant they used but could not find anything in their procedure that varied from the FAA's directions.

Finally, in desperation, they arranged for a technician with experience on the FAA windshield tests to come to the site and find the problem. The FAA technician repeated all the same inspections and tests, and he couldn't find anything the Brits were doing wrong, either. At last he asked them to repeat the test so he could watch it.

As they dropped the chicken into the cannon, the FAA technician leapt to his feet. "Hold it!" he shouted. "We thaw the chicken first."

Dave Elovitz

RISKY BUSINESS

The construction industry is a risky business. It's not like gambling, where the odds are mathematically calculated to favor the house. Rather, risk in construction comes from the fact that each building is one of a kind, a construction site has many more uncontrollable conditions than a factory, and construction is labor intensive with activities that cannot be completely standardized.

Successful projects manage risk, beginning with commitments by everyone involved to do their part to make the project a success. The owner, designer, and contractor must all recognize the risks, and each must accept the ones he is best able to manage and control.

The owner must communicate a basis of design. Engineers need to know what the owner expects the facility to do before they can design systems to meet those needs. The owner might need a consultant to develop the basis of design and translate it to terms the design team can understand. We call that function "mechanical programming", much like architectural programming that defines the size and shape of the building. Owners are more often disappointed because the designers did not know about needs that went unaddressed than because of design errors.

Engineers need to seek out the owner's needs and document those requirements as design criteria. They must help the owner understand how those criteria will affect the project, and they must create a design that meets the agreed criteria.

Design teams develop design concepts into plans and specifications a contractor can follow to implement the project. To make the project successful, engineers must strive to make the plans and specifications clear and understandable. Coordination is essential, beginning with design. Mechanical engineers need to consider how their systems affect the electrical system and vice versa. Engineers need to tell the architect how much room their equipment will need, and the architect needs to find space for it all.

Finally, the contractor on a successful project is committed to installing systems that work and implement the design intent. He does not wear blinders but tries to solve problems on his own or present workable solutions before running to the owner or the designer with every small problem.

A project must work on paper before it can work in real life. Yet plans and specifications can never show every minute detail. Even the best design cannot substitute for care and attention to detail by the individual trades people who actually build the project.

Ken Elovitz

CONTRACTORS and ENGINEERS

Our friend, Dan Holohan, recently published an article titled, "How to Tell an Engineer from a Contractor". We thought you would appreciate some of his insights:

Engineers will often look around a boiler room desperately trying to find the boiler, which they perceive as a square box labeled "BOILER" in block letters. Contractors will sometimes hold the blueprints upside down and, not realizing they are upside down, proceed to install the boiler on the ceiling.

Engineers have letters after their names on their business cards - PhD, PE, etc. Contractors look at those letters and decide this engineer has absolutely no knowledge of the real world. The more letters, the stronger the contractor's convictions.

An engineer is likely to say things like, "particles of iron oxide residue". A contractor asked to explain the same thing will simply say, "rust".

At lunch, engineers order iced tea. Contractors order beer.

Engineers and contractors both stare at the architects wondering what is going on in their heads.

Engineers love traditional methods because it means they can keep using the same standard details they've had in their files since 1955. Contractors love tradition because it means they don't have to learn anything new.

Engineers carry Swiss Army knives with all the attachments in their brief cases. They never use these tools, but they're content knowing they're prepared. Contractors carry Leatherman tools on their belts and use them to disassemble anything that looks even remotely mechanical.

Engineers look at the Leaning Tower of Pisa and figure out ways to straighten it out. Contractors look at the Leaning Tower of Pisa and start thinking of how many extras they can get.

Engineers write letters criticizing authors of controversial magazine articles. Contractors don't even read those articles because they are too busy trying to find ways around the plans and specs.

Engineers take a long time in a buffet line. They don't make a sandwich, they build one. The meat must align with the cheese, and the tomatoes can't overlap the crust by more than a millimeter. Contractors on a buffet line just eat as they go, wiping their mouths on their sleeves and smiling.

Ken Elovitz

July 2000

SOUNDS LIKE ...

Here's some fun with English words that look alike but have different sounds and meanings:

We polish the Polish furniture.

He could lead if he would get the lead out.

A farm can produce produce.

The dump was so full it had to refuse refuse.

The soldier decided to desert in the desert.

The present is a good time to present the present.

At the Army base, a bass was painted on the head of a base drum.

The dove dove into the bushes.

I did not object to the object.

The bandage was wound around the wound.

There was a row among the oars men about how to row the boats into a row.

They were too close to the door to close it.

The buck does funny things when the does are present.

The sewer lost her needle down the sewer.

To help with the planting, the farmer taught his sow to sow.

The wind was too strong to wind the sail.

After a number of Novocain injections, my jaw got number.

I shed a tear when I saw the tear in my clothes.

How can I intimate this to my most intimate friend?

I spent the last evening evening out a pile of dirt.

I watched the man in the bow tie bow down in the bow of the boat before fitting the arrow to his bow.

Her breath came in short pants as she struggled to squeeze into her tight new pants.

He wants the right to lean on his right side.

When he paid the last mortgage payment, he said, "This mine is now mine"

I'm going to fire the guard for not pulling the fire alarm.

Natives ground up the acorns they found on the ground.

Dave Elovitz

August 2000

URBAN LEGEND

Stricken with polio as a child, Itzhak Perlman has to struggle across the stage in leg braces and on crutches. Every Perlman concert starts with his progress one step at a time, painfully yet majestically, across the stage to his chair where he sits, tucks his crutches away, loosens the braces, and picks up his violin.

The audience is used to this ritual. They sit reverently silent while he makes his way to his chair, waiting for Perlman to nod to the conductor that he is ready to begin. But on November 18, 1995, in Avery Fisher Hall in NYC, something went wrong. After just a few bars, a violin string snapped with a report that echoed through the hall like a gunshot. It was clear what Perlman would have to do: Buckle on the braces, pick up the crutches and limp his way offstage to find another violin or find another string for this one. But he didn't.

Perlman closed his eyes for a moment, then signaled the conductor to begin again. He played with more passion, more power, and more purity than the audience had ever heard before. Any musician knows that it is impossible to playa symphonic work with just three strings, but that night Itzhak Perlman refused to know that. You could see him modulating, changing, recomposing the piece in his head. At one point it seemed he was detuning the strings to get new sounds from them that they had never made before.

When he finished, there was an awesome silence, then people rose and cheered. An extraordinary outburst of applause, the whole audience on its feet, cheering and screaming its admiration for what he had done.

He smiled, wiped his brow, and raised his bow for quiet. Then he said - not boastfully - but in a quiet, pensive, reverent tone, "You know, sometimes it is the artist's task to find out how much music you can make with what you have left."

Dave Elovitz

September 2000

DESIGN/BUILD

On a design/build project, the contractor is responsible for both designing and building the project. The design/build approach can be applied to an entire project or just a part of a project (like HVAC or electrical). When only part of the project is design/build, the remainder is plan and spec.

A successful design/build project requires detailed performance criteria. Otherwise, the owner does not know what to expect and has no way to determine objectively whether the design/build contractor has fulfilled the contract. Performance criteria usually go beyond design criteria to establish minimum required capacities or define how to achieve certain results.

Depending on the sophistication of the project, preparing design/build performance criteria might require architectural or engineering expertise. That consultant develops and documents the criteria in a performance specification. The owner might want to retain that same consultant for ongoing advice during the construction project. That type of professional engagement is becoming increasingly common. Since we often fulfill that role, we were pleased to learn that the industry has given it a name. The owner's design/build consultants are called "bridge consultants" because they bridge the gap between the owner and the contractor's design team.

Some people give design/build bad grades because of poor performance or unmet expectations. As often as not, those unhappy experiences resulted in large part from failure to establish or communicate appropriate performance criteria. In other cases, the process had basic deficiencies, like leaving the design part of design/build for trades people to perform in the field as they go along. On a proper design/build project, the documentation, in the form of design drawings and equipment submittals, is every bit as well developed as on a plan and spec job. The owner should insist that each design/build contractor submit a thorough design before that contractor begins its "build" activities.

Commissioning the final product is almost always a useful process. It can be particularly valuable on a design/build project as the owner's acceptance test. That way, the owner can determine clearly and fairly whether the final product meets the stated performance criteria.

The complexity and sophistication of a project determine the level of engineering services required. Design/build is a method of project delivery. It is not a substitute for the necessary engineering.

Ken Elovitz

October 2000

CHAPTER 13

If you are involved with commercial or high rise residential projects governed by the Massachusetts State Building Code, you should be interested in the new Chapter 13, Energy Conservation, effective January 1, 2001. While much of the new Energy Code will be familiar, section 1301.8.4 introduces an entirely new administrative concept to the Code. Under the new Chapter 13, every permit application for an HVAC, power, or lighting system will have to include a "design narrative" report. The design narrative must describe:

- Design intent
- Basis of design
- Sequences of operation and how each system interacts with other systems
- Systems and equipment capacities
- Acceptance testing requirements and pass/fail criteria
- Format and content of operation and maintenance manuals

The design narrative report may be new to the State Building Code, but it won't be new to many EEI clients. We have routinely prepared such documents for many years. We feel it is extremely important for the designer to explain and document what the new system can be expected to do as part of the early stages of design. In addition to improving communications among owner, designer, contractors, and building operators, including a design narrative helps expedite review and inspection by building and fire officials. It also improves understanding by the installing contractors.

Since the Code-required narrative must describe what is to be tested, who tests it, who witnesses the tests, what is to be measured, and what measured quantities represent acceptable performance, the new requirement should smooth the whole construction process. It should also reduce the number of last minute surprises at Certificate of Occupancy time. The report will be a valuable reference for building operators so the way the building is operated and future changes can be consistent with the original design intent.

Some may see this new requirement only as another burden of paperwork, but we have found that having the design narrative information at the start of a project produces a better project.

Dave Elovitz

November 2000

THE DEVIL'S DICTIONARY

American journalist and short story writer Ambrose Bierce (1842-1914) began publishing his own definitions for some English words in a weekly newspaper in 1881. These definitions were later published as *The Devil's Dictionary* in 1906. Here are some we like:

ARCHITECT, n.
One who drafts a plan of your house, and plans a draft of your money.

BAROMETER, n.
An ingenious instrument which indicates what kind of weather we are having.

BELLADONNA, n.
In Italian a beautiful lady; in English a deadly poison. A striking example of the essential identity of the two tongues.

CONFIDANT, CONFIDANTE, n.
One entrusted by A with the secrets of B, confided by him to C.

DENTIST, n.
A prestidigitator who, putting metal into your mouth, pulls coins out of your pocket.

EGOTIST, n.
A person of low taste, more interested in himself than in me.

ELECTRICITY, n.
The power that causes all natural phenomena not known to be caused by something else.

FIDDLE, n.
An instrument to tickle human ears by friction of a horse's tail on the entrails of a cat.

HASH, x.
There is no definition for this word - nobody knows what hash is.

HIPPOGRIFF, n.
An animal (now extinct) which was half horse and half griffin. The griffin was itself a compound creature, half lion and half eagle. The hippogriff was actually, therefore, a one quarter eagle, which is two dollars and fifty cents in gold. The study of zoology is full of surprises.

INSURANCE, n.
An ingenious modern game of chance in which the player is permitted to enjoy the comfortable conviction that he is beating the man who keeps the table.

KILT, n.
A costume sometimes worn by Scotchmen in America and Americans in Scotland.

LITIGATION, n.
A machine which you go into as a pig and come out of as a sausage.

POSITIVE, adj.
Mistaken at the top of one's voice.

Ken Elovitz

December 2000

DILBERT'S CHRISTMAS

There are about 1 billion children (under 18) in the world. Since Santa does not visit Muslim, Hindu, Jewish, or Buddhist children, his workload for Christmas Eve is only 378 million. Figuring 3½ children per household and assuming at least one good child in each, that comes to 108 million deliveries.

If Santa goes east to west, taking advantage of time zones and the earth's rotation, that *gives* him 31 hours, which comes to 967.7 visits per second: About 1/1000th of a second to park the sleigh, hop out, jump down the chimney, fill the stocking, eat whatever snack has been left for him, jump back up the chimney, get in the sleigh and get to the next house.

Even if we assume the 108 million stops were evenly distributed around the earth, that comes to 0.78 miles per household, 75.5 million miles altogether, not counting bathroom stops or breaks. Which would be 650 miles per second, 3000 times the speed of sound.

Assuming that each child gets nothing heavier than a medium LEGO set (2 lbs), the payload is 500,000 tons, plus Santa. On land a conventional reindeer can pull 300 pounds. Even if flying reindeer can pull 10 times as much, eight won't do the trick: You would need 360,000 of them, weighing another 54,000 tons. (About 7 times the weight of the Queen Elizabeth - that's the ship, not the monarch) 600,000 tons at 650 miles per second would create enormous air resistance, heating up the lead reindeer like a space ship re-entering earth's atmosphere. The lead pair of reindeer would absorb 14.3 quintillion joules per second each and would burst into flames instantaneously. Not to mention the deafening sonic booms. The entire reindeer team would be vaporized within 4.26 thousandths of a second, about the time Santa reached only the fifth house on his trip.

Not that it matters, however, since Santa, as a result of accelerating from a dead stop to 650 mps in 0.001 seconds would be subjected to acceleration forces of 17,000 g's. Even a skimpy 250 pound Santa (which seems ludicrously slim) would be pinned to the back of the sleigh by 4,315,015 pounds of force, instantly crushing his bone and organs and reducing him to a quivering blob of pink goo.

Therefore, any engineer can prove to you that, if Santa did exist, he doesn't any more.

Shamelessly plagiarized by Dave Elovitz

January 2001

RHYME

As year end approaches
    our brain cells start reeling
To think up some rhyming
    that you'll find appealing.

That look back and think of
    the year that is ending
While bearing the wishes
    to you that we're sending.

It's been our tradition
    for twenty some years
To send you in verse
    our good wish for next year.

Y - 2 - K got started
    without any meltdown;
No bugs and no crashes -
    'Twas sort of a let down.

Heating oil prices
    did take a big jump;
Gas followed suit
    with a similar bump.

Aside from some grumbling
    as weather turned cool,
Not much conservation -
    Low first cost still rules.

The weather went wacky:
    The West was aflame;
The South had a drought,
    and the East just saw rain.

Of work we've had plenty,
    sufficient and then some.
Tough to meet schedules,
    keep clients' momentum.

Whatever the workload
    we struggled to meet it.
Whatever the challenge
    we tried to defeat it.

We surely are grateful
    to clients and friends
For a year that was super.
    Our thanks we extend.

So, farewell Two Thousand
    And on to Oh-One.
We wish you a year
    of health and great fun.

Dave, Ken, and Gary Elovitz

February 2001

DRYING OUT

My phone rang one steam bath day last summer. The occupants of a new building complained about high humidity. Comfort wasn't the issue - the problem was paper curling in their copy machines. "Today's the day to check humidity," I said, knowing the forecast was for a break in the weather the next day. I rearranged my schedule and bopped over to the site, psychrometer in hand.

It was humid alright. The outdoor wet bulb was 78°F. We only rarely exceed 75°F or 76°F. But outdoor humidity alone didn't account for the high indoor dew points and relative humidity over 80% in one room.

The building was cool (low 70's). Most of the lights were off, and many rooms were empty. Even though it was a hot, muggy afternoon, the rooftop units cycled on and off, some running only 15 minutes at a time. When the units ran, the supply air temperature was in the low 60's - a little high. The higher the supply air temperature, the less moisture the units remove, though there was plenty of condensate flowing out of the units onto the roof. Still, the indoor dew point was higher than I could account for knowing the supply air temperature. There had to be a source of moisture in the building.

It wasn't people. There weren't enough of them to make a difference. The site wasn't wet and the carpet didn't feel damp, so it wasn't moisture coming up through the concrete slab. The walls were drywall and tape. Plaster releases copious amounts of moisture, but drywall and joint compound contain comparatively little water.

Quick calculations showed the units had some reserve capacity, at least the way this tenant used the building. I recommended slowing the supply fans to lower the supply air temperature and wring out more moisture. I made that recommendation knowing it would help but troubled that I was treating a symptom without identifying the underlying problem.

As I wrote my report, the answer struck me. The rooftop units had setback thermostats that shut off the cooling at night. The HVAC drawings called for the electrician to install a time clock on the toilet exhaust fan. The electrical drawings didn't show the time clock, and none was installed. The toilet exhaust ran all night long, sucking in humid outside air for make up. Furnishings absorbed moisture and released it during the day. I had found the moisture source.

A time clock for the toilet exhaust plus slowing the rooftop unit supply fans did the trick. After a week, humidity was back to normal. A follow up check another week later confirmed humidity was under control.

Ken Elovitz

March 2001

DEREGULATION

In 1998, the Massachusetts legislature deregulated the electric industry. A few years before, they also deregulated the sale of natural gas. A whole energy marketing industry sprung up almost over night with promises of saving 10%, 20%, or even more on the cost of electricity and natural gas.

The first question to ask when evaluating claimed savings is 10% or 20% of what. Quoted savings percentages apply only to the cost of generation for electricity or the cost of the gas itself. The savings do not apply to the local distribution portion of your bill, which makes up 1/2 to 2/3 of a customer's total cost. Actual savings, if any, will be a much smaller percentage of the total bill.

Another trap for the unwary is a common contract provision that obligates the customer to buy both electricity and gas from the same supplier. The energy marketer might come in with a seemingly attractive deal on electricity. Those savings could be more than offset by what they charge for gas when they start that program in a year or two. It's kind of like a loss leader at the supermarket. In addition, savings this year do not guarantee savings next year. Some customers who switched gas suppliers in 1998 did not save money in 1999 when energy prices continued to drop.

Switching electricity or gas suppliers means that utility supply is no longer automatic. The customer has another contract to negotiate and administer each year.

People have been understandably cautious about switching to alternate suppliers. In April 1999, when the program was new, just over 5000 of almost 2.5 million electricity customers in Massachusetts had switched to alternate suppliers. By September 2000, the numbers were still about the same. Total electricity purchased from alternate suppliers was about 220 million KWH out of almost 4 billion KWH delivered in Massachusetts that month. More than 90% of the electricity purchased from alternative suppliers was sold to large commercial and industrial users - customers who most likely have full time facility managers if not full time energy managers on staff.

Although you might not be able to be the first on your block to jump on the deregulation band wagon, there's no reason to rush in. The train has definitely not left the station. The numbers for 1999 and 2000 say there's plenty of time left to board, when and if you want to go. It's probably smarter to invest in using energy wisely and making systems work effectively than shopping for a lower cost supplier.

Ken Elovitz

April 2001

THE CASE OF THE MISSING kW

The winter cold snap brought a cold complaint in a new tenant's space. The service contractor said he thought the heating coil was too small, but the owner wanted to be sure that was the problem.

The base building HVAC drawings were rather sketchy, and there were no tenant fitout drawings, so I didn't really know what capacity was actually there. I had been working on a load study for the whole building, so I had enough of a feel to question whether a small heating coil was really the problem.

When I arrived at the building, room temperature in the problem area was around 69°F. It was 28°F outside, so a system designed for a 0°F day should have been able to heat the room to a lot more than 69°F. My load calculations said the zone needed 3.5 kW on a winter design day. At 28°F outside, the zone should need only 2.1 kW. The fan powered box was on full heat, and my amperage measurements confirmed it was delivering its full 6 kW. Where was the other 3.9 kW going?

As I climbed the ladder to trace out the ductwork, I found out: The ceiling plenum felt very cold. "That's odd!" I thought. "With conditioned space above, the plenum should be warm from heat from the lights. The rest of the plenum *was* much warmer - 73°F at the other end of the space. A quick calculation predicted an extra heating load of 3.3 kW to bring the plenum air from the 56°F I measured to 73°F like the other end of the suite. I had figured out where the extra heat was going, but why?

Back on the ladder, I used my new infrared thermometer to scan the ceiling plenum, looking for temperature differences. There! A cold spot at a column by the outside wall. Moving the ladder to that corner of the room, I could peer behind the soffit over the window to see some disturbed and missing insulation at the outside wall. I could even see the back side of the brick facing. Cold air was pouring in from outdoors, overloading the heating system. That is what made the room cold!

A little judicious stuffing with fiberglass and sealing of joints would solve a vexing problem that likely had existed from the day the building opened.

Ken Elovitz

May 2001

DEAR MR. ENGINEER

Please design and build me an HVAC system. I am not quite sure what I need, so use your discretion. My building will have somewhere between two and forty-five rooms. Make sure the plans are such that partitions can be easily added or deleted. When you bring me the blueprints, I will let you know what I want.

Also bring a cost breakdown for each type of system so I can pick one. Keep in mind that the system I ultimately choose must cost less than the one in my current office. Make sure, though, that you correct all the deficiencies that exist in my current system.

Please take care to use modern design practices and the latest materials. I want my building to be a showplace for the most up-to date ideas and methods. Also remember that I want to minimize maintenance costs.

To insure you build the correct system for all my tenants, make certain to contact each one. Our leasing agent will have very strong feelings about how the system should be designed, too. Make sure you weigh all the options carefully and come to the right decision. However, I retain the right to overrule any choices you make.

Please don't bother me with small details right now. Your job is to develop the overall plans. Your first priority is to prepare detailed plans and specifications. Once I approve those plans, I would expect the major ductwork to be in place within 48 hours.

While you are designing this system specifically for me, keep in mind that sooner or later I will sell the building. The HVAC system therefore should appeal to a wide variety of potential buyers.

Please prepare a complete set of blueprints. It is not necessary at this time to do the real design, since the prints will be used only for construction bids. Be advised, however, that you will be accountable for any increase in construction costs for later design changes.

You must be thrilled to work on as interesting a project as this! To be able to use the latest techniques and materials and to be given such freedom in your designs cannot happen very often. Contact me as soon as possible with your complete ideas and plans.

PS: My partner just told me that he disagrees with many of the instructions in this letter. As engineer, it is your responsibility to resolve these differences. I have tried in the past and have been unable to accomplish this. If you can't handle this responsibility, I will find another engineer.

Dear Reader: Do you think this is just a joke?

Ken Elovitz

SHAGGY PIRATE

Friend and manufacturer's rep extraordinaire Gerry Rosen tells this story:

A sailor had been sitting at a bar for several hours when this pirate walks in and sits down next to him. As they were sitting there drinking, the sailor looks over and sees the pirate is pretty much the worse for wear. "Pardon me," the sailor says, leaning over to the pirate, "I don't mean to be nosy, but I can't help wondering how you got that wooden leg."

"Aye, that's quite a tale," the pirate responds enthusiastically. "We were in this terrible heavy weather, the waves as high as the mast, and all of a sudden, me ship heels up so steep I slide right off the deck into the briny deep. Me mates threw me a line to haul me out right quick, but I had landed plumb in the middle of a bunch of sharks. Just as the line got me halfway out of the water, one of them sharks grabs me leg and chomps it right off."

"That's quite a story. I hope you won't think I'm out of line, but could I also ask about your hook?"

"Not atall me lad. That's another exciting tale. Y'see me and my mates had just o'ertaken a merchantman and were boarding her to capture her cargo, but her crew was puttin' up a terrible fight! One of 'em, a giant of a man, came at me swingin' the biggest cutlass I ever saw, and without thinkin' I put up me arm to ward off the blow. Sliced me hand clean off, he did. So after we had captured that ship and disposed of the crew, me mates made me this hook."

"Incredible. You sure have had some harrowing experiences. Could I just ask one more thing? How did you get the eye patch?"

"Oh! That's a strange tale, indeed. 'Twas a glorious day and we're racing toward shore to bury our treasure. I'm looking up at the sunny blue sky with not a cloud in sight, just checking for any sign of weather when this seagull flies over, and his droppings fell right in me eye!"

"Ohmigod! How awful. I never knew that seagull guano could destroy your sight."

"Oh it don't, me lad. But that was the first day after I got me new hook."

Dave Elovitz

WATER BEARER

Edie Rosenblum from N'Awlins sent me this story:

A water bearer in India had two large pots, one hung on each end of a pole he carried across his neck. One of the pots had a crack in it. While the other pot was perfect and always delivered a full portion of water at the end of the long walk from the stream, the cracked pot arrived only half full. For a full two years this went on daily, with the bearer delivering only one and a half potfuls of water.

Of course the perfect pot was proud of its accomplishments, perfect to the end for which it was made. But the poor cracked pot was ashamed of its imperfection and miserable that it was able to accomplish only half of what it had been made to do. After two years of what it perceived to be bitter failure, it spoke to the water bearer one day by the stream: "I am ashamed, and I want to apologize to you."

"Why?" asked the bearer. "What are you ashamed of?"

"I have been able, for these past two years, to deliver only half my load because this crack in my side causes water to leak out. Because of my flaw, you don't get full value for your efforts." The water bearer felt sorry for the pot, and said, "As we return today, I want you to notice the beautiful flowers along the path." Indeed, as they went up the hill, the cracked pot saw the beautiful flowers on the side of the path. The flowers cheered it some, but at the end of the path it still felt bad because it had again leaked out half its load. Again, the pot apologized to the bearer. "Didn't you notice," said the bearer, "that the flowers were only on your side of the path? That is because I have always known about your flaw, and I took advantage of it. I planted flower seeds on your side of the path, and every day while we walk back from the stream you water them. For two years, I have been able to have beautiful flowers for the table. Without you being just the way you are, this beauty would not be there to grace the house."

Moral: We are all cracked pots with our own unique flaws. But it is the cracks and flaws that make our lives together so interesting and rewarding. You just have to take each person for what he is, and look for the good in him.

Blessed are the flexible, for they will not be bent out of shape. Remember to appreciate all the different people in your life.

Dave Elovitz

August 2001

A LITTLE HUMOR

Two guys were walking their dogs. One had a Labrador, and the other had a Chihuahua. It was a hot day, and there was a bar on the next corner, so the guy with the Labrador suggested they go in for a beer.

"We can't go in there. What'll we do with the dogs?" the guy with the Chihuahua said.

"Just watch," his friend replied as he put on his sun glasses, stretched out his arm, and told the dog to lead on.

After several minutes the guy with the Chihuahua figured the ruse must have worked, so he tried the same thing. The bar tender stopped him at the door and told him animals are not allowed in the bar.

"This is my seeing eye dog," the guy protested.

"That's not a seeing eye dog," the bar tender retorted. "That's a Chihuahua."

"What?! They gave me a Chihuahua?!"

REST ASSURED

Just in case you thought you safely escaped the Y2K bug, here's something new to add to your computer-based worries.

If a nuclear device explodes in the atmosphere, the resulting electromagnetic pulse will destroy all semiconductor devices for hundreds of miles around. Suppose, for example, that a nuclear bomb hit Chicago. The only cars that would still work would be those '57 Chevys and the like located at least 300 miles away in Iowa.

Most military circuits are hardened against electromagnetic pulses, but almost no commercial or industrial computers or electronic systems have that protection.

Luckily, Micro Circuit Engineering, based in the UK, has come to the rescue. They have developed an early warning device that detects a nuclear explosion and shuts off the power to all devices connected to it. They believe that shutting off the power will prevent damage from the electromagnetic pulse and electrical surges.

So, if we all rush out and buy Micro Circuit's devices, we can rest assured that our computers will reboot successfully after the nuclear explosion.

Of course, who will be left to turn them back on?

Ken Elovitz

September 2001

FACTS ABOUT WATER

Less than 1% of the water treated by public water systems is used for drinking and cooking.

Kentucky Bluegrass uses 18 gallons of water per square foot each year. Tall fescue and wheat grasses use 10 and 7 gallons of water per square foot each year, respectively.

Producing a typical lunch -- hamburger, french fries, and a soft drink -- uses 1500 gallons of water. This includes the water needed to raise the potatoes, the grain for the bun and the grain needed to feed the cattle, and the production of the soda.

According to NASA, the 10 trillion tons of water stored in reservoirs have slightly altered the natural rotation of the Earth over the past 40 years.

The Nile perch has wiped out nearly all the other species of fish in Lake Victoria, Africa since its introduction in the 1960s.

In the United States, approximately 500,000 tons of pollutants pour into lakes and rivers each day.

Nationwide, about 11% of pollution in rivers comes from storm sewers and urban runoff.

Four quarts of oil can cause an eight-acre oil slick if spilled or dumped down a storm sewer.

The maximum 24-hour snowfall in the United States is 75.8 inches at Silver Lake in the mountains west of Boulder, CO. The storm occurred on April 14-15, 1921.

Snowflakes that have become small rounded pellets (usually two to five millimeters in diameter) are called graupel. It is sometimes mistaken for hail.

Rime is the ice that forms when supercooled water droplets freeze on contact with an object.

Fresh, uncompacted snow is usually 90-95 percent trapped' air.

Ozone gas is the only known disinfectant that works against the cryptosporidium parasite in drinking water.

Avalanches killed 914 people in the United States between 1900 and 1995.

Once evaporated, a water molecule spends ten days in the air.

from Colorado State University's water information web site, http://waterknowledge.colostate.edu/coolfact.htm

Ken Elovitz

October 2001

HAS COMMISSIONING CHANGED?

Many of you who have been reading these calendars for years know that we have long advocated a clear basis of design report, independent design review, well thought out acceptance testing, and meaningful/useful O&M manuals as essential elements of a successful HVAC or electrical system project. Some of you have watched the ASHRAE video on commissioning with Ken as one of the featured speakers and have read Ken's articles on commissioning in several publications. So you know this is a concept we feel important and valuable to the Owner. To us, the goals of the commissioning process are to provide:

> Information up front so the Owner can understand what he is getting and whether the systems are likely to meet his expectations;

> Information on whether the building systems can actually deliver the performance the design documents say the systems should deliver; and

> Information to help the Owner's staff operate and maintain those systems effectively.

The contractor is, and always has been, responsible to furnish and install properly operating systems that deliver the design capacities. I reflect back 30 years or so to the inception of the independent Test and Balance (TAB) concept. This new trade role was supposed to ensure that the Owner got a complete and properly working HVAC system. TAB promised to be a process that solved all the problems and complaints that resulted from systems not being started up and adjusted properly. Sound familiar? Isn't that just what we are now being told to expect from commissioning? Yet contracting practices are following the same path that added the TAB contractor to the circle of people standing around pointing fingers at each other. Will commissioning just add even one more guy to that circle?

In my view, for commissioning to add value, it must be something the Owner *does*, not something the Owner *buys*. It should not be left to a contractor or even a design consultant to confirm that the system does what the Owner needs. The Owner has to be involved in, and in charge of, the commissioning process. If the Owner needs technical assistance, he can hire it, but the role of that expertise has to be to support the Owner's involvement not to substitute for it.

Dave Elovitz

November 2001

DUAL ROLE

A recent call thanking me for a report got me thinking about the dual role of construction contract documents like design drawings. Most of the time, we focus on the need for drawings to convey sound technical concepts: an HVAC or electrical system that works.

It's easy to forget that the drawings also define the scope of the contractor's work. For that reason, they must communicate clear, enforceable contract requirements. That was the issue my report had addressed. The drawings contained lots of ambiguities that screamed opportunity for "change order artists" who might bid the job. Sure, it was obvious to me that the riser for the console heat pumps at the ends of the corridors had to be connected to the heat pump loop. Maybe the designer didn't notice it was missing because it was obvious to him. And it was probably obvious to the contractor, too, but it wasn't on the drawings. "That's extra," the contractor could demand, and he'd be right! He might even be right if he ignored the whole thing until it came to a head during balancing when those heat pumps wouldn't work. Then it would really cost extra to plumb them in.

On a plan and spec job, the contractor has no obligation to figure out how the system is supposed to work or to check the design. Under the AIA A201 General Conditions (section 3.2), the contractor must report errors to the architect promptly when discovered. The contractor cannot install something he knows to be wrong, but the contractor has no duty to perform an independent design review. And if the contractor does discover an error, the AIA A201 calls for him to request clarification from the architect and, if extra work is involved, allows him to submit a claim for extra.

Designs are made by people, so they are never perfect. Design review, whether internally within the design firm, by the owner's staff, or by an outside consultant, can't guarantee perfection, either. However, it can often find errors and inconsistencies while they are lines on paper before they become ducts and pipes in a building. Design review that finds everything checks out OK is not wasted. At the very least it confirms that the design presents a clear, consistent picture to an objective, independent reader.

"You not only saved me big money on the project," the client said at the end of the phone call, "but I also saved a pile of dough on you! It's a lot less expensive to have you analyze things while they're still on paper than when I wait until things don't work and need you to figure everything out at the site after the fact."

Ken Elovitz

DON'T TREAT THE SYMPTOMS

The topic was a clean room that was cold (61°F) and damp (60% RH). The rooftop units were reportedly running, pulling moisture out of the air. A quick calculation showed that the reheat coils had nowhere near enough capacity to bring the cold, dehumidified air leaving the rooftop units back up to room temperature.

"Before I size new reheat coils," I said, "let me analyze the problem."

When troubleshooting a problem, it's often tempting to fashion a solution that counters the observed symptoms. If a room is cold, add more heat, right? Not necessarily. True, recognizing and identifying symptoms is the first step in diagnosing a problem. But you need to find the underlying cause. A room can be just as cold from over cooling as from under heating.

More reheat would raise the room temperature, reducing relative humidity because warm air can hold more moisture than cold air. More reheat would address the symptoms, but it would not change the amount of moisture in the room air.

HVAC systems (and most other moisture related physical phenomena) are functions of dewpoint, not relative humidity. The clean room didn't need low RH; they needed low dew point. Analyzing performance ratings showed that the rooftop units could not produce air at a low enough dewpoint to achieve the stated design conditions. As a result, the real problem was not insufficient reheat but insufficient dehumidification.

The solution was to precondition the incoming outside air with a unit that could deliver the desired dewpoint without over cooling. With an outside air unit providing the dehumidification, the rooftop units could simply heat and cool in response to room temperature. The clean room not only wouldn't need more reheat, it wouldn't need reheat at all.

Almost every time I'm tempted to treat the symptoms before I understand the underlying cause, subsequent analysis proves I would have gotten it wrong. It takes some time to analyze the problem, and there might be no outward sign of activity while we crunch the numbers and scratch our heads. But in the end, we aim to address the underlying problem, not just mask the symptoms.

Ken Elovitz

January 2002

2001 REPRISE

This is the end of twenty oh-one
A year of shock, but a year of fun.
Started like a house afire
Backlogs climbing higher 'n' higher.

Then Dot Coms evaporated.
Greenspan soon the rates abated.
Cut them more, again, again.
One more cut will make it ten.

Congress passed a tax rebate
So the boom would procreate.
Jeffords moved across the aisle;
Democrats began to smile.

As we saw the boom relax
Some friends even got the axe.
Searched for new jobs anyplace
While others kept their frenzied pace.

March roared in with lots of snow.
On sailing days, no wind did blow.
Sun and rain did interweave
To bring us lots of colored leaves.

Stephen Flemmi's under lock:
Rifleman refused to talk.
Whitey Bulger hides someplace;
FBI is red of face.

Southie's hero, Moakley, died -
Hometown guy, but dignified.
Cellucci left, promoting Swift.
When Helms retires, will he be
    missed?

Bin Laden made the towers fall,
Crashed the Pentagon's west wall.
Then we got the anthrax scare.
Made us all much more aware.

Al Qaeda and the Taliban
Let Osama work his plan.
Rumsfeld, Bush, and Colin Powell
Bombed to make the Afghans howl.

Bombs and missiles ev'ry night,
Special forces to the fight.
Can't tell yet how it will end
But to terror we won't bend.

So hail! Farewell 2001.
I think we're glad this year is done.
From Ken and Gary, Dave and Fran
A great new year for your whole clan.

Dave, Ken, and Gary Elovitz

February 2002

REWARD

In the middle 1930s, construction unions had not yet made much impact outside the big cities. The union leaders decided it was time to try to organize workers on rural projects. They selected a small town with one dominant contractor as a trial target. The location was not nearby, so they recruited a handful of local rowdies to harass the contractor's job site. Each day, the ruffians would show up to jeer the workers and suppliers going to and from the site.

The owner was not pleased, and the situation was getting grim, but the contractor was ingenious. At the start of each day he would walk out to greet the hoodlums and give each one a quarter for his efforts. Delighted, they shouted their insults for a while and then moved on to a local watering hole.

After a week or so, the contractor apologized to the hoodlums and said he could now only afford to give each one a dime. They took their dimes, did their jeering, and left. The next week, the contractor said he could only afford to give each of the hoodlums a nickel. They reluctantly accepted the cut in pay, shouted a few threats, and adjourned to their favorite spot.

A few days later, the contractor was outside waiting for the hoodlums when they arrived. He apologized once again, held his hands up in the air, and announced sadly that he could only afford a penny.

The young toughs were indignant. They would certainly not spend their time sneering for a mere penny! So they didn't. And the contractor lived happily - more or less - ever after.

LYSISTRATA REVISITED

In the Greek comedy *Lysistrata,* the women barricaded themselves inside the Acropolis and went on a sex strike to persuade their husbands to stop the Peloponnesian War. The play is considered fantasy because Greek women had no political power in 411 BC.

Don't mock Aristophanes too loudly if you happen to be in Sirt, Turkey. Last July, the women in Sirt went on a sex strike to force the men to provide running water to the village. The women had grown tired of hauling water, sometimes for miles. By August, the government had agreed to the men's request for pipes so they could build a water system.

Ken Elovitz

March 2002

DIVINE INSPIRATION

The president of HighFlyer.com had not been sleeping well. All around him, he saw his friends' companies crashing and burning. That made him very anxious, so he called his accounting firm for advice. They could not offer any words of encouragement. Then he called HighFlyer's management consultants. They recommend an extensive study and report with a detailed market survey to evaluate the situation. He tried to call the investment banker who put together HighFlyer's IPO, but the banker didn't return his calls.

The president of HighFlyer.com was afraid and didn't know what to do. Finally, in desperation, he decided to talk over his worries with the swami at the local Ashram. (This is California, after all.)

The swami listened, nodded wisely, hummed quietly, and finally said, "Take a beach chair and a copy of Nostradamus and drive to the ocean. Set up the beach chair on the water's edge. Sit in the beach chair and open up the copy of Nostradamus. Close your eyes and let the wind riffle the pages for a while. Eventually the book will stay open at a particular page. Read the first words your eyes fall upon, and they will tell you what to do."

The president took the swami's advice: He put a beach chair and a copy of Nostradamus in his car and drove to the beach. He sat in the chair at the water's edge and opened the book. The wind riffled the pages and eventually stopped at a particular page. The president of HighFlyer.com looked down at the book, and his eyes fell on the words that told him exactly what to do.

Three months later the president went back to see the swami. The man was wearing a $1000 Italian suit, his wife was all decked out with a full-length mink coat, and their child was dressed in beautiful silk. The president handed the swami a thick envelope stuffed with cash and said he wanted to donate the money to the Ashram to thank the swami for his wonderful advice.

The swami was delighted. He recognized the president and asked him what words from the book of Nostradamus brought this good fortune.

The president replied, "Chapter 11."

Ken Elovitz

April 2002

HISTORY LESSON

The one dollar bill we use today first came off the press in its present design in 1957. This so-called paper money is actually a cotton and linen blend with minute red and blue silk fibers running through it. We've all washed it without it falling apart. A special blend of ink is used, the contents we will never know. It is overprinted with symbols, and then it is starched to make it water resistant and pressed to give it that nice crisp look.

Most of us know that George Washington's picture adorns the front of the bill. But how many of us know what's on the back? The back of the one dollar bill has two circles that comprise the Great Seal of the United States. The First Continental Congress asked Benjamin Franklin and a group of men to come up with a Seal. It took them four years to create the design and another two to get it approved.

The left hand circle encloses a Pyramid. The east face is lighted and the western side is dark. Our country was just beginning. We had not begun to explore the West. The Pyramid is uncapped, again signifying that we were not even close to being finished. Inside the capstone is the all-seeing eye, an ancient symbol for divinity. Franklin believed that one man couldn't do it alone, but a group of men, with the help of God, could do anything. "IN GOD WE TRUST" is on this currency. The Latin above the pyramid, ANNUIT COEPTIS, means "God has favored our undertaking." The Latin below the pyramid, NOVUS ORDO SECLORUM, means "a new order has begun". The Roman Numeral on the base is 1776.

The right-hand circle has a bald eagle holding an olive branch and arrows in his talons. This country wants peace, but we will never be afraid to fight to preserve peace. The Eagle wants to face the olive branch, but in time of war, his gaze turns toward the arrows.

The Bald Eagle was selected as a symbol of victory for two reasons: first, he is not afraid of a storm; he is strong and smart enough to soar above it. Second, he wears no material crown, symbolizing our break from the King of England. The shield is unsupported because the country can stand on its own. The top of the shield has a white bar signifying congress, a unifying factor. The banner in the Eagle's beak reads, "E PLURIBUS UNUM," meaning "one nation from many people." The 13 stars above the Eagle represent the thirteen original' colonies and any clouds of misunderstanding rolling away.

We all handle dollar bills every day and seldom take the time to notice the design. Next time, take a closer look at the back of a one dollar bill and think about what it stands for.

Ken Elovitz

May 2002

MONDEGREENS

Mondegreens are mishearings of familiar phrases, song lyrics, slogans, and sayings. The word never made it into the dictionary, but it provides a lot of fun:

For years, the Manhattan White Pages carried the phone listing: "Bonds and Noble, see Barnes and Noble".

How many people say their favorite Sousa march is "Tarzan Strikes Forever"?

A man met a lady for dinner at Taipei Restaurant, which she thought was for people with Type A personalities.

If you're ever in New York City, why not meet for a cup of coffee at "The World of Astoria"?

The Beatles song, "Eight Days a Week" has been sung, "Ain't they Louise", and "Norwegian Wood" has been interpreted as "Knowing she would".

An L.A. traffic report came across as, "Maniacs see dents on the 405 Freeway". The message was "Many accidents on the 405 freeway".

A computerized speech recognition program answered the question, "Can you recognize speech?" with "No I can't wreck a nice beach."

A teenager went to the library looking for a copy of "Catch Her in the Rhine" by JD Salinger.

The chorus to "Yellow Submarine" has been sung, "We all live in a tub of margarine" instead of "We all live in a yellow submarine".

Jose Feliciano's famous recording of "Feliz Navidad" has been called "Police naughty dog".

A pet shop clerk told her boss that her parents' wealth did them no good at all because they just sat around their backyard and "drank themselves to Bolivia".

A seventh grade English class had to write a paragraph about an exciting moment in their lives. One student wrote, "I'll never forget the time my brother choked at dinner and my father gave him the Hemlock Remover!"

The pledge of allegiance is a hotbed of Mondegreens: I pledge a lesion to the flag, of the United State of America, and to the republic for Richard Stans, one naked individual, with liver tea and just this for all.

Ken Elovitz

June 2002

DRIP, DRIP, DRIP

Water dripped from the ceiling of the indoor pool all winter. The same design had worked flawlessly on another project, so the architect couldn't figure out what was wrong. He wondered if it could be because the pool water was 82°F instead of the design 80°F.

The contractor had made a test cut through the roof. It showed condensation on the warm side of the vapor retarder and in the insulation behind the vapor retarder. Not understanding that a vapor retarder can only reduce moisture migration, not eliminate it, the contractor proposed to remove the entire roof, replace the vapor retarder and wet insulation, then add two more inches of roof insulation, all for the bargain basement price of $150,000. Apart from the cost was a practical question of how to redesign the roof and its connections to the rest of the building to accommodate an additional 2" of thickness. That's when the architect asked for help.

Calculating the temperature and dewpoint profile through the roof structure showed the temperature dropped below the dewpoint in the middle of the insulation. No wonder the insulation was wet! "You need more ventilation, not more insulation," I assured the architect, "And the fix won't cost $150,000!"

A site visit confirmed that the ridge vent had been installed as specified, but the eave vents had too little free area. Without enough airflow, any moisture that passed through the vapor retarder would be trapped under the plywood instead of flowing out through the ridge vent. Since the ridge vent had ample free area, the fix seemed simple: install eave vents with more free area.

The work was finished later that spring, and everyone waited for fall and cold weather to test the results. Guess what? The ceiling still dripped. A site visit on a cold day and a pair of binoculars revealed the vapor retarder did not wrap around the wood curb under the skylight. As a result, moist air from the pool was entering the roof structure. Removing the skylight panels showed us gaps big enough for mice to crawl through, never mind moist air to flow.

Adding insulation to the roof wouldn't have worked, and normal roof ventilation cannot overcome wholesale air leakage. The skylight curb had to be sealed. Patience and persistence found and solved the problems. They also knocked the cost of the work down from $150,000 to more like $5000.

Ken Elovitz

July 2002

HOT STUFF

The plastic plant's new 4000 amp switchgear was fine until they doubled the number of molding machines. The new machines routinely raised loads to 3000 to 3500 amps, and sometimes even over 4000. Then breakers started tripping, halting production. The electric room got so hot they had to prop open the outside door for extra cooling. That's when everything in the electric room started getting rusty and covered with an oily film.

"We've had a slew of so-called experts look at this situation - chemists, electrical engineers, you name it - and they all assure us that no corrosive vapors get into the electric room from the production area. But they can't explain where the corrosion and oily film came from or why the most corrosion and the thickest oil film is inside the switchgear," the client complained. "Then our lawyer said to call you guys."

Since the switchgear had been fine until the new molding machines increased the load, maybe the electric service was too small. That would make everything run hot. The service was a bank of 12 PVC conduits buried in sand. Each conduit had a set of 500 kcmil conductors. The standard tables in the electrical code rate those conductors for 380 amps each, so the contractor figured he had plenty of wire for 4000 amps.

When the contractor sized the underground conductors, he didn't realize that the standard tables in the code book did not apply. Dry sand around tightly spaced conduits dissipates much less heat than air in a building. Using the complicated Neher-McGrath equations, Ken calculated the temperature of the conductors in buried PVC conduit at various loads. Pushing 3500 amps through the underground service heated the cables to 270°C!

Ken's calculations made everything clear: Overheated cables made the PVC conduit so hot it started to decompose. Sure enough, lab tests of a piece of the conduit confirmed that it started to break down at 250°C, releasing hydrogen chloride and oily vapors from the plasticizer. Hydrogen chloride mixed with moisture in the air to form corrosive hydrochloric acid that attacked the switchgear. No wonder the heaviest corrosion and thickest oil film were inside the switchgear!

"Those chemists may have known chemistry, and the electrical guys may have known electricity," enthused the client, "But you guys were the only ones who looked at the WHOLE problem. Next time we'll call you first!"

Dave Elovitz

August 2002

MEMORY LANE

Years ago, when I worked for a design/build contractor, a big firm had a clean room that had never worked right. They hired us to fix it, which we did, to their delight. Six or eight months after I finished that project, the customer invited me to submit a proposal to build two new clean rooms, designed by the same guy as the clean room I had just fixed.

As I went through the drawings, I could see the design was not going to work. So when I priced the project, I included the changes to make the clean room work. My bid stated that we did not believe the clean room would work as designed, but our price included the additional work necessary so it would.

After the bids had been submitted, the customer who had been so pleased with my fix of the first clean room asked to meet to discuss my bid. "Your price is 5% more than Acme's," he started. Acme had built the first clean room and had left the customer hanging when it didn't work. "We appreciate what you did on the first clean room," he continued, "so we'll let you have this project if you meet Acme's price."

"Do you understand that I included money in my price to make up for the design problems, and Acme's price does not include that work?" "Oh, yes," he nodded, "That's why we're willing to let you do the job for Acme's price."

I don't know how red my face got, but I stood up and said, "If I am not worth a few thousand dollars more after what I did for you, give the job to Acme." I turned on my heel and walked out.

When I told my boss what happened, I thought he might fire me on the spot. I mean, this was a BIG job. He didn't hit me, but he sure yelled a lot.

About two years later, that same customer called again. The two new clean rooms were done. As I had predicted, they didn't work, and nobody could make them work. Would I like to bid on fixing them?

"Yes, Mel, I know how to fix them. But the only way I will take the job now is time and materials." He hemmed and he hawed, and he stammered and he stuttered. A few days later he called back and gave me the go ahead. It wasn't easy, but we made the clean rooms work. When we were done, they even exceeded the spec. The cost of the repair was about 2½ times our original bid to do the whole job! Why is there never enough money to do it right in the first place, but always enough money to do it over?

Dave Elovitz

September 2002

BIG TROUBLE

Think of the most mischievous, hyperactive, into-everything 11 year old boy you have ever known. Now double it, and that describes Jeb. This particular day, he had been even more hell-raising than usual, tormenting his 8 year old brother, taking off all the door knobs in the house, then chasing the dog around the parlor and tipping over a big vase of flowers. Jeb's Pa was absolutely beside himself. He just didn't know what to do to get some peace and quiet. Finally, he decided to take the boy over to the preacher to put the fear of God into him.

Pa put Jeb into the car, drove over to the preacher's office, marched Jeb inside, pushed Jeb into a chair, unloaded all his misery and frustration to the preacher, then went back out to the car to wait for Jeb.

Now, what Jeb's Pa didn't know is that the preacher had been sick with the flu all week, and this was his first day back in the office. He was regretting it even before Jeb and his Pa barged in. The preacher felt just awful, like death warmed over. His head still ached, and his stomach was still churning. He just sat there and glowered at Jeb across the desk for what seemed an eternity. Finally the preacher got up gingerly, walked around the desk, stood right in front of Jeb, stomped his foot, and growled, "Where's God?"

Jeb didn't know what to do. "Where's God?" he quavered back. The preacher just stood there glowering for a couple of minutes, then sat down behind his desk, picked up some papers, and started to read them. Jeb didn't waste any time. As soon as the preacher's eyes were off him, he skedaddled out of the office, jumped into the car and told his Pa, "Let's go home." No words were exchanged on the trip home. When they got home, Jeb went straight into the house, into his brother's bedroom, and closed the door.

"Boy, are we in big trouble!" he announced.

"What's up?"

"God's missing, and they figure we had something to do with it!"

Dave Elovitz

October 2002

ACROSS THE POND

From time to time, I come across some fascinating, but probably useless, information that I can't resist sharing. This batch comes from a column by Professor John Swaffield in one of the trade magazines.

Three water authorities in Scotland recently reported the results of a detailed 1999 study of water usage in Scotland. For instance:

Each Scot uses, on average, 37 US gallons per day, with 25% of them using 33.5 or less and 25% using more than 43.8 gallons/day. (The EPA says the average US household uses 400 gallons/day. That's well over 100 gallons per person.)

66% of Scottish dwellings have a shower, 10% have two. 10% of the showers have pressure boosters. Another study, 17 years earlier, found almost all dwelling units had a bath, but only 32% of dwellings had showers back then.

The increased number of showers installed may be why bath use declined from 0.71 baths per day in each household to 0.62, with a corresponding increase in shower use from 1.07 per day in 1982 to 1.55 in 1999.

Almost all dwellings have washing machines, and 97% of them are automatic front loading models. And contrary to the historic image of Monday wash days, households averaged almost one load of laundry per day! While 26% of households have dishwashers, they average only 0.7 loads per day.

Water closets were the major water users, swallowing 31% of all water used with 4.4 flushes per day. Washing machines came next at 22%, then baths (15%,) showers (10%,) and dishwashers only 8%.

Peak rate of water use is in the morning, 8-9 AM, except on weekends it is at 10-11 AM. Incidentally, overall water use is about 10% higher on weekend days than on weekdays, and the lowest daytime rate of use is 2-3 PM, weekdays or weekends. The second highest period of use - weekdays or weekends - is between 6 and 7 PM.

Ken says, "Why should anyone in the US care about how much water they use in Scotland?" I have to confess, I can't say why, but both Fran and I did find it fascinating. Fran wants to know why they all have front loading washers while most of us have top loaders. And Ken points out that he uses about 72 US gallons per day per person, about twice as much as the Scots. But then, who but Ken would even know how much water he uses?

Dave Elovitz

November 2002

THE VILLAGE HVAC-SMITH

Under a spreading chestnut tree
A cooling tower sits;
The water rushes o'er the fill,
The fan does roar and spit;
An air in-take was right nearby
And sucked in all that grit.

I found a ladder on the site
And gave myself a boost.
Got high enough to see the place
Where all the birds did roost.
My stomach turned as I surveyed
The guano they had loosed.

The water in the tower sump
Was thick with greenish goo.
The chiller tubes must be all clogged.
What water could go through?
The pumps inside the plant did groan.
This clearly would not do.

Legionella, it must thrive,
And Stachybotros, too.
Would Aspergillus show its face?
Cladosporium say boo?
What other fungi colonized
As filamentous algae grew?

And then it was upstairs to see
The unit and its coils.
My, oh my, what a big mess!
The fan did creak and toil.
I wondered what delicious crops
Would grow within that soil.

And then I walked into a room
That felt all damp and cold.
The ceiling tiles had spots of green:
Encrusted all with mold.
The maintenance here is really good.
At least that's what I'm told.

You have to clean it up I said,
As nasty as that seems.
For stuff like that dissolves the walls
And rots away the beams.
And not to mention what it does
To people on your team.

So first you must expel the birds
And get that louver clean.
The next task is the tower sump:
Add filters to the scene.
And water treatment must improve.
You need a guy who's keen.

The strainers all must be blown down
'Til water flow is clear.
And use some disinfectant soap
Instead of day old beer
To wash the walls and mop the floors.
Get maint'nance staff in gear!

The A-H-U must be fixed up.
The condensate must drain.
The filters must be changed on time.
E'en if it be a pain.
And if you do those things with haste,
The system will sustain.

Ken Elovitz

December 2002

DID YOU KNOW

Almost everyone knows that the Pilgrims landed in Plymouth, MA in 1620. The Pilgrims were not the first English settlers in this country. Jamestown had been established 13 years before. However, Massachusetts does claim a number of historic firsts:

In 1621, the Pilgrims were the first to celebrate Thanksgiving. It became a national holiday in 1863 courtesy of Abraham Lincoln.

In 1634, Boston Common became the first public park in America, only 4 years after William Blackstone moved from Dorchester to Beacon Hill to establish Boston.

Boston Latin School was the first public secondary school in America. It was founded in 1635, a year before Harvard College. The Mather School in Dorchester was the first free American public school. It was founded in 1639. Boston Public Library, established in 1653, was the first public library in America.

America's first lighthouse was Boston Light, built in Boston Harbor in 1716. It is still in service today.

In 1839 in Woburn, Charles Goodyear discovered that applying steam under pressure to a sulfur and gum mixture produced rubber as we know it today. Charles Thurber invented the first typewriter in Worcester in 1840. Elias Howe made the first sewing machine in Boston in 1845.

In 1846, William Morton, a Boston Dentist, demonstrated the first use of anesthesia in surgery at Mass. General Hospital. He used a specially designed glass inhaler containing a sponge soaked in ether. The famous "ether dome" where that operation occurred is open to the public as a museum.

In 1891, the Kennedy Biscuit Works (later Nabisco) used a machine to mass produce Fig Newton cookies, which were named for the town of Newton, MA.

In 1896, landscape architect Charles Eliot developed Revere Beach as America's first public beach. Eliot was a student of Frederick Law Olmstead, who often worked with architect HH Richardson. Olmstead is considered the father of landscape architecture. He designed Boston's Emerald Necklace and New York's Central Park.

On September 1, 1897, Boston Mayor Josiah Quincy VI inaugurated America's first subway system, riding from Park Street to Boylston station.

In 1947, Raytheon's Percy Spencer invented the microwave oven. It weighed 750 pounds and was 5½ feet tall.

Ken Elovitz

January 2003

with apologies to
PAUL REVERE and HENRY W. LONGFELLOW

Listen, dear friends, and you shall hear
Of the stories and exploits that came so near.
Of things that we saw
and that happened to us
Followed by greetings
That tell you thus:
"To you and yours, a Joyous New Year."

The Pats traveled south, to the Super Bowl.
Amazed us and thrilled us, with goal after goal.
Alas the poor Sox
Did not follow suit.
They started out great
then went down the chute!
The Bruins and Celtics weren't quite on a roll.

Ashcroft switched colors and "W" grinned.
OPEC cut output, would war soon begin?
We worked with Musharaff
for whom Afghans did vote.
We searched for Bin Laden
If he lives, he still gloats.
But we caught Richard Reid and
John Walker Lindh.

Shannon met Mitt, in ads and debate,
Fighting for who would be head of the State.
They strove for our votes
But no matter who won
The shots will be called
by Tom Finneran!
And we'll pay for the Dig at the
Mass Pike tollgate.

We heard tales of Enron and Anderson, Arthur,
And Faneuil implied hanky-panky by Martha.

Kozlowski used TYCO
as his own piggy bank.
For Jack Welch's lifestyle,
He had GE to thank.
Such scandals effected CEO departures.
The incredible shrinking 401ks
Were only one part of the Wall Street Malaise.
Polaroid bankrupt,
Ames Stores and more.
More white collar workers
Were shown to the door.
Have we hit the bottom? When will start better days?
In June we turned golden, that's
Franny and Dave.
Fifty great years. Few are happy as they.
And grateful to you,
for your years of support.
Adieu, 0 0 2,
and we leave with this thought:
Happy 2 0 0 3, may it bring all you crave.

Dave, Ken, and Gary Elovitz

February 2003

MICROSOFT HAIKU

The Japanese have replaced impersonal and unhelpful Microsoft error messages with Haiku poetry. Haiku has strict construction rules. Each poem has three lines, 17 syllables: five syllables in the first line, seven in the second, five in the third. Haikus are used to communicate a timeless message, often achieving a wistful, yearning and powerful insight through extreme brevity - the essence of Zen.

Your file was so big.
It might be very useful.
But now it is gone.

The Web site you seek
Cannot be located, but
Countless more exist.

Chaos reigns within.
Reflect, repent, and reboot.
Order shall return.

Program aborting:
Close all that you have worked on.
You ask far too much.

Windows NT crashed.
I am the Blue Screen of Death.
No one hears your screams.

Yesterday it worked.
Today it is not working.
Windows is like that.

First snow, then silence.
This thousand-dollar screen dies
So beautifully.

With searching comes loss
And the presence of absence:
"My Novel" not found.

The Tao that is seen
Is not the true Tao-until
You bring fresh toner.

Stay the patient course.
Of little worth is your ire.
The network is down.

A crash reduces
Your expensive computer
To a simple stone.

You step in the stream,
But the water has moved on.
This page is not here.

Out of memory.
We wish to hold the whole sky,
But we never will.

Having been erased,
The document you're seeking
Must now be retyped.

Serious error.
All shortcuts have disappeared.
Screen. Mind. Both are blank.

Dave Elovitz

March 2003

FOGGY, FOGGY, DEW

The new lab facility was running smoothly - the hoods passed all the tests, and the controls ran like a charm. But there was one problem: water dripping from the ceiling, clearly coming from the big make-up air unit on the roof. The final filters beyond the cooling coil were soaked - the coil was clearly shedding condensate. The contractor and the manufacturer had come back time and again and tried this and that. After two summers without improvement, my phone rang.

The facilities manager and I shut down the unit and looked inside. "It looks like someone sprayed the inside of the unit with a garden hose!" I marveled.

The "usual suspect" when a coil sheds condensate is excessive air velocity - either overall airflow is too high, or air is bypassing around the coil at a high velocity. Either way, high velocity air can pick up moisture and carry it away from the coil. But that was unlikely in this case - this was a variable air volume unit that usually ran at less than 70% of full speed. Face velocity had to be too low to blow condensate off the coil fins.

The only way to see what is going on is to watch inside the unit while it is running. So I climbed in, we closed the doors and started the unit up, and I began to check for air bypass and high air velocity. At first, I saw nothing unusual. Then I saw nothing at all - my glasses were all fogged up!

What in the world? That shouldn't happen inside an HVAC unit: condensation can only form on a surface that is colder than the surrounding air, but after the cooling coil, the coldest thing in the unit should be the air. How could my glasses be colder than the air coming off the coil?

As I carefully studied the surface of the cooling coil I noticed that in some places the air felt very cold, and in other places the air felt a warmer. Suddenly, the problem (if not my glasses) became clear: "Because of the way that this coil is built," I explained, "under part-load conditions the coil does not cool the air uniformly but makes alternating 'strips' of very cold (about 40°F) and relatively warm (65°F) air. My readings show that both airstreams are at 100% relative humidity. Just as when a mass of cool air settles into a valley after a hot, humid day, when these two air streams mix they make - fog! I could actually see fog in the beam of my flashlight as I shined it across the coil section!"

Fortunately, we could adjust the staging controls to make more uniform air temperatures across the coil face. With that change, the unit would stay nice and dry inside all summer.

Gary Elovitz

April 2003

NUMBERS

The Census Bureau recently issued *Demographic Trends in the 20th Century,* filled with fascinating (but probably useless) numbers contrasting 2000 with 1900. From 76 million Americans in 1900 we grew to 150 million in 1950, 220 million in 1980, and 281 million in 2000. The population not only got bigger, it got older: In 1900 one out of three Americans was under 15, and the median age was 23. Now it is 35, and only one in five is under 15. Life expectancy stretched from 47 in 1900 to over 54 in 1920, 64 in 1940, and 77 in 2000.

Families got smaller: From 45% of households having five or more members in 1900, now only 11 % are that large. 5% of households were singles, now 26%.

And we moved closer together: About 3/4 of the population lived on farms or in small towns in 1900 - what demographers call non-metropolitan areas with population under 200,000. By 1920 the rural population had fallen below 50% of the total and farm residents to less than 30%. In 2000, 80% of the population lived in metropolitan areas, half of them in suburbia. (It seems to me it would have made a lot more sense to live so close together back in 1900 when there were only about 8000 automobiles and less than 10 miles of paved roads. But there were about 18 million horses and mules providing transportation, plus 10 million bicycles!)

People like my parents who were born just before 1900 saw the first radios, movies, televisions, and airplanes. From only one telephone for every 66 people, now it seems like there are 66 phones for every person. Eastman Kodak introduced the Brownie Camera for a dollar, and a six shot roll of film cost 10 to 15 cents. 1903 saw the first commercially successful phonograph records. What major life changing inventions have my kids seen? Computers, the Internet, what else is to come?

The population not only got bigger, it got dramatically re-distributed in the process. Montana, Wyoming, Colorado, and New Mexico had 5% of the population in 1900, up to 23% at the turn of the millennium. From New York as the largest population state with 7 million people and California as 21st with 1.5 million people, California is now the largest (34 million), and New York has dropped to third with 19 million, behind second place Texas with 21 million. I am not sure whether our industry should claim credit or admit responsibility, but that population shift would never have happened without air conditioning.

Dave Elovitz

May 2003

SHARE AND SHARE ALIKE

"The electrical contractor just hit us with a huge 'extra' for our tenant's renovation project," the building owner told me. "It sounds like they are going way overboard. Can you help sort this out?" "Sure," I replied, "what's up?" We 'conferenced in' the electrical contractor:

"We've already roughed in new wiring for the workstation cubicles exactly as we were told - three circuits to each group of nine cubes," he explained, "but there's a problem. The tenant just gave us new load information: every cubicle will have 3 PCs, at 4 amps per PC. That's 12 amps per cube - more than twice what our wiring can handle. One circuit for 3 cubes is not enough. We need a dedicated circuit for every cubicle! The existing panels don't have enough circuits or capacity for that load, so we need new transformers and new panels. There's no space in the electrical room for more equipment, so you'll need to add a new electrical room for the new transformers and panels."

"Hmm," I mused, "everything you say makes sense - if the loads are right - but doesn't 4 amps seem high for a PC? I would expect a PC to be under 2 amps."

"I thought so, too," the contractor replied, "but I asked the tenant, and he said he got the data directly from the equipment nameplates. How can I design for less?"

"You're right. But maybe some hard data will help us pin down the real loads. Do you have a 'True RMS' ammeter - a meter that gives accurate readings even with non-linear loads such as PCs? No? Well, get one, and then measure the actual load of the 'worst case' cubicle setup."

The measured load of a "worst case" cubicle - with 3 PCs - was only 4.6 amps - almost 2/3 less than the tenant thought. Faced with the test data - and the prospect of a huge 'extra' for all of the new wiring and equipment - the tenant quickly agreed that 3 cubicles per circuit would work as originally designed. With the reduced loads, the existing transformers and panels could handle the entire load - in the end, the electrician only needed to run a handful of additional circuits. There was still an 'extra: but it was manageable.

Gary Elovitz

June 2003

WOERTENDYKE vs. BUCK

When I first got out of college I worked as a facilities engineer for a guy named Bill Woertendyke at Pratt & Whitney. Bill, in turn, worked for a guy named Dick Buck.

One of my first projects for Bill was air conditioning a conference room. This was a big deal, because at that time Pratt & Whitney did not believe in air conditioning for comfort, so this was one of the few air conditioned conference rooms in all of P&W. It was a pretty simple system by today's standards, a vertical package unit with an air cooled condenser on the roof. What we used to call a "drug store unit." It sat in a corridor just outside the conference room with a return grille through the partition and a supply duct off the top and over the conference room to a couple of ceiling diffusers. You've seen that type of unit: Built in thermostat under a little metal door in the middle of the front of the unit.

Now, whenever our department had a meeting with one of the other departments who were our "customers," we would meet in that conference room. And if the meeting was about a fairly big project, Dick Buck would sit in. Invariably, 15 or 20 minutes into the meeting, Dick would start running his finger around inside his shirt collar, then get up and go out the door. You could hear the clang as he closed the little door on the unit after he changed the thermostat setting. Then he'd return to his seat and - after a few more minutes - announce, "Ahh! That's much better now!" For some reason, this just drove Woertendyke nuts.

So after the fourth or fifth episode, Bill called me aside and told me to disconnect the wires to the thermostat inside the unit and install a ductstat out of sight in the return air duct to control the compressor.

Sure enough, at the start of the next meeting Dick Buck runs his finger around inside his shirt collar, leaves the room, the little door goes clang, and he returns to announce shortly thereafter, "Ahh! That's much better now!" Woertendyke just grinned at me across the table, immensely pleased with himself.

I wonder if Dick Buck ever found out that I had taken the wires off that thermostat and his twisting the knob didn't change anything at all.

Dave Elovitz

July 2003

FIRE & ICE

"We have a serious problem in our new school," the town administrator said. "The classrooms are too hot all winter - it's hard enough to keep the kids awake when it isn't hot! On top of that, the heating coils keep freezing and bursting. We lost over 20 last winter! How can coils freeze when the rooms are so hot? The engineer and the contractor just point fingers at each other."

The HVAC system was typical for the 1970s - unit ventilators with pneumatic controls - except for the coils. The design called for 4-row coils for future air conditioning. Sounds clever enough: change from hot to chilled water in warm weather, and voila - air conditioning in every classroom.

Cooling coils need to be much bigger than heating coils. My calculations showed that coils sized for cooling were 3 - 4 times too big for heating - that explained the overheating. But with excess heat, why did coils freeze?

The standard unit ventilator control sequence includes the ability to use outside air for cooling. This is very helpful on cool, sunny days, but it can be a disaster with oversized heating coils. Even when the room needs just a little heat, the control valve opens and fills the coil with hot water, delivering much too much heat to the room. By the time the thermostat closes the control valve, it's too late: the room is already several degrees above setpoint. Then the control opens the outside air damper to cool the room. Cold air hits the standing water in the coil, and it's "Ice Age".

Changing the coils would solve the problem, but that would be expensive and would foreclose future air conditioning. Reducing the heating water temperature would also work, but other heating equipment on the same loop needed the hotter water. Also, it would be hard to avoid opening the outside air dampers in cold weather. Repiping to create a separate low temperature loop for the unit ventilators was a non-starter.

But there was a way to reduce over heating and maintain flow through the coil to reduce the risk of freezing when the outside air damper opened: Install an inexpensive pump at each coil and install smaller control valves that would meter the water more precisely. When the valve opens, a small amount of hot water from the boiler loop mixes with water that the small pumps recirculate in the coil, making water warm enough to heat the room, but not enough to overheat it.

Temperature control is A-OK, and there hasn't been a single coil freeze in two winters.

Gary Elovitz

August 2003

PIED PIPER

A young Irish immigrant worked hard, served his customers well, and over the years built up a successful general contracting business. As part of his climb to respectability and wealth, he became a noted collector of art. Finally the business reached the size where it did not need his continuous attention, and he decided to make a pilgrimage to the county in Ireland where he grew up. Wandering around a picturesque little country town, he peered into the window of a dusty curio shop and spied a very lifelike, life-sized bronze statue of a rat.

It had no price tag, but it was so striking he decided he must add it to his collection.

He went inside and asked the owner, 'How much for the bronze rat in the window?'

"12 Euros for the rat; 100 Euros for the story," responded the owner.

The contractor handed the owner 12 Euros, "I'll just take the rat, you can keep the story." As he walked down the street carrying his bronze rat, he noticed that a few real rats had crawled out of the alleys and sewers and had begun to follow him down the street. This was disconcerting, and he began to walk faster.

Within a couple of blocks, the herd of rats behind him had grown to hundreds, and they were squealing. He began to trot toward the bay, looking around to see that there now seemed to be MILLIONS of rats, squealing and coming at him faster and faster.

Terrified, he ran to the edge of the bay and hurled the bronze rat as far out into the bay as he could.

Amazingly, the millions of rats jumped into the bay after it, and were all drowned.
The contractor walked back to the curio shop.

"Aha!" smiled the owner, "You have come back for the story?"

"No," said the contractor. "I came back to see if you have a bronze Architect."

Dave Elovitz

September 2003

TALE FROM A CHEM LAB

I've been working in a fan room
All the live long day.
I've been working in a fan room
Air won't flow where the prints say.

Fan won't push air down the
    ductwork
It won't heed the call.
Fan won't build up static pressure
It seems to be in stall.

    Fan why won't you blow
    Fan why won't you blow
    Fan why won't you blow
    And fill that duct?

    Fan why won't you blow
    Fan why won't you blow
    If you don't blow
    Then I'll be stuck.

I spent days and nights a-plotting fan
    curves
I spent days and nights on
    calculations.
Then I ran another round of field
    tests.
Fan won't do what fan curves show.

So I said, Fe fi fiddle e i o
Flow is high but pressure's way too
    low.
Fe fi fiddle e i o
All that work and nothing yet to
    show.

(melody restarts at the beginning)

So then I called the manufacturer
And explained my tale of woe.
But when I called the manufacturer
He just told me where to go.

He said the problem's in the system
And spouted other flack.
But I wish he'd admit the problem
And take those damn fans back.

    Motor's in the way
    Motor's in the way
    That's why fan won't do
    What fan curves say.

    Motor's in the way
    Motor's in the way
    Move it back and
    Fan will work OK.

Motor at the inlet blocks the air flow
Motor takes up lots of space where
    air should go
Motor at the inlet restricts air flow
That's why pressure is so low.

So I said, Fe fi fiddle e i o
Move the motor back and watch the
    pressure grow
Fe fi fiddle e i o
Problem solved so homeward I did
    go.

Ken Elovitz

October 2003

IF THINGS GO WRONG

No one likes to think about problems, but construction projects are prone to encounter some bumps in the road. The difference between success and failure sometimes depends on how the parties respond to and resolve problems.

Disputes between the owner and the designer usually involve unanticipated extra costs or disappointment with the end results of the project. To the extent the problems are due to design errors, the owner might be able to recover from the designer. Once the problem is fixed, the owner has the benefit of the corrected or improved system. Therefore, the owner's recovery is limited to the additional cost to fix the problem over what the work would have cost if the designer had not made the error. This cost can sometimes be substantial but often is small.

Disappointment is more often due to unmet expectations than to design error. Diligent efforts to develop a solid program and clear design criteria go a long way to defining expectations and heading off disappointment.

Cost reductions are one of the most common sources of disappointment or problems on construction projects. After the bids come in, the owner sometimes wants to cut the cost and pressures the designer to take out some feature the designer thought was important. When the end product is deficient in some way, the designer attributes the problem to the owner's pressure to cut costs. Rightly or wrongly, the upset designer sometimes reasons that the cost reduction eliminated the part that would have made the project work.

Contractors are notorious for offering "proposals for cost reduction". Owners are sometimes quick to accept them. If the designer has reservations about the contractor's proposal, the project becomes a mix between plan and spec and design/build. The designer does not want to adopt the contractor's "design", yet the contractor might not have put enough design effort into the proposed cost reduction to coordinate with the overall project design. The result can be a system that does not work. There is nothing wrong with contractor proposals as long as they include the necessary engineering.

Not every problem attributed to design rises to the level of an actionable claim. Unless designers promise to deliver a certain result, their liability is measured by what a similarly situated professional would have done. If that professional might have made the same mistake, the owner has to pay the extra.

Ken Elovitz

November 2003

ENGINEER JOKES

How many engineers does it take to change a light bulb? None, if the contractor installed it properly.

How many contractors does it take to change a light bulb? None. It's a design error.

* * *

A physician, a civil engineer, and a computer scientist were arguing about what was the oldest profession in the world.

The physician remarked, "The Bible says God created Eve from Adam's rib. This clearly required surgery, so I can rightly claim that mine is the oldest profession in the world."

Then the civil engineer said, "But even earlier in the book of Genesis states that God created the order of the heavens and the earth from out of the chaos. That was the first and certainly the most spectacular application of civil engineering. Therefore, fair doctor, you are wrong: mine is the oldest profession in the world."

The computer scientist leaned back in her chair, smiled, and then said confidently, "Ah, but who do you think created the chaos?"

* * *

What is the difference between Mechanical Engineers and Civil Engineers?

Mechanical Engineers build weapons, Civil Engineers build targets.

A mathematician, a physicist, and an engineer are told to find the volume of a red rubber ball.

The mathematician carefully measures the diameter with a caliper and evaluates the resulting triple integral.

The physicist fills a beaker with water, puts the ball in the water, and measures the amount of displaced water.

The engineer laughs and looks up the model and serial numbers in his red-rubber-ball table.

* * *

If you want to practice engineering and wind up with a little money at the end of your career, you have to start out with a lot of money.

Ken Elovitz

December 2003

LEARN TO ADJUST

Our friend Dan Int-Hout, who is chief engineer for Krueger diffusers, asked if this story qualifies him as an honorary "energy guy". Dan was in an engineer's conference room giving a talk on air distribution. The room was long and narrow, with a window at one end and a pull down screen at the other. Dan's talks are usually both entertaining and informative, so it is no surprise the room was packed to its limits.

The host called the building maintenance staff twice during the talk to have them lower the thermostat setting, but the room remained stuffy. Meanwhile, the secretary outside wondered what was going on in the room because she said "you can hang meat" out where she was sitting.

During the talk, Dan noticed that the room had a single 2-slot linear ceiling diffuser just in front of the screen, at the opposite end of the room from the outside wall. He also noticed that both slots were directed down.

Desperate to make the room comfortable, Dan' took out his pocket screwdriver, removed his shoes, jumped up on a credenza under the diffuser, and adjusted both slots to direct air along the ceiling toward the outside wall. The response was immediate. One lady sitting 2/3 of the way toward the window immediately said "Oh". Within a few minutes they had to call the maintenance guy to reset the room temperature to where it had been earlier. Even the secretary remarked that it was far better outside the room where she sat. There were no complaints for the remainder of the meeting.

Dan was back at the same office about six months later for a follow up talk. It was in the same room. Many of the same people attended. Before starting the talk, Dan asked what people thought about comfort in that conference room. They said it had become the firm's most popular conference room. Out of curiosity, Dan asked how long they had been in that building. They said 20 years!

In all that time, no one ever thought there might be something wrong with the conference room air distribution. Even after Dan's simple adjustment made such a big difference, no one considered checking how the diffusers were set in the other conference rooms. They just accepted poor HVAC system performance.

Dan left me with this moral: "Adjustable diffusers must be adjusted. It only takes a few seconds. It's not rocket science, but the result can be astounding."

Ken Elovitz

January 2004

apologies to
CLEMENT MOORE

"'Tis the month before New Years,"
    I groaned to my spouse.
"Can't finish this poem
    just by clicking my mouse.
"We always remind them,
    our clients and friends,
"Of all that has happened,
    and how the year ends."

"Just tell them we thank them,"
    she said with good cheer,
"For making '03
    Such a tremendous year.
"For friendship, for kindness,
    for sending new clients;
"From small local Owners
    to national giants."

"And wish them," adds Ken,
    "All the best for '04.
"Health, wealth, and happiness;
    joys by the score!"
"From our house to yours,"
    chimed in Dena and Gary,
"Bon Annee, and Salud:
    All the toasts customary."

The one hundredth birthday
    of the Kitty Hawk Flight.
There was even a stamp
    for the two Brothers Wright!
The Spring found us speechless
    at the victims of SARS.
In August stared skyward:
    Close encounter with Mars.

June was so rainy
    we drew plans for an ark.
Then a blackout put much
    of the East in the dark.
Annika struck out, but
    Not our heroes from Saugus.
The Red Sox did choke,
    but those kids made our August.

When you look back and ponder
    on the year just gone by
Recall projects and schedules
    to be met, do or die.
Some with budgets too small and
    ambitions too high
We're pleased that you called
    on an Energyguy!

May two thousand and four
    be a year filled with peace.
A year when you prosper
    through our expertise!
A year of excitement
    from projects well done.
A year of contentment,
    A year filled with fun.

Dave, Ken, and Gary Elovitz

February 2004

NEW TECH

One response to the oil crisis of the '70's was a great deal of research and publicity about solar water heaters. By the early '80's we no longer stopped to stare at an array of glass topped solar panels on a neighbor's roof. Then - as energy prices stabilized and even relaxed a bit - and as the panels aged and tax credits disappeared, the panels began to disappear, too. Because I had always thought this was new-fangled high tech stuff, I was amazed and amused to read in a recent article in *Invention and Technology* that a patent was granted to Clarence Kemp of Baltimore for the design of a solar hot water heater in 1891! Kemp called it the Climax Solar Water Heater. The article included a copy of a Climax ad announcing a special $15 price in 1892 for his $25 Model 1 and claiming "This Size will Supply sufficient for 3 to 8 baths."

In the 1800's the alternative was chopping wood or lifting heavy hods of coal into a cookstove, kindling the fuel, and stoking the fire. Of course, the wealthy had heaters that burned gas made from coal, but it didn't burn clean, and the heater had to be lit every time you wanted hot water. And if you forgot to turn off the flame, the heater might well blow up. Wood, coal, and coal gas were expensive in many places and often hard to find. So a solar-heated alternative was very appealing.

Southern California had lots of sunshine and expensive fuel, so by 1897 1/3 of the residences in Pasadena had Kemp solar water heaters. The Kemp heater had a solar collector integral to the storage tank. In 1909 William J. Bailey designed a split system with an insulated indoor storage tank that kept water hot overnight. Climax went out of business but Bailey sold 4000 "Day and Night" heaters between 1909 and 1918. Natural gas discoveries in the '20's and '30's killed the business in southern California, but a Florida firm bought Bailey's patent rights just in time for the Florida building boom in the '20's. By 1941 they had installed over 60,000 units in Florida. Over half the population of Miami got hot water from the sun. After World War II, electricity prices dropped, and local utilities offered low cost electric water heaters to increase electric load. The solar hot water industry faded away even in Florida.

Although we don't see many solar hot water heaters here nowadays, American solar hot water technology has been enthusiastically adopted elsewhere. According to the article, 10 million Japanese homes heat their water with the sun, as do 90% of Israelis and Cypriots. At a time when much of the world criticizes American actions, it's rewarding to know they still enjoy our Yankee ingenuity.

Dave Elovitz

March 2004

LOW FLOW WOE

"We can't get enough chilled water to the penthouse units on one wing of our new luxury apartment building," the client complained. "The design engineer says he designed for diversity and the problem is balancing. We think he's covering his tracks because we don't see how 860 GPM of pump is enough for a system with 1200 GPM of connected load."

I convinced the client to send me drawings so I could calculate pressure drops before going to the site. My calculations said there should be plenty of pump, so maybe there was a cross connection, some other installation problem, or even a pump running backwards.

At the site, I started with the pump. The head and motor amps were on the pump curve, and the flow estimated that way was consistent with the pressure drop through the chiller. The pumps performed as advertised. The piping looked OK, too - no improper bypasses or cross connects. The control system said the bypass valve was closed, and it was. We moved upstairs to check the risers.

First I checked a branch that worked OK. Measured pressures confirmed my calculations. Then I checked the problem riser. The supply pressure was right on, but the return was 18 psi too high. That extra 18 psi meant 18 psi of extra pressure drop in the return line. "I'll need access to every riser on this leg so we can trace the pressure back toward the main tee and locate the obstruction."

A few weeks later the client had the ladders and access panels ready. I rechecked the pump pressures to be sure nothing had changed. Then we started checking risers. Riser 6 - restricted. Riser 5 - restricted. They were all restricted on that leg. "We have to get at the main tee," I said. "I can't imagine what the restriction might be, but now we know where it is. We just have to find it."

Two of us steadied a very tall ladder as the balancer climbed up. "I can see a bulge in the pipe," he said. "It looks like a flange. Wait! It's a valve! There's one on the supply and one on the return. The handle for the return is behind that duct. I think it might be closed!"

"Then open it," I said. He did, and we measured the pressure at Riser 6A and on the 21st floor. The pressure difference immediately went from 2 psi to 12 psi - more than enough to get design water flow in the $12,000/month penthouse apartment.

"If you hadn't figured out where to look, I never would have found that valve, and we never could have solved this problem!"

Ken Elovitz

April 2004

DEMONOLOGY

It had seemed like a pretty routine boiler replacement. That's why I was surprised when the property manager called over a year' later about water hammer complaints in some basement apartments, especially when the burner was on high fire. "We had some complaints last year, but the contractor tweaked a few things and it seemed to go away. I guess it was just the end of the heating season, because the problem is back."

I went to the site to investigate. First a few little plinks, but then real solid banging like a bunch of demons with sledge hammers. The demons seemed to hang out in a condensate main a foot or so above the floor along the far wall. "How can that be?" I asked. "Water hammer comes from steam and condensate in the same pipe: The steam pushes the condensate along until it bangs into something. But this pipe is the wet return of a one pipe steam system. It should be full of just condensate with no steam in it at all. The line rises a foot above the condensate tank then drops into the top of the tank. Water seeks its own level, so the line should be full of condensate all the way up into all the risers."

My infra-red thermometer showed that the pipe was too hot to hold just condensate. As I was measuring at the condensate tank, the property manager said, "There used to be a whole kluge of traps stacked up on the connection to the tank, but the contractor took them out." Light dawned. I knew what they had tried to do: create a false water line higher than the water level in the boiler.

The boiler was originally installed with a gravity return. The condensate tank and feedwater pump had been added later. When the wet return was connected directly to the boiler, steam pressure in the boiler pushed back against the condensate to keep the wet return full of liquid. But when condensate empties into a vented condensate tank, steam supply pressure can push the condensate into the tank. At start up, steam condensed and filled the return line OK until the condensate rose to the level of the tank connection: Then the return was a wet return, filled full with condensate up to about 3½ feet above the floor. Thus, steam met the condensate only in the vertical risers that carried condensate away. But once the steam pressure built up enough to more than balance that 3½ foot head, steam pushed all the condensate out of the return line into the condensate tank, and the wet return became a dry return where steam and condensate could meet violently.

Raising the height of the tank connection further above the tank, and lowering the boiler steam pressure let the return always stay full of condensate and banished the demons!

Dave Elovitz

May 2004

PERSONNEL REFERENCES

The English language is full of ambiguity. Many employers find that ambiguity helpful when asked to provide a recommendation on a former employee:

"I can't recommend the candidate too highly."
Does this mean the candidate is not recommended at all?

"I assure you no person would be better for the job."
Does this mean you'd be better off leaving the position vacant than hiring this person?

"I am pleased to say he is a former colleague."
Does this mean we are glad to be rid of him?

"I enthusiastically recommend this candidate with no qualifications whatsoever."
Does this mean the person is completely unqualified?

"I urge you to waste no time in making this candidate an offer of employment."
Does this mean the applicant is not worth considering?

"A man like him is hard to find."
Does this mean the employee is always absent or hiding?

"I would say his real talent is getting wasted at his current job."
Does this mean the person gets drunk on the job?

"I can't begin to tell you what a fine person he is."
Does this mean he's not a fine person at all?

"You'll be lucky if you can get this person to work for you."
Does this mean the person is not very industrious?

"This candidate is solid, competent, and has good work habits."
Does this mean the person is a plodding dullard who will never do anything original?

"I'd say his accomplishments are unheard of."
Does this mean he never did anything?

"I recommend this person warmly, strongly, to any department with a job in her area."
Does this mean do not hire this person?

For more examples of sentences and phrases that say one thing but often mean just the opposite, get a copy of *Lexicon* of *Intentionally Ambiguous Recommendations (LIAR)* by Bob Thornton, professor of economics at Lehigh University.

Ken Elovitz

June 2004

BIGGER IS NOT BETTER

The air handling unit always shuddered. A millwright had balanced the fan wheel and aligned the sheaves, but it still shook. The fan was in a roomy fan compartment and discharged to a plenum, so this was not a system effect problem. That was all good, but it meant the answer would not be obvious.

The balancing report showed the installer sped up the fan to try to get design CFM. Even at the higher speed, the fan was 25% low on airflow. On top of that, the balancing report listed 3.3" wg total static pressure. The fan curve said the fan would develop almost 5" wg of static pressure at the reported speed and airflow.

High RPM, low CFM, and low static. That sounds just like a fan running backwards. Watching the fan coast to a stop ruled out that possibility. The fan blades were clean, and there were no obstructions or damage to the fan wheel, so the fan was physically OK.

Measuring the pressure drop across the coil showed that airflow had not changed much since the balancing report. But plotting the data on the fan curve explained the pulsing and shuddering: The fan was operating well to the left of the peak efficiency point in a region with unstable performance. How did that happen? This fan must have been way too big for the job from day one. The real puzzle was what to do about it.

The air handling unit was built in place, so changing fans would mean taking the building apart. I needed a way to make this fan behave like a smaller fan. Then I remembered years ago seeing some constant volume air handling unit fans with fixed inlet vanes. I learned that old time engineers used inlet vanes to trim fan capacity and shape the fan curve. As inlet vanes close down, the fan curve gets steeper, and the unstable region of the fan curve gets smaller.

The shuddering fan had been converted from variable inlet vanes to variable speed drive years ago. The contractor was probably supposed to remove the old inlet vanes, but luckily he didn't. The fan curve predicted that closing the inlet vanes to 60° would steepen the fan curve enough to provide stable operation and still deliver the design capacity.

"I adjusted the inlet vanes last Saturday, just like you said. The fan runs as smooth as a baby's bottom, and there's no problem meeting the static pressure setpoint!"

Ken Elovitz

CREATIVE THINKING

An architect, an engineer, and a contractor were going on a partnering retreat. The contractor offered to host the retreat at his hunting lodge in the northern wilderness. The architect and engineer agreed, figuring the retreat would let them see the quality of the contractor's work.

At the last minute, the contractor had to work late to finish up a job. So he gave the architect and engineer driving directions to the lodge and told them to help themselves to anything they wanted while they waited. The contractor expected to get to the lodge in time for a late dinner.

The lodge was a simple cabin. It was just two rooms, with minimal furniture and household equipment. There was nothing unusual except the cast iron pot bellied stove that heated the place. The stove was suspended in mid air by wires attached to the ceiling beams.

"Wow!" the architect exclaimed. "It's obvious that this contractor has a deep sense of the relationship between humans and structures. Mounting the stove up high like that lets him curl up underneath it and make this simple hunting lodge a vicarious return to the womb."

"This guy is much more sophisticated than I would ever have guessed," the engineer announced. "By mounting the stove up high, he found a way to distribute heat throughout the cabin more evenly."

They could hardly wait for the contractor to arrive so they could compliment him on the understanding and insight he seemed to bring to the construction team.

"I appreciate your compliments," the contractor said, feigning a blush, "but it's a lot simpler than you think. I had plenty of left over wire but not much stove pipe."

***

A man was convinced that his aging wife was losing her hearing but was too vain to get a hearing aid. One day he decided to conduct an experiment to prove his point. He walked into a room where she was sitting with her back to the door. Standing just inside the doorway he said, "Darling, can you hear me?" When he didn't get an answer he moved into the middle of the room and said again, in a normal voice, "Darling, can you hear me?" Still no answer. Finally he walked just behind her chair, bent over, and said, "Darling, can you hear me?" She looked at him with an irritated expression and said, "for the third time, YES!"

Ken Elovitz

August 2004

DON'T JUMP TO CONCLUSIONS

*Shamelessly plagiarized by
Dave Elovitz*

September 2004

BLOWING HOT AIR

The luxury "garden apartment" had it all: "Sub-Zero" refrigerator, high ceilings, rich woodwork, elevator, prestigious Back-Bay address, everything - except enough heat. It was never warm enough, and on really cold days some rooms were simply unusable. The furnace looked big enough (and I confirmed that with load calculations), but still, it just couldn't make the place warm.

The first thing that caught my attention was the return air duct: starting at the furnace air inlet (near the floor), the duct rose straight above the ceiling and extended about 10 feet to a grille in the hallway ceiling. The ductwork and intake grille looked a bit too small for a furnace of that size. Also, the duct had several sharp bends that created a lot of pressure loss. Sure enough, my pressure measurements showed that half the fan's power was expended drawing air through the return air duct even though it was relatively short. Restrictions in the return air ductwork kept the furnace from moving as much air as it should.

The combination of reduced airflow and a ceiling-mounted return air grille spelled disaster - this was a textbook case of stratification. The air at the ceiling was more than 10°F warmer than at floor level.

Here's how it happens: There is, of course, a natural tendency for warm air to rise, so air at the ceiling is almost always a bit warmer than the air at floor level. Delivering warm air from grilles in the ceiling always presents a challenge to get the warm air to mix in the room rather than hang near the ceiling. With low airflow because of duct restrictions, the supply air will be warmer - the same amount of heat applied to a smaller volume of air; that alone increases the risk of stratification. When warm supply air hangs near the ceiling, the furnace draws in the warmest air in the apartment, making the supply air even warmer and more buoyant - a vicious cycle. Meanwhile, air near the floor just sits there and gets cold.

If we could put in a new return air duct - a "straight shot" to a grille in the hallway wall closer to floor level - it would draw in the cool air near the floor, and there would be less pressure loss, which would increase our airflow. Unfortunately, the water heater blocked the only path between the furnace and the hallway wall. Happily, there was just enough room to raise the water heater onto a platform and squeeze in a new return duct below it. The new return duct was "short and sweet", the furnace breathed easier, and the homeowners no longer got cold feet.

Gary Elovitz

October 2004

THE TRUTH ABOUT CATS AND DOGS

A dachshund was out frolicking in the forest and wandered into a dark, unfamiliar area. Soon he noticed out of the corner of his eye that he was being stalked from afar by a leopard. Running was obviously futile, but as he looked around for an escape route he happened to see a bone on the ground and was inspiration struck. He dropped to the ground and began gnawing on the bone, paying no attention to the approaching leopard.

Just as the leopard was about to pounce, the dachshund sat up and said, "Boy, that was the best leopard I've ever eaten! I wonder where I can find another one?" The leopard froze in midpounce, turned tail, and ran.

Unfortunately for the dachshund, a monkey sitting in a nearby tree saw the whole thing and decided he could turn the situation to his advantage. The monkey sought out the leopard, explained how the dachshund had tricked him, and struck a deal: He would lead the leopard back to the dachshund, and the leopard would promise never to harm the monkey from then on. The monkey jumped on the leopard's back and began to direct him to where the dachshund was hiding.

The poor dachshund caught sight of the angry leopard approaching with the monkey on his back and immediately understood the gravity of his plight. "This time I'm really done for," the dachshund said to himself.

But once again inspiration struck: the dachshund sat down with his back to the approaching leopard and pretended he didn't have a care in the world. Just as the leopard came within earshot, the dachshund stood up, looked at his watch impatiently, and loudly declared: "Where is that monkey? I sent him half an hour ago to bring me another leopard!"

from "Click & Clack", the Car Guys)

Shamelessly plagiarized by
Gary Elovitz

OUT OF THE MOUTHS

The Community Farm in Natick is a great place to take kids. Maple sugaring in the late winter and lots of farm animals. This past spring the farm had a new calf. Franny often takes our 5½ year old twin grandsons to visit the farm, and, of course, they had to go see the new calf. The twins were duly impressed, and a good time was had by all.

On the way home, one asked, "Where was the bull?" Franny replied, "That's right, I didn't see a bull there, did you? Maybe they don't have one." "Oh, no, Bubby," explained the other soberly, "If there's a calf, there has to be a bull!"

Dave Elovitz

November 2004

SPEED DEMONS

BANG! CLANG! CLUNK! These are not sounds that a property manager likes to hear on her home phone at 2 AM, accompanied by grumbling from the red-eyed tenant of a luxury apartment.

The service contractor must have adjusted the controls for the motorized valves a thousand times. He checked the pitch of the piping to make sure none of the piping had sagged - he knew that condensate (i.e., water) could collect in a sagging pipe, and that steam could pick up that condensate and slam it against the end of the pipe. But the banging went on, night after night.

I met the service contractor and the tenant at the building. The contractor fired up the boiler and opened the motorized valve for one half of the building. (No need to send steam heat to both sides of the building in mid-June.) For a while, nothing. Then, all of a sudden, BANG! CLANG! KA-CHUNK! I couldn't sleep through that either. We sent the tenant off with assurances that we would move heaven and earth to solve this problem.

I am always leery of motorized valves on a steam system - they often cause problems without improving comfort - so I wanted to fire the boiler from a "cold start" again, this time with both motorized valves open. While waiting for the pipes to cool down I traced out and measured the piping in the basement, counted and measured radiators, and looked for clues. But the system looked good - I didn't see any of the usual water hammer suspects.

We fired up the boiler with both valves open, and waited. And waited. No banging at all for a full half hour - at least 10 minutes longer than with one valve open. I was just about to congratulate myself, but then - Bang! Definitely quieter and less frequent than before, but still, hold the applause. Nonetheless, I knew I was on the right track.

Back in the office, I pored over reference data, and the answer came into focus. This boiler was twice as large as it needed to be! When the boiler builds up a good head of steam, it makes too much steam for the pipes to handle. With both motorized valves open, steam in the main piping zooms along at over 70 miles per hour. With only one valve open - imagine steam shooting through the piping at 150 mph and picking up a bit of water along the way ...

The solution was gloriously simple - adjust the burner to fire at half its original firing rate, and keep both motorized valves open all the time. Still plenty of steam to heat the building, but no more midnight percussion ensemble.

Gary Elovitz

December 2004

WHEN?

"Boy! How am I supposed to defend my guy against this?" the lawyer groaned. "He pulled the oil burner permit and started up the burner when the house was built in 1995, but the plumber had already put in the copper oil line from the tank to the burner, running under the slab. My guy never saw the line before the slab was poured, but he could see the copper line coming up out of the slab on both ends inside a corrugated plastic sleeve, just like the code requires. He hooked it up, started the burner, and sold them oil until July 2000.

"The next thing he knows he's being sued. The new oil dealer fills the tank September 8th, and it's empty again on September 27th. He sees oil stains on a patch in the basement floor near the boiler, digs it up, and finds the plastic sleeve has been cut and a compression connector installed in the copper line. The connector was over tightened and it leaked. He puts in a new overhead oil line, fills the tank, and gets the burner back on. But all that oil has contaminated the ground. Now the homeowner's insurance company says that connector was always there and always leaking, and since my guy had the permit, they want him to pay for all the environmental remediation. How am I supposed to show that the connector was put in later, and not by my guy? How can anyone tell when the leak occurred?"

I thought about that for a while, and then analyzed the oil delivery records, the service records, and the weather records. Oil use for heating should be proportional to heating degree days. After all, that's how the oil company knows when your tank needs to be filled. This was a little more complicated because the same boiler also used oil to make domestic hot water. But looking at average oil use per day during the summer when there was no heating let me make a pretty good estimate of how much oil was used to make hot water. Subtracting the hot water use from each oil delivery and comparing it to the weather data gave me this graph of gallons per degree day versus time:

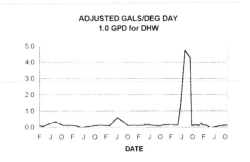

Any question when the leak occurred?

Dave Elovitz

January 2005

'TWAS THE NIGHT BEFORE

'Twas a cold winter night,
    and all 'round the block,
Not a burner was running,
    lest all be in hock.
The children were shivering,
    all huddled in bed,
And dreaming of 70,
    not 50 instead.
Mama in her blanket,
    and chilled to the liver,
Sat shivering there,
    while the windows did quiver.

Then out on the lawn,
    there arose such a clatter,
I sprang from my bed
    to see what was the matter.
There what to my wondering
    eyes did appear,
But the Energy Guys,
    Those three great engineers!

More rapid than eagles,
    their remedies came.
They whistled and shouted,
    and called them by name:
"Weatherstrip! Insulate!
    Efficiency Test!
Add timers, computers
    Get maintenance best!
Turn off extra lights!
    Caulk up all the cracks!
A new hybrid car
    earns credit on tax."
And then from the boiler,
    I heard one of them roar:
"Your flue is 600!
    CO-2 should be more!"

A wink of the eye,
    and a twist of the head,
"Don't shiver, stop wasting!
    Conserve fuel," he said.
"'Lectricity, too
    is too precious to fling.
If you use it all wisely,
    you'll live like a king."
They spoke not a word,
    but went straight to their work.
They audited everywhere,
    then turned with a jerk,
Made out a report,
    in clear, lucid prose,
Said, "Do all of these.
    End your energy woes!"

Then they called out to all,
    "Don't be OPEC-reliant."
"HAPPY NEW YEAR TO ALL,
    EV'RY COLLEAGUE AND
    CLIENT!"

From our homes to yours
    for each family member,
"A wonderful year,
    from Jan. through to December!"

Dave, Ken, and Gary Elovitz

February 2005

FREEZE DISEASE

"The make up air unit in our new dormitory keeps shutting down on freezestat. We did some testing on our own, and we found uneven temperatures across the heating coil. The control sub says the problem is uneven airflow. The mechanical sub says the problem is a control valve that hunts. Help!"

Heat transfer calculations showed that the coil was too big for the job. Running the design 55 gpm of 180F water from the heating system through the coil would heat air from -10F all the way to 120F. The make up air unit was supposed to introduce "temperature neutral" air, so the control valve choked the water flow way down to achieve the desired 70F supply air temperature. The water flow was so low that the coil pulled enough heat out of the water to approach the danger point for freezing.

With little more than a trickle of water flowing through the coil, the water would cool off significantly before it completed its first pass across the coil. The hot supply water heated the air crossing the first part of the coil to 100F or more. Sub-freezing air could suck enough heat out of the water that air on the other side of the coil was hardly heated at all. That explained the uneven air temperatures.

100F air from one side of the coil mixed with 40F air from the other to meet the 70F supply air setpoint.

That averaging effect might sound OK, but the 40F part of the airstream was cold enough to trip the freezestat. At 20F outdoors, when freezing is still a danger, the control valve would have to throttle the water flow even more - the perfect recipe for freezing a coil.

"You probably should change out that 4-row coil for a 2-row coil," I told the contractor, "but I might be able to find a way for you to make this one work." A few more heat transfer calculations showed that this coil only needed 110F supply water to heat -10F outside air to the required 70F.

"You could keep using this coil if you add a three way valve in the boiler room so the make up air unit can have a lower supply water temperature than the rest of the system. With lower temperature water, the water flow through the coil will stay more nearly constant, and the water temperature drop through the coil will be more a more reasonable 30 degrees. The leaving air temperature across the coil will be more even, too."

"I'll take your word for it," the contractor said, "but it would be cheaper to change the coil." So that's what he did.

Ken Elovitz

March 2005

ICE DAM DILEMMA

"You saved us so much money on that pool ceiling condensation problem a couple years ago, I was hoping you could help us get rid of some ice dams. The waterproofing maven said the building has design and construction defects. He wants us to take the whole roof apart. But we haven't had a single rain leak, so I question whether he's on the right track and whether we need to do all that work."

The design drawings showed insulation tucked all the way into the corner where the sloping roof met the wall. There were gable vents but no overhang or soffit vent. This building would function like an unventilated attic.

Using the R-values for each element of the roof structure, I calculated the roof sheathing temperature at various outdoor temperatures. If the sheathing is warm, snow on the roof starts to melt and form little rivers between the snow cover and the roof. If the weather is cold enough or the snow on the roof is deep enough, that little bit of water can refreeze before it gets to the ground. The gutters or the eaves fill up with ice, and ice dams proliferate. My calculations showed there was enough insulation to keep the sheathing temperature below the danger point, but the ice dams in the pictures proved that it did not.

Crawling through the attic revealed why there were ice dams - it was shirt sleeve weather in the attic when it was freezing cold outside. The warmth in the attic made the sheathing warm enough to melt snow and provide a water source for forming ice dams.

The real question was why the attic was so warm. Uninsulated or leaky HVAC ducts are common culprits, but not in this case. The ducts were completely buried in the blown-in insulation. And there were no voids in the insulation that would account for heat transfer from the occupied space below. But there was a clue. Food odors in the corridor were just as strong up in the attic. Air had to be leaking from the occupied space into the attic.

Sniffing my way through the attic identified the suspects. Toilet and dryer exhaust ducts were not safed off at the attic floor. Those penetrations provided a chimney for warm air from below to rise into the attic. There were also locations where a plastic vapor retarder ran across the unused portion of a shaft. The plastic was not protected with drywall, and it was punched through in several places. Warm air from below flowed easily up into the attic. The contractor was pretty happy I had saved him from replacing another roof, but I doubt he relished having to seal all those air leaks.

Ken Elovitz

April 2005

WRENCH RELIEF

"The VFD on the air handling unit for the right side of the building maxes out every afternoon," the client complained. "It happened again yesterday. Even though it was sunny, the load should have been well below design. The temperature was only in the 50's. I had my drive guy raise the maximum speed to 72 Hz, but we still have the problem," he continued. "I need you to tell me if we should change sheaves or maybe replace the whole unit."

"It sounds like a big duct leak," was my first reaction. With a big duct leak, speeding up the fan sends more air out the leak without much increase in CFM at the boxes. Whether it was a duct leak or not, raising the VFD maximum to 72 Hz was clearly treating symptoms without understanding the underlying cause. After all, the system had run OK for years before the VFD was installed.

The first step was to check fan rotation. The problem supposedly began shortly after the VFDs were installed, so maybe the electrician had reversed some leads. Nope. The fan was turning in the right direction, and the inlet vanes, which had been left in place, were wide open. Everything looked OK at the fan.

Next victim was the static pressure sensor. Maybe something was out of calibration, making it read lower than the actual pressure. A sensor problem that over-speeds the fan could even cause excess pressure that eventually busts a seam, creating a duct leak.

As I walked over to the control cabinet, I noticed an access panel in the main duct. "Let's take a look in there," I suggested, more out of curiosity than anything else. Curiosity may have killed the cat, but it sure made my day. A damper inside the duct was almost half way closed. "No wonder you don't have enough pressure downstairs," I beamed. "You're chewing it all up in that damper."

"That's our morning warm up bypass damper," the manager protested. "It closes to push air through the furnace during morning warm up. If you want it to open more, we need a bigger actuator."

"That's not how it works," I explained. "The damper is supposed to close only part way in morning warm up because the furnace is sized for only part of the system airflow. That damper has to open wide in the cooling mode. Here, take my wrench, readjust that linkage, and your problem will be solved."

The now proud building manager called late that afternoon. "The static pressure was OK all day for the first time in a long time, and the drive was running at only 47 Hz."

Ken Elovitz

May 2005

I BLEW IT

Yep, I blew it! What kind of businessman passes up pointing out his company's 25th anniversary? But that is what I did. I was just too involved with doing what I do every day and every week, and January 1, 2005 flew right by without a single trumpet fanfare!

Well, better late than never. Sure, I had been an independent consultant on energy systems before that, but Energy Economics, Inc. officially came into this world on January 1, 1980, and January 1980 was the first of this series of monthly calendars and messages. The April 1980 calendar was the first to bear the corporate name. It included this message:

> INCORPORATION and PHILOSOPHY
> A few people have asked me about the long term significance of my new corporate form of business. Energy Economics, Inc. incorporating doesn't have to change my personal philosophy or my personal relationship with my clients. I just want to assure you that I continue to adhere to the belief that a client retains me instead of some big firm because he wants my own judgment - not the judgment of a faceless helper the client doesn't know and who doesn't know the client or understand his operations. I know I could do a lot more business and respond more quickly to your needs if I built up a large staff, but I am not going to pass your trust in me along to someone else unless I can have confidence that every piece of work that goes out with my name on it is as good, and as responsive to your individual needs, as I could make it myself. Keeping the books as Energy Economics, Inc. won't change that.

Many of you reading this will have read that message in 1980. I sincerely hope that I have fully lived up to that promise in the 25+ years since I wrote it. I know it is a commitment to you that Ken and Gary both share. I am so proud to see what good engineers they have both become and to see the confidence so many of you have placed in them.

I still remember vividly sitting at a client's plan table looking over some drawings when he turned to me and said, "Boy! That kid of yours. I wouldn't have a kid like that working for me!" "Why, what do you mean, Ed?" I stammered. The reply came back, "He's so good, he makes you look like a bum!"

25 years is a long time, and looking back on it, it is clear to me that, while we might have made more money doing things differently, we could never be richer in relationships with clients than we are today. And for that I most sincerely thank you.

Dave Elovitz

June 2005

QUOTES

A graceful taunt is worth a thousand insults.
Louis Nizer

He has all the virtues I dislike and none of the vices I admire.
Winston Churchill

A modest little person, with much to be modest about.
Winston Churchill

I've just learned about his illness. Let's hope it's nothing trivial.
Irvin S. Cobb

I have never killed a man, but I have read many obituaries with great pleasure.
Clarence Darrow

Thank you for sending me a copy of your book; I'll waste no time reading it.
Moses Hadas

He is not only dull himself,. he is the cause of dullness in others.
Samuel Johnson

He had delusions of adequacy.
Walter Kerr

He can compress the most words into the smallest idea of any man I know.
Abraham Lincoln

I've had a perfectly wonderful evening. But this wasn't it.
Groucho Marx

They never open their mouths without subtracting from the sum of human knowledge.
Thomas Brackett Reed

He inherited some good instincts from his Quaker forebears, but by diligent hard work, he overcame them.
James Reston, about Richard Nixon.

I didn't attend the funeral, but I sent a nice letter saying I approved of it.
Mark Twain

She is a peacock in everything but beauty.
Oscar Wilde

He has no enemies but is intensely disliked by his friends.
Oscar Wilde

He has Van Gogh's ear for music.
Billy Wilder

Dave Elovitz

July 2005

OUT OF HOT WATER

The residents had complained about insufficient heat and hot water from the day the building opened. Two other consultants had said that if the boilers don't provide enough heat and hot water, they must be too small. They recommended adding separate gas-fired hot water heaters. I figured insufficient boiler capacity was an unlikely cause: The problem occurred in the summer when there was little or no space heating load. I knew I would have to dig into the entire system.

The piping was definitely unconventional, but calculating pressure relationships throughout the system showed me it should have worked. Then I noticed the maintenance man's log. The boilers locked out on manual reset high limit almost every day. "Bigger boilers won't help if the boilers lock out," I pointed out. "You can't make hot water when the boilers are off."

The maintenance man didn't care. He wanted new boilers. I needed to find out why the boilers kept shutting down, and he wasn't going to help, so I started tracing out the control wiring. Nothing was labeled. The junction boxes were tangled rats' nests. Unused leads were wrapped with electrical tape instead of wire nuts - an invitation for an accidental or intermittent short to ground. Despite all that, there were no fundamental flaws, and it should have worked.

So I made myself comfortable in the boiler room and waited for a lock out. I put thermocouples on the pipes so I could watch the temperatures. I put my ammeter on a pump motor. A noise caught my attention. Whrrrrr, click. Whrrrrr, click. It sounded like a pump motor trying unsuccessfully to start. The boiler water temperature display started to climb. 193F. 209F. 224F. Click. The boilers shut down on high limit.

I scrambled to find the pump that was having trouble. The shaft turned freely, so it wasn't a mechanical overload. It must have been electrical. Raising and lowering the domestic water heater thermostat duplicated the problem. Sometimes the pump motor whirred but did not start. When the boiler made heat with no water flow, the operating control could not always back the burner down before the boiler temperature reached the high limit setpoint.

The next day the contractor replaced the motor. A lot less expensive than adding a boiler! And the owner didn't get a single boiler shutdown or "no heat" call for the rest of the winter.

Ken Elovitz

August 2005

HIDDEN HEAT LOSS

"These heating bills are killing me! I replaced the windows, added insulation, and installed a new heating system with a fancy, high efficiency boiler, and I'm spending more now than before I made all the improvements."

"First let me analyze your fuel bills. That lets me separate out rate escalation and adjust for weather so we can see what the system is doing." The results indicated excellent automatic temperature control. It was almost too good - as if the building was taking in a lot of outside air.

The fuel use analysis also let me determine the actual heating load for the building. The peak heat loss was more than twice what it should have been compared to similar buildings - another pointer towards excessive outside air. Then I found out the system is perimeter baseboard. There are no fancoils or air handling units. It couldn't bring in too much outside air.

The fuel use analysis showed that the building heated consistently to the same temperature, but it did not tell whether that temperature was 60F or 80F. Heating to 80 would waste a lot of fuel and induce people to open the windows to offset the over heating. But indoor temperatures were normal or even cool, and all the windows were closed.

Could a grossly inefficient boiler be the culprit? Nope. A combustion efficiency test showed a normal 82%, so the problem wasn't inefficient combustion.

The combustion efficiency test itself didn't show anything unusual, but watching the boiler run gave some clues. The boiler did not connect to a chimney. It was power vented through the outside wall. A fan and duct pulled products of combustion away from the boiler and discharged them outdoors. When any zone called for heat, the fan would come on, and the boiler would fire. But wait a minute! The power vent fan continued to run long after the boiler shut off. Any time a zone called for heat, the power vent fan would run, regardless of whether the boiler fired. Running the power vent when the burner was off pulled cold basement air through the boiler, heated up that air, and promptly dumped it outdoors. Pulling air from the basement also sucked in cold outside air to replace it. That extra air flow through the boiler was the outside air load I had been looking for!

A simple wiring change interlocked the call for the power vent fan with the burner instead of the zone thermostats and brought the fuel use back to normal.

Ken Elovitz

September 2005

SIGNS OF OLD TIMES

Before the Interstate Highway system, driving from city to city was a long, slow, tedious experience. Some of you might remember the groups of small red signs with white letters that sprouted along the roadside. The signs were in groups of five, about 100 feet apart, each with one line of a four line couplet, with the last proclaiming

BURMA SHAVE

Most of the messages related to safe driving. Here are a few:

Don't lose your head
To gain a minute
You need your head
Your brains are in it ... Burma Shave

Brother Speeder,
Let's rehearse.
All together:
"Good Morning, Nurse!"... Burma Shave

Speed was high
Weather was not
Tires were thin
X marks the spot ... Burma Shave

Around the curve
Lickety-split
Beautiful car
Wasn't it? ... Burma Shave

Passing school zone
Take it slow
Let our little
Shavers grow ... Burma Shave

Trains don't wander
All over the map
'Cause nobody sits
In the engineer's lap ... Burma Shave

The one who drives
When he's been drinking
Depends on you
To do his thinking ... Burma Shave

Both hands on the wheel
Eyes on the road
That's the skillful
Driver's code ... Burma Shave

At intersections
Look each way
A harp sounds nice
But it's hard to play ... Burma Shave

No matter the price
No matter how new
Best safety device
In the car is you ... Burma Shave

It's great to be able to hurtle along the superhighway, but I miss those signs, both their humor and their messages.

Dave Elovitz

October 2005

INSTRUCTIONS FOR LIFE
(From the Dalai Lama)

(1) Take into account that great love and great achievements involve great risk.

(2) When you lose, don't Jose the lesson.

(3) Follow the three Rs;
   Respect for self
   Respect for others
   Responsibility for all your actions.

(4) Remember that not getting what you want is sometimes a wonderful stroke of luck.

(5) Learn the rules so you know how to break them properly.

(6) Don't let a little dispute injure a great friendship.

(7) When you realize you've made a mistake, take immediate steps to correct it.

(8) Spend some time alone every day.

(9) Open your arms to change, but don't let go of your values.

(10) Remember that silence is sometimes the best answer.

(11) Live a good, honorable life. Then when you get older and think back, you'll be able to enjoy it a second time.

(12) A loving atmosphere in your home is the foundation for your life.

(13) In disagreements with loved ones, deal only with the current situation. Don't bring up the past.

(14) Share your knowledge. It's a way to achieve immortality.

(15) Be gentle with the earth.

(16) Once a year, go someplace you've never been before.

(17) Remember that the best relationship is one in which your love for each other exceeds your need for each other.

(18) Judge your success by what you had to give up in order to get it.

(19) Approach love and cooking with reckless abandon.

Ken Elovitz

November 2005

DUCT TAPE

Many people think the term "duck tape" is a corruption of the real name of that sticky, silver colored product called "duct tape". But according to *Wikipedia,* the free on-line encyclopedia that anyone can edit, the original "duck tape" got its name because it was made of cotton duck fabric and because it repelled moisture like "water off a duck's back". The tape was developed for the military in 1942 by Permacel, then a division of Johnson & Johnson, as a waterproof sealing tape for ammunition casings. Because its water resistant, rubber-based adhesive stuck to everything and its fabric backing made it easy to rip to the desired length, "duck tape" soon became a favorite way to repair almost anything. It remains a near universal repair kit today.

Wikipedia says "duck tape" became "duct tape" during the post World War II housing boom when people started using "duck tape" to connect heating and air conditioning ducts together. Whether the *Wikipedia* story of the origins of duct tape is historical fact or urban legend, duct tape apparently doesn't stick so well to ductwork. Under section 1305.3.8 of the Massachusetts State Building Code, duct tape is not permitted as a sealant on any metal ducts in commercial and industrial buildings. Duct tape is not permitted as a sealant on ducts of any kind in one- and two family dwellings (section J4.4.8.2).

For once, what might look like a baseless provision in the building code actually has some scientific support. In a series of research reports and magazine articles, researchers Iain Walker and Max Sherman of the Lawrence Berkeley National Laboratory reported the results of their study of the durability and longevity of duct sealants for more than a decade. They found that duct tape was the worst performer of all the commonly used duct sealants. Duct tape seems to be most prone to failure at high temperature (like heating ducts) and when used to seal round duct "stab in" connections to a plenum (where the geometry makes sealing particularly difficult for any product).

Research is ongoing, but for now, duct tape retains its honored status as a universal sticking agent for joining almost anything together except HVAC ductwork. Even though the adhesive in duct tape apparently does not stick well to ductwork, it sticks well enough to everything else that removing duct tape leaves behind a troublesome residue. Those situations call for applying the universal "unsticking" agent, better known as WD-40.

Ken Elovitz

December 2005

THE ECSTASY AND THE AGONY

The grand old Victorian was so lovingly preserved it nearly sold itself. The previous owner had lovingly restored so many period details - lush woodwork, built-in cabinets, window seats, clawfoot tubs - and the original windows opened and closed "like butter". Of course some things had been updated - the kitchen had all the latest appliances and the boiler was "state of the art" - but this house screamed "class".

Just sitting in the formal drawing room that fall sent shivers up and down the new owners' spines. But then early fall turned to late fall, and those shivers of excitement turned to shivers of ... cold! The whole first floor just wouldn't stay warm! By mid-December the front parlor was an ice-box, and the family had to huddle together in blankets to be on the first floor at all.

My first thought was to make sure everything was working properly - no need to go through lengthy analysis if it's just a wrong control setting or failed circulator. I found a few minor glitches - the owners had shut off the kickspace heaters under the kitchen counters because they thought they were electric heat (actually, they only used a few watts to blow air over a hot water coil from the boiler) - but everything else was working just fine. So I measured wall and window areas, calculated the heat loss for every room, and figured the heat output of the cast iron baseboard. Sure enough, some of the rooms had much less baseboard than they needed!

That was a surprise - usually we find more than enough heating capacity in the rooms. But in this house, the combination of barely insulated walls and single-pane windows meant high heating loads, and the built-in cabinets and fancy woodwork didn't leave much wall space for baseboard. On closer examination I discovered that the cast iron baseboard was new - whoever renovated the house had probably removed some monster radiators, put in as much baseboard as would fit, and hoped for the best.

Replacing windows or adding storms was a non-starter: the owners couldn't abide the thought of changing the appearance of the house. We needed more heat in the house.

Those kickspace heaters in the kitchen were the answer: Drop a couple more of those heaters in the front parlor, recessed into the floor and covered with fancy brass grilles, and conceal another in the custom-built cabinet in the master bath. As a bonus, warm air blowing gently down onto the tile floor would also make a luxuriously warm floor! Agony into ecstasy at low cost, and with a dash of style!

Gary Elovitz

January 2006

TO THE TUNE OF ....

Frosty the snowman
Was a fixture in '05
    Through the winter snow
    We yelled "Go, Pats, Gal"
And the S-bowl they survived.

The snow it did continue,
So we shoveled it away.
    With near seven feet
    We were all quite beat
Until Spring and its mild days.

    Then summer simmered over
    All the days were mighty hot
        Cooling made the meter spin
        'Twas worth every kilowatt.

Along came Katrina
And her mighty winds did blow
    When the levees broke
    New Orleans got soaked
But that did not end the show.

Soon there was Rita
Kept the oil production low
    As fuel prices rose
    We just froze our toes
And watched Exxon's profits grow.

    There was fuss and turmoil
    When Chief Justice Rehnquist
        died
    And when John Roberts donned
        the robe,
    All the Democrats just cried.

Real estate markets
Were just headed for the sky
    Housing prices soared
    How can we afford
What we'd really like to buy?

Despite those disasters
We remain with spirits high
    And we wish the best
    For the year that's next
Bon Anee! '06 is nigh!

    Once a year we write a rhyme
    To bring a smile or two
    And send a wish from all of us:
    'Happy New Year to your crew!'

Energyguys are
Engineers you need to know,
    Make your systems work,
    Find the flaws that lurk,
No more song, we gotta go!

A HEALTHY AND HAPPY NEW YEAR!

Dave, Ken, and Gary Elovitz

February 2006

RAIN FOREST

"When they first ran the chiller in our new school last summer, it rained everywhere in the building. We shut down the chiller right away, but mold started growing anyway. We spent a fortune to replace ceiling tiles and get rid of the mold. We want to run our air conditioning this summer, but we don't want to start with mold again."

The drawings showed a 2-pipe system that switched between boiler and chiller depending on the season. That would work OK for the classroom units. They were designed for heating or cooling and had condensate drain pans. But it could be a problem for the baseboard and the vestibule heaters. They don't have drain pans so should not be piped to chilled water lines. The designer had thought about that problem: The drawings showed automatic valves to isolate the heating-only loads from the distribution pipes when the chiller is on.

With no explanation for the "rain" apparent from the drawings, the only way to find and solve the problem was to put the system through its paces. I convinced a reluctant maintenance supervisor to start the chiller. When he did, the pump and chiller ran normally. The main pipes were insulated, and there was no sign of sweating. Spot checking some classroom units showed the supply air temperature was low enough to dehumidify. The classrooms might get muggy on a muggy day, but there should not be rain clouds.

Our test had run all day without incident. We were about to pack up when the custodian yelled down the hall. "There's a puddle on the floor in the dietitian's office. The heater in the ceiling sprung a leak!"

There was water on the floor alright, but it wasn't a leak. The little heating coil was wringing moisture out of the air. Maybe the contractor had forgotten to install the isolation valve. Nope. Maybe the valve was faulty. As we started to check it out, the custodian called us over to the main vestibule. There was a puddle on the floor there, too. Something was wrong with ALL the isolation valves. We traced the wiring back to the panel to figure out why the valves were not closing. When we opened the panel, the problem was obvious. "See this loose wire?" I asked. "It's supposed to be connected to this relay. Hook it up, and the rain will stop."

Ken Elovitz

March 2006

WHAT GOES AROUND

"One of our board members told me to ask you to review our heating contractor's proposal to replace our pumps with bigger pumps and variable speed drives. The contractor says that's what we need to get heat at the ends of the system. At $9800 plus tax, we want to be sure it will work."

So the client sent me the design drawings, and I calculated the system pressure drop. The project was 22 years old, and the designer had picked the same pumps I would have chosen. The installed pumps were even bigger, so the pumps on the job should have worked. Installing still bigger pumps with variable speed drives would not solve the problem. Unless they had lived with this problem for 20 years, something must have changed. I had to find it.

The system had 2 in-line pumps, one above the other, supported from a wall. As luck would have it, the pumps cycled off as soon as I got to the mechanical room. I ran over to watch the lower pump coast down and check the direction of rotation. It was fine.

I figured out how to override the controls and got the pumps to run again so I could measure pressures and motor amps and compare them to the pump curve. I checked the lower pump first. Amps were high, and the pump was running way out on its curve, so there was much more than design flow. As I checked out the lower pump, I noticed the upper pump spinning, too. "Oh yeah," the maintenance man said. "We always run both pumps. And we still don't have enough flow."

I couldn't figure out how the contractor had wired the pumps so both ran at once. I was really stumped when I measured zero current in the wire to the upper pump, and my voltage detector said there was no voltage. Finally, I shut the system down so I could poke around in the wiring compartment. That's when I got a good look at the upper pump. It wasn't running at all. It was wind milling backwards! A faulty check valve let water recirculate back to the inlet of the operating lower pump. Closing the isolation valve at the upper pump stopped the bypass flow and let the operating pump push that water out into the system where it belonged.

They didn't need to spend $9800 on pumps or variable speed drives. All they needed was a new check valve. Closing the isolation valve solved the problem temporarily. It also reminded me that you can't determine pump rotation when the pump is on. You have to shut it down, watch it coast to a stop, and verify by watching the shaft when the pump starts up again.

Ken Elovitz

April 2006

HISTORY

What a difference a century makes! In the year 1905:

The average life expectancy in the U.S. was 47 years. More than 95 percent of all births in the U.S. took place at home. Five leading causes of death in the U.S. were:

1. Pneumonia and influenza
2. Tuberculosis
3. Diarrhea
4. Heart disease
5. Stroke

Ninety percent of all U.S. doctors had no college education. Marijuana, heroin, and morphine were all available over the counter at the local corner drugstores. Back then a pharmacist said, "Heroin clears the complexion, gives buoyancy to the mind, regulates the stomach and bowels, and is, in fact, a perfect guardian of health!"

There were only 8,000 cars in the U.S., and the maximum speed limit in most cities was 10 mph.

Only 14 percent of the homes in the U.S. had a bathtub. Most women only washed their hair once a month and used borax or egg yolks for shampoo.

Only 8 percent of the homes had a telephone. A three-minute call from Denver to New York City cost eleven dollars.

The average wage in the U.S. was 22 cents per hour, and the average U.S. worker made between $200 and $400 per year. A competent accountant could expect to earn $2,000 per year, a dentist $2,500, a veterinarian from $1,500 to $4,000, and a mechanical engineer about $5,000 per year!

Sugar cost four cents a pound; eggs were fourteen cents a dozen; and coffee was fifteen cents a pound.

18% of households in the U.S. had at least one full-time servant or domestic help.

Only 6 percent of all Americans had graduated from high school.

It staggers the mind to think what life might be like in another hundred years. But extrapolating from the data above, in another hundred years maybe we can expect:

> The average lifespan will be 109 years.
>
> The average wage will be $2,000/hour.
>
> Speed limits in cities will be 129 mph,
>
> and women will wash their hair every 3 hours!

Gary Elovitz

May 2006

WORDS FROM POPPY

My sister Poppy sent me this: Apparently she came across the phrase "FENDER SKIRTS" in a book. A term she hadn't heard in a long time, and thinking about "fender skirts" started her thinking about other words that quietly disappear from our language with hardly a notice. Like "curb feelers" and "steering knobs." Any kids will probably have to find some elderly person over 50 to explain some of these terms.

Remember "Continental kits?" They were rear bumper extenders and spare tire covers that were supposed to make any car as cool as a Lincoln Continental.

When did we quit calling them "emergency brakes?" At some point "parking brake" became the proper term. But I miss the hint of drama that went with "emergency brake."

Did you ever wait at the street for your daddy to come home, so you could ride the "running board" up to the house?

Here's a phrase I heard all the time in my youth but never anymore - "store-bought." Of course, just about everything is store bought these days. But once it was bragging material to have a store-bought dress or a store-bought bag of candy.

On a smaller scale, "wall-to-wall" was once a magical term in our homes. When we were first married, everyone covered his or her hardwood floors with, wow, wall-to-wall carpeting! Now my wife wants to replace our wall-to-wall carpeting with hardwood floors. Go figure.

And when's the last time you heard the quaint phrase "in a family way?" It's hard to imagine that the word "pregnant" was once considered a little too graphic, a little too clinical for use in polite company. So we had all that talk about stork visits and "being in a family way" or simply "expecting."

Apparently "brassiere" is another word no longer in usage. I said it the other day and my daughter cracked up. I guess it's just "bra" now. "Unmentionables" probably wouldn't be understood at all.

Here's a word I miss - "percolator." That was just a fun word to say. And what was it replaced with? "Coffee maker." How dull. Mr. Coffee, I blame you for this.

Some words aren't gone, but are definitely on the endangered list. The one that grieves me most is "supper." Now everybody says "dinner." Save a great word. Invite someone to supper. Discuss fender skirts.

Dave Elovitz

June 2006

FLUSHED WITH PRIDE

When I bought my oid Victorian 10 years ago, it had one barely functioning bathroom. Our plans called for four new bathrooms (and a whole lot more), so I was on the lookout for bargains. Fortunately, just as we closed on the house, a large home center store declared bankruptcy and held a close-out sale. Our renovation plans were far from complete - so I didn't know exactly what we would need - but I knew we needed toilets. So I bought a couple - I think I paid about $40 each and felt pretty good about it. That is, until our plumbah, Mahk, saw the unmarked boxes.

"Oh boy!" he chuckled, "Wheah did you get these? Oh, rnan - they're made in Mexico did you pick 'em up south of the bordah?"

It took a bit of persuasion, but Mahk did install my bargain toilets.

"You get what you pay for, ya know," Mahk counseled as he tightened the flange bolts. "Next time, get a cola!"

I still needed two toilets, so I took Mahk's advice and bought "name brand" fixtures. I even bought one from the famous (and expensive) Wisconsin manufacturer whose name - in Boston - can easily be confused with a popular soft drink. Mahk was pleased.

Over the next ten years I had a lot of opportunity to comparison test name-brand vs. no-name toilets. (I'm not sure how to put this delicately - let's just say that when manufacturers design and test their toilets, they don't have me in mind.) My two "no-name" toilets have been by far the best performers! They clog much less frequently than the others and are much easier to unclog when they do. And between the two "name-brand" toilets, the fancy Wisconsin throne was by far the more troublesome! In fact, I recently gave up on it completely (after spending well over an hour unsuccessfully trying to unclog it).

I've learned a lot about toilets over the last decade -I hesitate to call it "hands on" experience - and have thought a lot about why some toilets are more prone to clogging than others. The shape of the waterway (the path from the bowl to the outlet connection) seems to be the key factor: some toilet waterways have sharp turns that seem to invite clogs, while others are much smoother. So when I replaced my troublesome toilet I looked for a fixture whose waterway had no sharp bends. So far my theory seems to hold up.

There are two morals to this story: first some low flow toilets perform much better than others; and second - the fancy name doesn't guarantee quality design.

Gary Elovitz

July 2006

BORN TO GOOGLE

"Oh, no, not again!" My usually boringly dependable car had become intermittently exciting in hot weather. Several times it failed to restart after a short stop, but it always started right up after "burping" the gas cap. (Memo to self: buy a new gas cap some day.)

Then one hot day, after sitting outside at a construction site all day (far from home, of course), the engine cranked and cranked but just would not turn over. I finally gave up, had it towed to a garage, and left the mechanic with instructions to fix whatever was wrong. The next morning the mechanic called to say the car started right up, and he couldn't find anything wrong at all. For the next few days I limited myself to tentative local trips, and the car performed flawlessly.

Then, just as I started my car after a short meeting (on my first long trip), I remembered that I needed to check a couple of dimensions at the pad-mount transformer. So I pulled over to the loading dock, got out and made my two measurements, and ... the car would not start! Within five minutes every mechanic in the plant was trying his hand at getting my car started. After trying every folk remedy known to man for about 10 minutes, the engine finally turned over and I was on my embarrassed way.

It was time to take action. But I had already struck out with a mechanic, and who knew if I would have any better luck with the dealer? Maybe I could see if anyone else has dealt with a similar problem. So I "Googled" "Honda Accord intermittent starting problem", and within seconds dozens of "hits" popped up describing my exact problem - a corroded terminal in the "main relay that expands just enough when heated to break the circuit. In fact, a couple of sites included pretty detailed instructions - with photos - describing how to fix it! It took less than an hour to find and remove the relay, resolder the bad terminal, and put it all back together - probably less time than I would have spent ferrying the car to the dealer!

The World Wide Web has become an indispensable tool. Once you master the free-form poetry of the search engine (entering just the right combination of words to capture useful search results), useful information on almost any subject is at your fingertips.

Gary Elovitz

August 2006

DIFFERENTIAL DIAGNOSIS

20 below is really cold, even in January. But I'm talking about INSIDE. With the wind chill from the cooler fans on top of the temperature, I couldn't stay inside the big freezer for more than a few minutes at a time. As cold as it was, it wasn't cold enough -- the -20F freezer only got down to 12 below.

My nearly frozen fingers couldn't tell the difference between 12 and 20 below, but the client's product could. And if trying to move my fingers inside fat gloves wasn't bad enough, the unit coolers where I needed to take measurements were 30 feet up in the air! Steadying myself atop a wobbly lift and squinting against the cold blast from the fans, I saw that the bottom quarter of each coil was clogged with ice. That didn't help. It effectively made the cooler only 3/4 size.

Fortunately I could measure refrigerant pressures and temperatures on the roof (a relatively balmy 30F outside) and in the machine room. Adjusting the defrost controls helped a lot, keeping ice off the coils so the coolers could deliver their full capacity. Fine tuning the thermal expansion valve and making the refrigerant sub-cooler operate properly helped some more. But not enough: The room only cooled to minus 16F. My refrigerant heat balance calculations said the system was delivering more than 19 tons. The design called for only 18 tons, so the system should have walked away with the job. Could the designer have undersized the whole system?

My own load calculations said 18 tons should have been enough, and the system was delivering 19. Either the calculations were wrong or the room wasn't built according to the specs. Checking the outdoor surface temperature of the walls said they were OK, but a rough survey with my infra-red thermometer showed there might be some big heat leaks at the edge of the roof.

"I think there's a problem where the roof meets the wall," I told the owner. "We need to survey with an infra-red camera to check for heat and air leaks at the construction joints." Sure enough, the infra-red survey showed insulation breaks at the edge of the roof all along the perimeter of the freezer. And some big leaks around the gaskets on the doors glowed brightly in the infra-red images.

The phone rang a month later: "We peeled back the roof and filled in lots of missing foam at the top of the wall. Then we adjusted the gaskets on all the doors. It's warmer outside than any time since we started the system, and we're finally down to -20F in the freezer! Now we can start accepting our shipments of ice cream."

Ken Elovitz

September 2006

INVISIBLE HISTORY

How hard could it be to replace an old steam heating system? When I saw the inside of the magnificent theater, I understood. It was like one of the elegant rococo movie palaces I remembered from my high school years, when you dressed up to take a girl to the movies.

The radiators were concealed within elaborately paneled walls with sill grilles about 11 feet above the floor. Cutting and patching was out of the question. "We knew we would need to replace the old steam system pretty soon," the Business Manager explained, "and it would be great to add air conditioning. We brought in several engineers and contractors, but no one has been able to figure out even how to redo the heat and keep the character of the space, never mind add air conditioning." "Even if we could get the new heat in," the Director of Facilities added, "I don't know how you could get those huge old radiators out. The only access is through little holes from the pipe tunnel below, just barely big enough to change a trap." "Pipe tunnels below?" My ears perked up. "How big?" "Pretty good size. They used to bring return air from the back of the theater back to the ventilating fan, but that fan hasn't run since probably World War II."

"I think I have an idea how to solve your problem. Let's look at that fan, and then at the old drawings." We climbed up, and up, and up to a mechanical loft 35 or 40 feet above the stage. The fan was magnificent, obviously idle for years, but the wheel still spun with the touch of a finger. It was part of a built up air handling unit, with a full size outside air connection, steam preheat coils, steam reheat coils, and supply ducts out over the auditorium ceiling. With a new motor, new heating/cooling coils, and new economizer dampers and controls, it would make quite a nice system. And the drawings showed me how to deliver the air with an invisible system.

We could reverse the duct connections so what was now the return duct system, including the big ducts (and pipe tunnels) under the floor became the supply. Abandon the radiators in place, blank off the floor level wall intake grilles, and cut big slots in the concrete floor to let conditioned supply air from the rebuilt fan system flow over the dead radiators and out the grilles beneath the windows. And the old ventilation supply grilles could become the returns. There was even some existing ductwork up to a cupola over the entry which would provide economizer relief!

Even before they had money for a chiller, they would have economizer cooling much of the year. And invisible!

Dave Elovitz

October 2006

FLOOD!

Friends who had retired to Florida were in the area, so we asked them and another couple of mutual friends to dinner. We had a delightful evening - the first hot weather of the season, and the air conditioning was especially welcome. As part of cleaning up after our guests left, I brought the tablecloth and napkins down to the washing machine and was dismayed to find the washer sitting in the midst of an immense puddle.

What the .... ??? Where did all this water come from? Did a washing machine hose let go? Something fail inside the washer? That's when I noticed that the little plastic line from the air conditioner's condensate pump was not sticking into the standpipe along with the washer's drain hose. Aha! Mystery solved, that's where the water came from. But the plastic line was nowhere in sight. I started tracing from the condensate pump - that's odd! Leaning against the basement wall, pulling down the condensate line, was the jackpost I had installed to stop the big display cabinet with Franny's giraffe collection from jiggling when you walk by it. I remembered I had been meaning to investigate why the giraffes started rattling again lately when someone walked by the cabinet.

Sure enough, I found the end of the condensate line up in the joists back over some storage cabinets. I pulled the slack out and stuffed it back into the standpipe. With some mopping up, that solved the flood. Then I carefully re-installed the jackpost, trying to figure out what I had done wrong originally that allowed it to come loose. This time, I adjusted the post so it was realty pushing very solidly against the joist under the cabinet. It still bothered me how the post could have come loose, but there is, after all, no fastener between the top of the post and the joist: All that holds them together is the weight of the floor. I finally gave up, went to bed, and turned out the light.

Forty minutes later, I suddenly sat bolt upright in bed. The end of May we had the wall-to-wall carpet in the living room taken up and the wood floors sanded and refinished. Of course, everything had to come out of the living room before the floor guys could do their work. We had moved all the small stuff and the floor guys took the big furniture, including that big heavy giraffe cabinet, out of the living room to do the floor. They moved it back when they were done.

With all that weight temporarily off it, the floor must have sprung up enough to let the jackpost fall loose!

Dave Elovitz

November 2006

THE TELLTALE STAIN

After scoping out a new project on Newbury Street, Mark Goldstone and I entered the very "chi-chi" boutique next door. When he marketed that space last fall, Mark needed to have the space ready for almost any kind of tenant, so I had selected two HVAC units sized to allow plenty of flexibility. The tenant's contractor would design and install the ductwork to suit the tenant's needs as part of the tenant fitout. Now the new tenant had been open a while, and we stopped in to see how everything was going. The store felt pretty good to us - but then we had just spent the last hour in a non-airconditioned building.

"How's the AC been?" Mark asked the sales clerk. "Fine," she answered.

"That's good," I said, "because if it wasn't, I'd be in trouble!"

"Well, actually," the manager piped up as she joined us, "it has not been so good. Especially in the afternoon when the sun shines through those windows it really gets too warm in here." (Gee, I thought, it does suddenly feel kind of warm in here.)

The temperature in the space was 73F, which was not bad, but the thermostat was set at 70 and it wasn't that sunny out - the system should not have had much trouble making 70! I took a quick look around, and saw that the tenant had reused existing sill-height grilles along the storefront windows (which had been left from a previous tenant) in addition to putting in new ceiling outlets. But there didn't seem to be very much air coming out of those perimeter grilles, and it wasn't even very cold.

A quick trip to the basement solved the mystery. The space had two 10-ton air-conditioning systems. One of the systems fed ductwork overhead, and it was working fine. The other system had been connected to feed only the perimeter grilles. That system's fan was running, but the refrigeration was not. (No wonder the store was struggling - only half their cooling was working!) Considering the weak airflow from the perimeter grilles, I would bet that the system wasn't moving enough air to support 10 tons of cooling; the cooling coil might have iced up and cut out weeks ago! Sure enough, there was a water stain on the floor beneath the refrigeration piping where the melting ice must have dripped. I explained to Mark that the solution could be as simple as speeding up the fan, or they might need to rework some ductwork to get full airflow to those perimeter grilles. The tenant had an airflow problem to solve, but I knew they would be fine once both their units were working properly.

Gary Elovitz

December 2006

SWITCH HITTER

We finally finished getting Patti's condo into ship shape, so we were confident the buyer's home inspector would not find anything the buyer could use to walk away from the deal.

So we were completely surprised by the inspector's report that he didn't see a ground for the electric service (did he look in the right place?) and his "recommendation" that we install a "common ground". I was sure the building and the electric service were grounded. I figured the remark was likely the result of the frequent confusion between services and feeders in a multifamily dwelling. Nevertheless, I had to check it out.

First I went outside to the meter bank. That's the service where power enters the building and where the electric system should be grounded. I saw a heavy wire connected to a rod in the ground and to the meter bank, so I knew the electric service was grounded.

Then I went to our basement, removed the panel cover, and saw that the feeder to our unit was wired to the neutral and the ground in our panel. As I expected, our panel was grounded. There was no need to add a "common ground". Satisfied that everything was OK, we went upstairs to read.

Soon the bedroom felt warm. When I saw 81F on the thermometer, I trundled downstairs to check the air conditioner. The thermostat was set for 74F and the indoor fan was on, but the condensing unit was silent. We had just spent $500 to get the A/C repaired, and it had run well for at least a week. Had the thing failed again? Did the tech do a lousy job? It was late and it was dark. I couldn't do anything until morning. I slept; Patti worried.

At first light I went outside to check things out. I didn't have gages or meters, but I could look for something simple like a loose wire. I opened the access panel, and everything looked OK. I "bumped" the compressor contactor and nothing happened. We might have a big problem here. I went inside and made sure the thermostat was calling for cooling. Back at the condensing unit, I heard a relay click, so I knew the controls were OK. One last hope - could the breaker have tripped?

I rushed downstairs to the breaker panel. No, the A/C breaker wasn't tripped. But it was OFF! I must have nudged the handle when I put the panel cover back on last night. I threw the breaker, and the A/C sprung to life. Maybe those silly warnings to make sure your appliance is plugged in and turned on before you call for service aren't so foolish after all.

Ken Elovitz

January 2007

APOLOGIES TO CASEY

Think backwards to last New Year
    Jan One, 2-0-0-6
Prices were still suff'rin'
    from Miss Katrina's tricks
The outlook wasn't pleasant
    if you drove an SUV
Pretty hard to find gas
    less than two-twenty-three

"It's just a passing bubble,"
    the government declared,
"Until the gulf refineries and
    platforms are repaired."
Alas! Alack! and Woe to me.
    We found that wasn't so.
The months went by and soon we saw
    That gas cost mo' and mo'!

My nifty little Prius goes
    three hundred fifty miles
On less than twenty bucks worth,
    while friends were in denial.
The months went by
    As each went past
We found that we were paying
    More and more for gas!

Pipeline leak in 'Laska,
    refineries shut down
"Nothing we can do!" they claimed
    "It's all just circumstance."
Three dollars loomed,
    by mid July
It seemed 'twould never stop
    Would four bucks be the top?

Suddenly in September.
    The price began to plunge
Two-eighty-four, two forty nine,
    and now two seventeen.
How can this be? What's going on?
    Its lower every day!
Could it be the elections
    are only weeks away?

We scratch our heads
    What shall we write?
'Tween now and when you read
    Gas may zoom out of sight!
If we say up
    and gas down,
We will be regarded
    as just a bunch of clowns.

Gasoline, we cannot fathom,
    But do remember, please:
When bills for the electric,
    are not at which to sneeze,
When heat by oil or gas does cost
    a ransom that's king-size
We're here to help you tame them,
    Get help from the energy guys.

From all of our houses to all of yours, the happiest and healthiest of New Years.

Dave, Ken, and Gary Elovitz

February 2007

BOILER ODDITY

I recently came across this old boiler with its unusual logo in the casting. Based on info from the casting, it was a Burnham, from Irvington, NY, originally coal fired, but converted to gas. I didn't need any technical information on the boiler, which obviously dated from well before the Nazi era - it will have been replaced with a shiny new boiler before you gel to read this - but I was curious to know anything about the manufacturer's use of that logo, so I contacted Yvon Blais, Burnham rep and boiler guru extraordinaire, who replied:

"Manufactured by Lord & Burnham, Greenhouse manufacturers for the well-to-do. This was during a time well before the war, before the Evil, -- at a time when Lord & Burnham were manufacturing their own boilers. They even made a boiler called the Swasteeka -- the swastika logo was seen as a symbol of good luck and prosperity.

"A very ancient logo - Unfortunately the Nazis ruined this symbol which existed many years before in different cultures.

"There are several brands out there that have used the swastika symbol. These were all prior to Adolf Hitler changing the symbolism it represented. Oddly enough you will also see this symbol in just about every one of the old Federal style US Post Offices that were built during the early 1900's. If you look closely at the moldings, floor tiles, and columns, you will see the swastika in the molded patterns in these buildings. They are not as obvious because they are interconnecting in chain pattern, but they are there!"

A very interesting bit of history.

Gary Elovitz

March 2007

HAPPINESS AND SUCCESS

An ancient legend (modified for modern times) tells about a time long ago when society so abused wisdom that the wisemen decided to take the secret of happiness and success away and hide it where mankind would never find it again. The big question was where to hide it. The chief wiseman assembled a team of experts and called a meeting to discuss the problem.

First the Civil Engineer said, "We can bury the secret of happiness and success in as subterranean structure in the dark depths of the earth."

The general contractor responded, "I like that idea, for my excavation contractor can dig deep into the earth and find the secret for me to keep."

Not wanting the general contractor to have that advantage over everyone else, the Architect said, 'Then we will design an undersea building and hide the secret of happiness and success in the darkest depths of the deepest ocean."

At that point the mechanical contractor replied, "That's great because we will be able to provide gigantic pumps and pump the water away until we can recover the secret."

So the structural engineer chimed in, "We will design the highest skyscraper and take the secret into the stratosphere where the air will be so thin that no one will be able to climb to the top."

But the chief wiseman said, 'NASA has already gone to the moon, and the Russians have built the Space Station. Humankind will easily be able to climb even the highest structure, find the secret, and take it up for themselves."

Finally, the EnergyGuys said, "Here is what we will do. We will hide the secret of happiness and success deep inside every individual, for they will never think to look for it there."

To this day, people have been running back and forth across the earth digging, pumping, and climbing, in search of something they already possess within themselves.

Ken Elovitz

April 2007

GLOWING SUCCESS

When I got there, the maintenance man was running garden sprinklers to keep the remote air cooled condensers cool. He was trying to keep the chiller from "overheating", but the real problem was a condenser that kept shutting down. When electrical equipment shuts down, measuring current is a good place to start. The disconnect on the condenser didn't have a "cheater", so I would have to shut the unit down to open the cover and take my measurements.

"We can't shut it down," the maintenance man insisted. "We have a function going on, and we don't dare interrupt their cooling. We've had so many outages this summer we can't tolerate another."

So I implemented "plan B", which was to measure voltage and current for each fan and for the unit as a whole at the control panel in the condenser. All the currents were in line with the ratings, though I noticed that Phase B was always half an amp higher than the other phases. Small amperage differences are not unusual because voltage and wiring are never exactly the same. The voltage caught my eye. On every fan, Phase C volts to ground was 3-4 volts less than the other phases.

"I'm curious about the voltage," I explained. "Let's go to the main switchboard and see if the difference exists on the incoming service." There was no good place to measure at the main, but the disconnect for the idle chiller would do. It was fed directly from the main panel, and with no load on it, the voltage would be the same as at the main. Voltage was the same across all 3 phases, so the problem was not an imbalance on the service.

Next we went to the panel to check voltage leaving the breaker that feeds the condenser. Once again, all 3 voltages were equal. "We have no choice," I pleaded. "I don't see a problem on the line side of the disconnect, but there's clearly a problem on the load side. We have to shut the condenser down so we can open that disconnect."

Wow! We didn't beat the fire department by much! The Phase C fuse ferrule was discolored, the wire had melted insulation, and there was a little red glow where a damaged line side lug had arced through an insulator to the ground lug. Current leaking off Phase C to ground was like a large single phase load. It not only accounted for the lower voltage on Phase C, the heat it generated was a likely reason for the blown fuses that plagued the system. Not much longer and it would become a real fire! A new fused disconnect would put them back in business and save rolling the fire engines.

Ken Elovitz

May 2007

FIVE STRANDS

In his most recent book, *Overcoming Life's Disappointments,* Rabbi Harold Kushner cites the five strands of a good life: family, friends, faith, work, and the satisfaction of making a difference. You read these words months after I wrote them on Thanksgiving afternoon, following a marvelous holiday dinner, truly thankful for the good life which Ken, Gary, and I, with our families, are privileged to enjoy. We are fortunate that we share all five strands:

Family -- Certainly, the warmth and camaraderie not only of Thanksgiving's festive meal, but of so many other occasions when the Elovitz clan has gathered. is an important source of strength and energy as we set out in the wider world each day.

Friends -- Family is not the only source. So many of you reading these thoughts have become, over the years, not only clients and colleagues, but also real friends, and it is your cheerful greetings and sincere collaboration that lubricate the stresses and strains of everyday life.

Faith -- maybe not just "brand-name" religion, the book notes, continuing "but .... a way of making sense of the world and our place in it, helping us to understand the bad things that happen to us without giving in to despair, and having reason to believe that tomorrow can be better than today.'

Work -- Paid or unpaid, we need something that calls on our abilities and lets us feel competent. What tremendous satisfaction I feel when I work out a neat solution to a knotty problem in your building -- and when my 3rd grade tutorees at Wilson School solve the flash cards faster than I can flip them.

Making a difference -- Contributing to making the world a better place not just by helping you use a bit less of limited energy resources and release a little fewer harmful emissions, but by the things we do for our communities as volunteers and lay leaders.

We are lucky. The book points out a number of famous people, people most of us regard with awe and admiration, who only managed four out of the five, and here we are, three fairly ordinary people -- albeit extraordinary engineers -- who manage to have all five. Sometimes people with whom I am involved in volunteer activities ask me, 'How do you do it? How do you find the time to work full time and do all these things for others?" Sometimes I just flip them off with something like, "I don't sleep much." Other times I am more thoughtful and share a quote from the book: "You can have it all, not just all at the same time."

Dave Elovitz

June 2007

JUST A MOM

The Clerk was obviously a career woman, poised, efficient, and possessed of a high sounding title like, 'Official Interrogator" or "Town Registrar."

"What is your occupation?" she asked the young mother applying for some official paperwork or other. What made her say it? Maybe she had just been dismissed as "just a housewife" once too often. The words simply popped out. "I'm a Research Associate in the field of Child Development and Human Relations."

The clerk paused, ball-point pen frozen in midair, and looked up as though she had not heard right. Our heroine repeated the title slowly emphasizing the most significant words. Then she stared with wonder as her pronouncement was written, in bold, black ink on the official questionnaire. "Might I ask," said the clerk with new interest, "just what you do in your field?"

Coolly, without any trace of fluster in her voice, "I have a continuing program of research, (what mother doesn't) in the laboratory and in the field, (normally she would have said indoors and out). I'm working for my Masters, (first the Lord and then the whole family) and already have four credits (all daughters). Of course, the job is one of the most demanding in the humanities, (any mother care to disagree?) and I often work 14 hours a day, (24 is more like it). But the job is more challenging than most run-of-the-mill careers and the rewards are more of a satisfaction rather than just money."

There was an increasing note of respect in the clerk's voice as she completed the form, stood up, and personally ushered our friend to the door.

As she drove into her driveway, buoyed up by her glamorous new career, the young mother was greeted by her lab assistants -- ages 13, 7, and 3. Upstairs you could hear the new experimental model, (a 6 month old baby) in the child development program, testing out a new vocal pattern.

She felt she had scored a beat on bureaucracy! And had gone on the official records as someone more distinguished and indispensable to mankind than "just another Mom."

Motherhood! What a glorious career! Actually they do deserve a title on the door!

Thanks and a tip of the hat for this one to our young Pittsburgh friend Dana.

Dave Elovitz

July 2007

SLEEPLESS IN SEATTLE

"We just can't sleep!" the young couple exclaimed. "This is the second winter we have been living in our brand new condominium, and the contractor keeps coming back and doing something, then telling us it is fixed, but it is not. We think it is even worse now. You could safely store butler and eggs in our living room, then bake a cake from them without an oven in our bedroom upstairs!"

I sensed a certain amount of hyperbole in that cake baking claim, but clearly these nice young people were very frustrated. I mean, it really was chilly in the living room, but as I walked up their stairs I wasn't so sure the cake baking was such an exaggeration after all. When I hit about the fourth step, my head felt like it had plunged into a sauna, and the whole second floor was unbearable.

Fortunately, I had an opportunity to review the design drawings and do load calculations before this meeting. The drawings showed adequate heating and cooling capacity and about the correct distribution of airflow to each room, but the furnace on the specified unit was big compared to the heat loss, which would make the supply air about 130F. Kind of high for two stories with an open stairwell, because such warm supply air would tend to roll along the ceiling and rise up the open stairwell. At the site, I could feel the supply air temperature was much higher than 130F, and you could actually feel the 6" deep layer of supply air stratifying at the living room ceiling and rolling up to the second floor at the open stair.

During the meeting, I learned that the unit that was installed was not the one on the drawings but another brand with an even bigger furnace. The contractor had figured the larger furnace was not that much more money, so why not play it safe and get the extra capacity? I explained how the bigger furnace made a higher supply air temperature, and then how lower airflow made it even higher. I projected that even if there were no supply airflow to the second floor, the supply air stratification alone would over heat the second floor. And the stratification also kept any warm supply air from getting down to the occupied level of the first floor, which is why the living room was so cold and the thermostat kept calling for more heat.

The contractor finally agreed to reduce the capacity of the furnace and boost the airflow. That brought the supply air temperature down to 110F, and *voila!* only a couple of degrees difference between first and second floor. This contractor now believes there is such a thing as too much of a good thing!

Dave Elovitz

August 2007

TRUE OR FALSE?

Many myths are represented as scientific fact. According to "The Most Popular Myths in Science" published on *LiveScience.com*, some are true and some are not. Here's a sampling:

A chicken can live without its head

True. A chicken can stagger around without its head because its brain stem is often left partially intact after the butcher cuts the chicken's head off.

Eating poppy seed bagels makes you fail a drug test

True. Eating two poppy seed bagels can produce a positive result for opiates in a drug screening test. Have them with a gin and tonic, and you can get a false positive for cocaine use, too.

A penny dropped from the top of a tall building could kill a pedestrian

False. Because of its shape and wind friction, a penny dropped even from a building as high as the Empire State Building (1230 feet) would only build up enough speed to sting an unlucky pedestrian below.

Chicken soup cures the common cold

True (partially). Chicken soup has anti-inflammatory properties that help reduce congestion.

The five-second rule

False. There is no scientific rule that food dropped on the floor is safe to eat if you pick it up within a certain time. Germs stick to most foods on contact.

A failing cat always lands on its feet

True (sort of). Cats dropped from most heights will land on their feet, though cats dropped upside down from a height of a foot or less might not. Don't try this at home.

You get less wet by running in the rain

True. Complex mathematical analysis of this question includes factors like the number of rain drops that hit a walker's head vs. the runner's chest. However, the analysis does not include the risk that running increases your chance of slipping and falling into a puddle.

Water drains backward In the Southern hemisphere due to the earth's rotation

False. The earth's rotation is too weak to affect the direction of water flowing down a drain. The direction that water whirlpools in a sink depends on the sink's structure, not the hemisphere.

Ken Elovitz

September 2007

IN FROM THE COLD

It was a really spiffy, big, new condominium building in a posh suburb. Each unit had its own efficient and economical heating system. A sealed combustion, very high efficiency gas water heater connected to outdoors by PVC flue and combustion air intake pipes up through the roof provided domestic hot water and hot water for a heating coil in the air conditioning unit. Very compact and really elegant! But not a concept that this town's officials had seen before.

Residents had moved in just a few days ago, but today the building stood eerily empty, by order of the Fire Department. Carbon Monoxide detectors had gone off in a third floor unit late the previous afternoon. Two residents were hospitalized, and the Fire Department detected carbon monoxide in two other apartments directly above and below, so the Fire Department ordered the building evacuated until the cause was identified and corrected, and they could be sure it would never happen again. The new residents were distinctly not happy.

All the flue and combustion air intake piping looked OK, but it didn't take very long to find the cause. Opening up the ceiling and taking down the flue piping from the offending unit in order to inspect it released about a half gallon of water! The vent pipe made a long horizontal offset before going up through the roof and there was no pipe hanger supporting it, just the friction of the high hat flashing. Snow and water must have accumulated inside the pipe during winter storms when the heater was off, and the piping sagged down so that the horizontal pipe formed a trap at the elbow. Eventually that trap filled with enough water to block it off so the flue gases could no longer vent from the heater and they escaped into the equipment closet.

That explained the third floor unit, but the second and fourth floor flues were fine. Those heaters worked like a charm. But the overflow pans under the water heaters were connected by an open vertical drain pipe that provided a direct path for the escaped flue gas from the third floor to migrate to the second and the fourth floors. That's how CO got to detectable levels in those units. That narrowed the problem to just one water heater. Or did it?

When I went up on the roof and pushed down hard on each flue and combustion air intake pipe, three others moved down several inches and could conceivably form traps to hold water. A few pipe hangers in strategic locations, and the residents could come back in from the cold -- and with their confidence restored.

Dave Elovitz

October 2007

SHOEMAKERS' CHILDREN

One year ago, I was just finishing a construction project at home. I experienced the same frustrations our clients face -- cost increases, disruption and chaos, and fretting over whether the contractors and suppliers would be on time and coordinate their efforts.

I took comfort thinking the HVAC work would go smoothly. After all, I designed it. Kenny Ekstrom of Ducts, Inc. did a great job on the ductwork. Bernie and Bobby from Phoenix Air Services installed the refrigeration piping and started up the unit. I knew the unit would run right. They voluntarily took all the refrigerant temperature and pressure data I often struggle to get the contractors to take to show that a system works properly. I tried to do on my own job all the things I want done on yours.

So imagine my surprise when I went to the attic one fine, hot day and saw a puddle of water under the fancoil unit. The ducts were lined, but the attic was hot and humid enough (and the supply air temperature was low enough) for moisture to condense on the outside of the supply air duct at the mouth of the fancoil unit. Some peel and stick foam rubber insulation cured the problem. It would have looked a lot nicer if we had installed that extra insulation before the duct went in, but who expected an insulated duct to sweat?

Then one cold winter day took me into the attic where I saw a new wet spot on the plywood under the fancoil unit. The unit has no outside air connection, so it couldn't be rain or snow. The unit had been off for months. Where was the water coming from? I put a pan under the unit and checked the next day to see if water kept appearing. It did. When I opened the filter door, I found a soaking wet filter!

How does water get inside the filter box of an idle, 100% return air fancoil unit? I finally figured it out. New construction releases a huge amount of moisture - wood and plaster drying out kept the upstairs well humidified all winter. When the air conditioner is off, air in the room mixes freely with air in the return duct and in the fancoil. The moisture level equalizes. That moisture condensed inside the fancoil unit in the cold attic. The answer was simple: Blank off the upstairs return grille as part of the winter shutdown procedure.

What's the moral of the story? Every construction job is unique. No one anticipates every problem or mishap, no matter how many jobs you've done. When problems appear, we try to solve them and make the surprise an opportunity to learn instead of a reason to start a fight.

Ken Elovitz

November 2007

OUT WITH THE OLD ...?

Converting the 1980's-vintage wood-framed office building to a private school was no "slam dunk", but it could be done. Money was tight - of course - but it looked like most of the building's existing mechanical and electrical systems could be reused, and keeping the four main sets of toilet rooms intact would be a big cost savings.

On the first set of the Architect's floor plans, all the existing mechanical space and all the existing toilet rooms were gone. New mechanical space and new toilet rooms were shown elsewhere on the plans. That approach was clearly not consistent with the concept of building to a budget: It was going to cost a lot of money to demolish all that equipment and those toilet rooms and then rebuild essentially the same stuff a few yards away.

A few e-mails and phone calls later I received revised floor plans. This time about half the existing infrastructure remained in place, but the plan still called for demolishing one "core" with two sets of toilet rooms and two mechanical/electrical rooms, only to rebuild two new sets of toilet rooms about 30 feet away.

I asked the Architect, "Why can't we reuse the existing toilet rooms and mechanical space in area 'B'? That would save a lot of money and you know that money is tight for this project."

"Gee, I wish we could, but we can't use those toilet rooms because they don't meet the code for handicapped accessibility."

"You're right - those existing toilet rooms aren't 'accessible', but they don't need to be! You don't have to make every toilet room accessible as long as there are accessible toilets nearby. We can meet all the fixture requirements and the accessibility requirements by adding a new toilet room. Besides, we can design the new toilet rooms with enough space for fully accessible toilets without having to squeeze them into the existing toilet rooms. We need additional toilet fixtures anyway (a school needs more toilets than an office building), so we can use the new toilet rooms to meet accessibility requirements."

A few weeks later the final floor plans came through with the existing mechanical space and toilet rooms reused in place and some added, accessible toilets. The school might never realize how much money we saved them, but I was pleased that we found a way to save all that money without sacrificing function.

Gary Elovitz

December 2007

EATS, SHOOTS, AND LEAVES

An "RFI' (request for information) came from the plumbing contractor shortly after the construction contract for the renovated school was signed: "All water piping is undersized per Mass, Code and must be replaced. Please advise." A quick review of my design reassured me that I hadn't screwed up, and I responded that the pipe sizing was fine.

The plumber apparently wasn't going to give up a potential fat extra so easily. At a meeting a few weeks later, the plumbing inspector explained that the Mass, Code uses different "demand factors" for office buildings than for schools. Piping that had been OK for an office would be undersized for a school. The demand factors made the existing piping to the toilet rooms, and even the service pipe coming in from the street, too small.

"Oh," I replied, "I did **not** use the demand factors in the Mass. Code. I calculated all the pressure drops. The Code allows the piping to be designed by an engineer's calculations."

"Actually," the inspector said, "I don't think that's what the Code says." So we took out the book and read as follows:

The methods used to determine pipe sizes shall be the procedure in ... publication #1038, or a system designed by a *registered professional engineer, using the computation outlined in Tables* 1, 2, *and 3.*

"See", the inspector said, "the Code says an engineer has to use the tables in the Code."

"That wouldn't make any sense," I replied, "If that's what it says, **only** an engineer could design piping using those tables. That would mean that plumbers would never be allowed to size any piping. That can't be the intent of the Code - on most small projects plumbers size all of the piping, and they use those tables. There must be a "typo" in this new edition. The previous edition of the Code says "designed by a registered professional engineer **or** using the tables".

The inspector became thoughtful. "I see your point. I never read it that way before, but I guess that makes sense. I'll have to call the State Board and ask for an interpretation."

A few days later I got the following message on my voice-mail: "Hi, this is the Plumbing Inspector. I called the State Board about that code question. They say it's not actually a typo in the Code. That comma between "engineer" and 'using' means "or". So I guess your pipe sizing is OK as long as your stamp is on it."

I wonder what Lynn Truss would say about that use of a comma?

Gary Elovitz

January 2008

ENERGY GUYS ARE COMING TO TOWN

You don't have to doubt.
You don't have to cry.
There's no need to pout.
I'm tellin' you why:
    Energyguys are coming to town.

As mortgages failed,
The stock market tanked
'Til the Fed came along
And rescued the banks.
    Energyguys are coming to town.

While Clinton and Obama
Debate about who'll rule.
The rest of us just shake our heads
And try to pay for fuel.

Oh, You don't have to doubt ...
    (refrain)

Karl Rove said good-bye;
Gonzales did, too.
We're still in Iraq.
What are we to do?
    Energyguys are coming to town.

And then came Harry Potter
With his last block buster book.
The I-3-5 bridge fell apart
In the Mississippi brook.

Oh, You don't have to doubt ...
    (refrain)

Turn off the loads
That are not in use.
Tighten up windows and
Wall that are loose.
    Energyguys are coming to town.

Consider T-5 lighting.
Recover exhaust heat.
Control your fans with V-F-Ds,
And the savings can't be beat.

Oh, You don't have to doubt ...
    (refrain)

Your job might be big
Or maybe it's small,
Whichever it is,
Just give us a call.
    Energyguys are coming to town.

Investigate the problems.
Find out how things work.
Run the tests and analyze --
That's where most solutions lurk!

Oh, You don't have to doubt ...
    (refrain)

Patti and Ken
David and Fran
Dena and
Gary all do what we can
To bring to you our annual wish.

We hope you all will prosper.
A year of health and joy.
We hope the year's a happy one
For every girl and boy.

Oh, You don't have to doubt ...
    (refrain)

Dave, Ken, and Gary Elovitz

February 2008

THE BEAUTY OF NUMBERS

My nephew Bruce, who is not a mathematician, sent this to me with the following thought:

Poetry is not just for words,
There is also poetry in numbers.

Even if you are not a mathematician, you cannot fail to appreciate the beauty of numbers. Look at this symmetry. Brilliant, isn't it?

Dave Elovitz

$$1 \times 8 + 1 = 9$$
$$12 \times 8 + 2 = 98$$
$$123 \times 8 + 3 = 987$$
$$1234 \times 8 + 4 = 9876$$
$$12345 \times 8 + 5 = 98765$$
$$123456 \times 8 + 6 = 987654$$
$$1234567 \times 8 + 7 = 9876543$$
$$12345678 \times 8 + 8 = 98765432$$
$$123456769 \times 8 + 9 = 987654321$$

$$1 \times 9 + 2 = 11$$
$$12 \times 9 + 3 = 111$$
$$123 \times 9 + 4 = 1111$$
$$1234 \times 9 + 5 = 11111$$
$$12345 \times 9 + 6 = 111111$$
$$123456 \times 9 + 7 = 1111111$$
$$1234567 \times 9 + 8 = 11111111$$
$$12345678 \times 9 + 9 = 111111111$$
$$123456789 \times 9 + 10 = 1111111111$$

$$9 \times 9 + 7 = 88$$
$$98 \times 9 + 6 = 888$$
$$987 \times 9 + 5 = 8888$$
$$9876 \times 9 + 4 = 88888$$
$$98765 \times 9 + 3 = 888888$$
$$987654 \times 9 + 2 = 8888888$$
$$9876543 \times 9 + 1 = 88888888$$
$$98765432 \times 9 + 0 = 888888888$$

$$1 \times 1 = 1$$
$$11 \times 11 = 121$$
$$111 \times 111 = 12321$$
$$1111 \times 1111 = 1234321$$
$$11111 \times 11111 = 123454321$$
$$111111 \times 111111 = 12345654321$$
$$1111111 \times 1111111 = 1234567654321$$

THE NOSE KNOWS

The 30-plus story hotel had four separate kitchen exhaust systems, but they all discharged through a mechanical room wall Just above the 5th floor low roof. The hotel was built in the early '70's, and there had not been a problem until a year or so ago. Then the hotel started getting complaints of cooking odors in the 7th floor conference rooms.

I looked around as the hotel's team outlined the history. The room was conditioned by high-rise fan coil units with no outside air connections. The windows did not open, and I didn't feel air movement around them. There was outside air ventilation to the corridor, but it was from the same system that fed the guest room corridors above. If the corridor system were the source, odors would appear on every floor, but the 7th was the only floor that ever got odors.

Odorous outdoor air might be drawn in by stack effect, but odors occurred even in hot weather when the air conditioning would reverse the stack effect and make air flow out at the 7th floor, not in. There seemed to be no way the kitchen exhaust odors could be coming back into these conference rooms. Then I thought, "Maybe they aren't leaving the building at all."

"Let's go look at the mechanical room beneath us," I invited. The four kitchen exhaust fans were right near the outside wall, and there was no doubt about cooking odors down here. The hotel's engineer confirmed that odors were a regular occurrence, and when the chef grilled steaks, the mechanical room got so much smoke it sometimes set off the smoke alarm. So at least one of those systems was leaking kitchen exhaust inside the building, but which one? And how could I be sure?

"Can we get a chef to saute some garlic in each kitchen, one at a time?" I requested. They started with the 1st floor kitchen as we waited in the mechanical room. No odors. We stepped out onto the 5th floor roof and the odor was strong. Second floor kitchen next. Again no odor in the mechanical room, but plenty out on the roof. I was beginning to wonder if this was such a great test. Then the chef went to the 3rd floor kitchen, and odor filled the mechanical room. The 3rd floor kitchen system was clearly the culprit, but we decided to test the 4th floor kitchen anyway. It's a good thing we did. Once the odor from the third floor cleared, we gave the chef the go ahead. Garlic odor soon followed. Now we could not only be sure the odors were leaking from the ductwork, but we knew which ducts. Sure enough, later careful examination found leaking spots, and sealing the ducts contained the odors. That made it fair to say we actually sniffed out the problem.

Dave Elovitz

April 2008

RANDOM KINDS OF FACTNESS*

Nine American presidents had facial hair: five had beards and four had mustaches but no beards. All were Republicans.

Lots of us know that a group of geese is called a gaggle and that a group of lions is called a pride. How many of you know that a group of frogs is called a chorus and that a group of toads is called a knot?

Alligators have 11 broad, square snout. Crocodiles have a more narrow snout, and their bottom teeth stick out when their mouths are shut. Now you'll know which is which if one is ever chasing you.

Novelist Ian Fleming, famous for his James Bond series, also wrote the children's classic, *Chitty Chitty Bang Bang*.

The Sherman brothers, who wrote the classic hit, ·You're Sixteen (You're Beautiful, and You're Mine) also wrote many Disney hits including "It's a Small World" and the music for *Mary Poppins* and *The Jungle Book*.

Dr. Seuss !Theodore Geisel) created the first animated color commercial. It was for Ford in 1949.

Manhole covers are round because a circle is the only common shape that won't fall through the hole if tilted sideways.

In the United States, a billion means 1,000,000,000. In England it means 1,000,000,000,000.

It takes 822 peanuts to make an 18 ounce jar of peanut butter.

In the late 1960's Douglas Engelbart created the X-Y Position Indicator for Display Systems. 'XYPIDS' did not catch on as a name for the device. Some users called it a turtle. Others called it a rodent. Today we call it a mouse.

Richard Reynolds invented metal foil wrap, but not for food. His uncle, cigarette magnate RJ Reynolds, wanted a way to keep tobacco fresh.

The Incas all had the same blood type - O positive. (How do we know this?)

Before she was a master chef, Julia Child was a spy. She served in India and China during World War II.

* From a book of the same title by Erin Barrett and Jack Mingo.

Ken Elovitz

May 2008

LIFE SUCKS

'I need your help on this one.' came the call from ace HVAC guy Bobby Neilan of Tech Mechanical. 'I did a tenant fitup on the 7th floor of an office tower this spring. All I did was ductwork and diffusers, and I installed it exactly the way the engineer showed it on his plans. Now everything's going wrong and they're blaming me! The air-handler won't make as much static pressure any more, and condensate doesn't drain from the cooling coil, so they get a flood in the mechanical room.

'I didn't touch the air-handler: Bobby continued, "so this really shouldn't be my hunt, but I've already spent two days up there trying to figure out what's going on, and I'm coming up blank. The only thing I see that's different from the old design is that the engineer increased the airflow from 16,000 CFM to 20,000 CFM. I don't see how that would cause these problems."

At the site, I took as many measurements at the air handler as I could - static pressures before plus after the filters, static pressure at the coil, and fan intake and discharge pressures, fan speed, and airflow through the unit.

The static pressure leaving the fan was low (as Bobby said), and there was enough suction pressure at the fan to prevent the condensate from draining. I measured the condensate drain carefully, consulted the manufacturer's fan performance chart, and rendered my verdict:

"Bobby - you're off the hook. Here's what's happening: When I measured the airflow I came up with just about 20,000 CFM, so you're right at the design. The low static pressure issue is simple - when you increase airflow from a fan, the static pressure goes down. If you cut the airflow back to 16,000 CFM like it was before, the static pressure would go right back up. As long as you're getting enough airflow, no one should care that the static pressure is low. If they really want static pressure, they can speed up the fan.

"On the condensate issue, the condensate trap is kind of shallow to begin with. That wasn't a problem with 16,000 CFM. But when you increased the airflow to 20,000 CFM, that 25% increase in airflow made the suction pressure at the fan intake 50% greater. That means you need a 50% deeper condensate trap. Voila! A marginal trap becomes inadequate. Make the condensate trap an inch and a half deeper, and the condensate will drain like it's supposed to."

Gary Elovitz

June 2008

A PERSONAL MESSAGE

As I write this. Fran and I are tens of thousands of feet above the Atlantic, winging our way toward Zagreb, Croatia, formerly part of Yugoslavia. It is the end of a stressful week scrambling to finish everything I promised to finish before I left, a week of little sleep, tossing and turning over how to do that. I am exhausted. but now I am too exhilarated to take advantage of the few hours of darkness between America and Europe to sleep.

As many of you know, we usually try to take two big trips every year, and those trips, more often than not, tend to be "off the beaten path". A week or so before every trip I send a note to every client with an active project to let them know I will be away and will be out of touch by normal means of communication. Invariably those notes trigger a number of responses: About half are "Oh, oh, oh! Before you go could you just" (which is why the week before we leave is so stressful) and about half are along the lines of "What, again? Where to this time?" with wishes for a great trip, some good natured banter about the unusual places we pick, often ending up with "Why in the world there?"

It is to answer that last question that I am writing this personal message in a space that is usually devoted to machinery and energy and, sometimes, words you have to go look up.

Some people like to sit on the beach and soak up the sun, but we like to feel like we are Lowell Thomas. Those few of you who are old enough to have been in grammar school in the late 1930's and early 1940's will understand what that means: Exploring new places, meeting new people, learning their history, and sharing their unique cultures. Not to ignore the indigenous foods.

For ten days or two weeks, we not only do not talk on the phone or read an E-mail, but we generally do not watch any TV or even read a newspaper - except (from my mouth to God's ear) if the Red Sox are in contention for the pennant. Don't look at a single set of prints; or even get caught up on my technical reading. Just complete immersion in the place we are visiting and the people who live there. And the people who lived there hundreds, if not thousands, of years ago. Not actually relaxing - we walk a lot and are on the go all day for long days, but it is a wonderfully stimulating and enriching experience: Completely emptying your mind of all the stuff we deal with every day, and stuffing it full with new - and quite different kinds of information.

As the kids say, "It is awesome!"

Dave Elovitz

July 2008

IDEAS TO PONDER

Keep your eyes on the stars, and your feet on the ground. - *Theodore Roosevelt*

Never ever ever give up! - *Charles Schultz*

To improve is to change; to be perfect is to change often. - *Winston Churchill*

Diversity is the one true thing we all have in common.

The greatest glory in living lies not in never falling, but in rising every time we fall. - *Nelson Mandela*

Teamwork is the ability to work together toward a common vision. It is the fuel that allows common people to attain uncommon results.

No one can defeat us unless we first defeat ourselves. - *Dwight Eisenhower*

One's mind, once stretched by a new idea, never regains its original dimensions. - *Oliver Wendell Holmes*

Love and compassion are necessities, not luxuries. Without them humanity cannot survive. - *The Dalai Lama*

No act of kindness, no matter how small, is ever wasted. - *Aesop*

There are people of spirit and there are people of passion ... rarest of all is a passionate spirit. - *Martin Buber*

Challenges are what makes life interesting; overcoming them is what makes life meaningful. - *Ralph Waldo Emerson*

It isn't being busy that counts - it's what you get done. Don't confuse activity with results.

A foot is a device for finding furniture in the dark.

Many of the things you can count don't count. Many of the things you can't count really count. - *Albert Einstein*

Tact is the art of making a point without making an enemy.

You cannot argue a man into liking a beer. - *Oliver Wendell Holmes*

Following conventional wisdom is sure to be conventional, but it's not always wise.

It is a capital mistake to theorise before one has data. Insensibly one begins to twist facts to suit theories, instead of theories to suit facts. - *Sherlock Holmes* in A Scandal in Bohemia

Ken Elovitz

August 2008

STARVING UNITS

The residential condo owners moved in before the first floor retail spaces were done. By June, the piping had been extended to the retail space, and more than half of them had tenants. That's when the residential heat pumps started locking out on high head. As more retail tenants moved in, the problems occurred more frequently. The property manager was not happy. and she was even more concerned about what would happen when all the retail spaces were running.

The first suspicion that came to my mind was low water flow. At light load, no problem. But as more units tied into the system, the less flow there was for each unit. They might get by with low water flow during cool weather when the water from the cooling tower was cooler than design. But as the weather got warmer, so did the water from the cooling tower. A few degrees warmer from the cooling tower coupled with a few degrees extra temperature rise because of reduced water flow, and soon heat pumps started tripping out.

The balancing report was done before any retail heat pumps were installed. It showed 1300 GPM delivered compared to 1400 GPM design. The measured head of 122 feet compared to 110 feet design explained why the flow was low. Did the retail units increase the resistance even more, cutting flow to even less than 1300 GPM? And more important. what could they do right away to get through the summer? This was not the time of year to shut down for major work.

The first step was to add up how much flow the system would need once all the heat pumps were connected. That came to 1530 GPM. So the system actually needed more flow than the design provided, and it delivered less than design. Finding a solution started with analyzing the pump curves. The pump could deliver the needed 1530 GPM, but only against a head of 103 feet. The balancing report showed the system needed 122 feet of head to push 1300 GPM. Moving 1530 GPM would need more than 122 feet and certainly more than the 103 feet the pump could develop at 1530 GPM. We needed more pump or less restrictive piping.

The system had two identical pumps, one active and one standby. The pump curve showed that each pump could deliver 765 GPM against 130 feet of head. Running the two pumps in parallel would make the needed 1530 GPM available at a healthy increase in head over what one pump could do alone. Running both pumps did increase horsepower and energy use significantly, but it kept the system operating reliably through the summer while we looked for a long term cure.

Dave Elovitz

September 2008

FEEDING THE HUNGRY

Last month I told you about a mixed use condominium with low heat pump loop flow. The residential heat pumps started locking out on high head as soon as the retail tenants and the warm weather arrived. Analyzing the pump and system curves showed that we could get through the summer by running both the active and standby pumps In parallel. Running both pumps increased horsepower and energy use, and it left the system with no standby pump, so that approach was not a permanent solution.

The system was sized for 1400 GPM with 110 feet of pressure drop, but actual flow was low at 1300 GPM and actual pressure drop was high at 122 feet. There were more retail space heat pumps still to go in, so required flow would eventually be 1530 GPM. We needed more pump, less restrictive piping, or both. The question was how to get it without breaking the bank.

First step was to trace out all the piping. That turned out to be much more complicated than it sounds. Piping disappeared above units and down through floors into soffits in the garage. We had no clue where it came out or what was connected inside those soffits. So the building engineer and I spent two or three days with a Sawzall, finding all the pipes and determining how they were connected. The drawings weren't much help. We discovered there were actually three distribution loops, even though the drawings showed only two.

Once we knew where the pipes went, we measured pressures at key locations and compared the readings to my calculations to figure out where the problems were. That's how we found out the cooling tower heat exchanger (for summer cooling) and the boiler heat exchanger (for winter heating) were piped in series. Water had to go through both heat exchangers all the time, even when the loop did not need heat. By piping a bypass around the boiler heat exchanger, we saved 23 feet of pressure drop in summer, when the loop needed the highest flows but didn't need heat.

Repiping the heat exchanger was not quite enough. Ideally, we would replace the existing 60 HP pumps with 75 HP pumps, but that would involve serious money. Replacing the impellers would be fairly easy and would let us get more flow and head out of the existing pumps, but bigger impellers would put more load on the motors. Luckily, the pump curve showed that once we added the heat exchanger bypass, we could just squeeze the 1530 GPM flow needed for summer peak out of the biggest impeller the 60 HP motor could support. And we could do it without spending a fortune.

Dave Elovitz

October 2008

IF YOU WANT SOMETHING DONE RIGHT ...

The new retail tenant had no room for his own boiler, so he asked to tie into the building's utility steam service that heated the apartments upstairs. The Landlord insisted that the tenant provide a sub-meter and reimburse the Landlord for the tenant's steam use.

Steam turns back to water (condensate) as it goes through a heating system, so the tenant used a condensate meter to measure his steam use. Condensate meters are simple, accurate, and require very little maintenance.

After the first month, the tenant's meter showed no usage at all, and the Landlord reported that there was live steam pouring from the seams of the tenant's meter. The tenant was told to fix his meter and steam traps. Months went by, with e-mails back and forth - mostly the tenant claiming that their contractor had fixed everything, and the Landlord complaining that there was still live steam pouring from the meter. Meanwhile, the tenant's meter still didn't move, and the Landlord was not getting any reimbursement.

The Landlord was fed up. "I need you to dig into this: he finally asked me. "Can you meet my plumber at the building and get this resolved?"

The plumber and I met in the basement. By shutting various steam valves we easily isolated two steam traps that were passing live steam.

The plumber headed off to get rebuild kits for the traps, and I looked more closely at the meter.

The live steam from the faulty traps had killed a little microswltch in the steam meter. Replacing the microswltch would be easy, but I wanted to be sure the meter would work once the new switch was in. So I pumped some water through the meter and watched. A cog was supposed to trip the microswitch once for every gallon of condensate that passed through the meter. But the cog did not turn! The microswitch was not the only problem.

"I thought these meters were tested at the factory!" I grumbled as I unbolted the cover. The internal roller drum, which was supposed to turn like a water-wheel as condensate flowed through it, had a distinct metal-on-metal scraping as I tried to move it. Then I saw that the drum had been dislodged from its track - this meter must have gotten bumped around quite a bit somewhere between the factory and the job site - probably in the back of the contractor's truck! After reseating the drum and reassembling the meter, it passed a second flow test with flying colors.

Gary Elovitz

November 2008

BLAME THE CROATIANS

According to *La Grande Histoire de Cravat* (by Francois Chaille, 1994) around 1635, some 6000 soldiers and knights assembled in Paris to support King Louis XIII and Cardinal Richlieu. Among them were a great number of Croatian mercenaries, in their traditional garb, including unusual and picturesque scarves tied about their necks. The scarves were made of various cloths, ranging from coarse material for common soldiers to fine cotton and silk for officers.

The French quickly fell in love with this elegant "Croatian style", especially the gallant French officers in the Thirty Years War. Unlike the traditional lace collar that had to be kept white and carefully starched, the brightly colored scarf was simply and loosely tied around the neck and remained visible under the soldiers' thick long hair.

Around 1650, the Court of Louis XIV, where military ornamentation was much admired, adopted the Croatian scarf. It was soon accepted throughout France. The fashionable, *"a la Croate"* soon evolved into a new French word that symbolized the height of culture and elegance and still exists today: *"la cravate"*.

When Charles II returned to England from exile in France, he brought with him the new word and the new fashion. Within ten years, it had spread throughout Europe, and across the Atlantic and throughout the American colonies.

So, when you have to struggle to tie your necktie in the morning, now you know who to blame!

RED TAPE AND PIPE

An American firm got a nice order for threaded pipe from Australia. The firm applied to Washington for a permit to ship, but the permit was refused. Was it because we were at War (World War I)? Was it because we needed the pipe for home consumption? No. The permit to ship was refused because the application did not state whether the pipe thread was to be of linen, silk, or cotton.

from Compressed Air, *November 1920*

Dave Elovitz

December 2008

COLD IN GEORGIA

It gets cold in January in Georgia over near the Mississippi line. This particular weekend the temperature dropped to 13 degrees during the early morning hours. That caused a big problem in a nursing home that had been shut down and vacant since November: Sprinkler heads in the utility room and a sprinkler pipe in the attic froze and burst. Water cascaded down through the building for hours before the watchman discovered the leak and got someone to close the main sprinkler valve. The insurance company had to shell out serious money to repair the damage. Naturally, the insurance company went looking for someone to sue so they could get their money back. When they discovered the design engineer had a good sized insurance policy, they decided the loss was the result of bad design. But when we got a thorough look at the facts, our calculations and analysis showed it was not a design flaw at all.

The room where the sprinkler heads froze did not have a heating unit, but it housed two 100 gallon gas-fired water heaters. Those heaters leaked 12,000 BTUH of waste heat to the room. The calculated heat loss from the utility room at 70F inside and 13F outside was only 10,564 BTUH, so the room temperature should stay well above 70F. Examining the gas bills explained what happened: Not only had the owners shut off the gas the previous November, they did not install any alternative heat in the utility room or even close up the big outside air louvers that brought in combustion air for the gas water heaters. No wonder the room was literally freezing cold!

The attic was vented and unheated, but the design called for R-30 insulation to be installed above the sprinkler pipes in the attic. Calculations showed the attic air would be below freezing at 13F outside, but the sprinkler pipe, snug under the insulation blanket and warmed by heat from the 70F occupied space below, would be about 50F - well above freezing. The owners claimed the building heat was turned on and working all the time the building was vacant. The building was electrically heated, and a careful analysis of the electric bills gave lie to that claim: Analyzing prior years' electric bills against historic weather data showed the relationship between electrical use and outdoor temperature when the building was occupied. The analysis also showed electric use from November through January was only about half what was needed to keep the building at 70F, so the heat could not have been turned on. It didn't hurt the credibility of our conclusions that the watchman testified he had not heard the sprinkler flow alarm because he was out in his car trying to get warm!

Dave Elovitz

January 2009

DIRGE ON 2008

From the great Charles River Valley
To the shores of the Merrimack
Gas shot up about four dollars
But it dropped a dollar back.

When the markets hurtled downward
And the banks' outlook turned black
But the cost of heat kept soaring
'Twas like give the mole a whack.

"They" blamed the sub-prime crisis
For why things looked so bleak.
McCain, he blamed Obama,
And vice-versa. so to speak.

Crude's well below a hundred,
But heat costs didn't shrink
Now OPEC's getting nervous
Set Jan. for a group think.

It's not a cheerful view right now
Best button up, and play it cool.
Cash or energy's the same:
Conserve must be your golden rule.

So work together to cut waste.
Reduce our costs a lot, plus
Not send dollars overseas
To folks who just don't like us.

Our houses are worth less today,
Don't even mention stocks.
We must rebuild our savings:
Buckle down despite the shocks.

November 4 we got to choose
Who'll fix whatever ails us.
No matter how we voted
The winner mustn't fail us.

This year we all in Boston
Had other things to mind.
While the Dow was plunging
    earthward
Some things - for us - were fine.

For God blessed us with the Red Sox
'Tho they didn't Win the fight.
The Pats, they sure miss Brady,
But Cassell will be all right!

Despite the sports distractions
Right now things loom real scary.
We hope that rescue will arrive
On twenty January.

From all of us to you and yours,
Our fervent prayer for thine:
Good health, good luck, and better
    times
For all two thousand nine.

Dave, Ken, and Gary Elovitz

February 2009

A FREEZIN' IN THE SUN

The AC in the tanning salon hadn't worked right all summer. That was not good - tanning customers want to look like they've spent all day at the beach - not like they sweated it out in a sauna.

'Oh, and one thing more," the manager said, "We had a flood in here a couple of weeks ago, and we had to turn the AC off altogether since then."

I quickly figured out that the flood was due to a split "U"-bend in the hot water heating coil in the salon's small air-handler. The only time water coils split like that is when they freeze. Freezing cold air flowing over the coil will freeze the water in two adjacent tubes, trapping unfrozen water in the "U"-shaped bend that connects the tubes together. When the water in the tubes freezes, it expands, but there is no space to absorb that expansion. Tremendous pressure builds up, and the "U"-bend splits open. But this was only October - there hadn't been nearly cold enough weather to freeze a coil yet.

Well, I thought, I'll have to figure out why that coil split, but meanwhile these people need air conditioning, and a split heating coil is no reason to shut off the AC completely. So I turned the AC back on, but nothing happened. Was it the thermostat? The wiring? Or the controls? (The little HVAC system was a super-high efficiency system - two-speed air conditioning compressor, variable speed fan control, and even a zoning control system; the tangle of controls wiring looked like something out of "Star Wars".)

As I sat in front of the unit tracing wiring and poring over the controls schematics, I felt a slight cool breeze coming through the open door of the fan section. (Must be outdoor air coming in, I thought, as I concentrated on the wiring schematics.) But the longer I sat, the colder and stronger the breeze felt, and I suddenly realized what had happened: The air-conditioning was working fine - it was the fan motor that failed!

The air conditioning cooled the AC coil (which is right on top of the hot water coil), but with no fan to circulate the cooling, the AC coil got very cold - icy cold, in fact. So much cold air dropped down from the AC coil and flowed over the hot water coil that it froze the water inside the hot water coil! That explained the other mystery, too. The reason the AC hadn't worked too well that summer was that the fan hadn't been running.

Gary Elovitz

March 2009

DON'T BE SO SURE

Ken, Gary and I have been at this business for a long time, so we often have a pretty good idea what the answer will be before we go through the calculations, but we never know for sure until we run the numbers. We don't want to get left out in the cold like the owner's engineer on a recent project. He had decided that the large supply duct running down the corridor didn't need to be insulated. "You don't need it," he said, "and we'll save a ton of money."

Now the corridor was freezing! When the contractor claimed the long narrow corridor was so cold because of the big uninsulated duct, I wondered if he was trying to cover up for excessive duct leakage.

I knew an uninsulated duct would absorb a fair amount of heat from the space. There were no exterior walls or windows to add to the cooling load, but there were a lot of lights putting heat into the space: Twenty-two 60 watt incandescent bulbs. That's enough to need a half ton of air conditioning! I had never seen the corridor or the duct, but my first reaction was that something else had to be making the corridor so cold.

I wasn't anxious to do the long, tedious calculations to determine the duct heat gain and surface temperatures. In a long duct (this one was over 100 feet long), the temperature inside the duct changes as the air going down it absorbs heat. As a result you either have to use calculus or calculate the duct heat gain in a series of short sections. Then I remembered that Ken has a nifty computer program that could calculate in a just few hours what would take me a whole day by hand. When I described the situation to him, his first reaction was the same as mine: it seemed unlikely that the surface of the duct could absorb the heat from all those lights and make the corridor cold. But he also agreed that we needed to do the arithmetic.

So he did. I gave him the duct dimensions, airflows, and supply air temperatures, and he set up a spreadsheet so I could plug in different supply air temperatures and airflows.

When I saw the results, I was reminded once again that experience helps you figure out where to look and what questions to ask, but you have to crunch the numbers if you want to know what's really going on: At the design supply air temperature, the duct would suck over 11,000 BTUH out of the corridor. That was a lot more than I had expected, and more than twice the amount of heat that the lighting was putting in. No wonder it was so cold in there!

Dave Elovitz

April 2009

HOW THE STEAM GOT ITS HISS

Long ago, O best beloved, in the hoary precincts of the Big City, there lived a Precocious Property Principal, the kind with the large bank account), who owned six buildings in a row - each of which had an ancient hissy, wheezy steam boiler. So the Precocious Property Principal summoned a Clever Comfort Contractor (the kind with all of the answers - even if they don't match the questions).

"Wot we got 'ere: quoth the Clever Comfort Contractor, "is six boilers wot mighta' been installed by me granddad." He removed his cap reverently. "It pains me poor 'eart to see 'em reduced to this state - they're for the scrapper, they are. Six new boilers ain't cheap, mind you. But I've a mind to work up just one boiler 'ere, wit' a bunch o' pipin' wot connects all these 'ere buildns and give each buildin' its own ee-lectronic valve control - it'll work like a charm, and save you a pretty penny!"

Now this was an answer the Precocious Property Principal (always with an eye on that bank account, O best beloved) could quite cotton to. So before you could say "float and thermostatic steam trap", there were furlongs of new pipe criss-crossing from building to building, all from the one Brand New Big Boiler. Now this was in the days of "rent control", O best beloved, so the banging and the clanging, the water spurting from the air vents, and the "Places the Heat Never Got To" were but minor inconveniences. But the years passed and the Precocious Property Principal repainted all those buildings (and tripled the rent), and repopulated his buildings with the "upscale".

But attend and hear, O best beloved - the "upscale" definitely do not cherish sound effects whilst they sleep. So the Precocious Property Principal called upon the Jinn of Good Heating, who arrived in a whirlwind of pencils and tape measures and levels, and spent some good long hours amongst the furlongs of basement piping. At length he delivered his decree: "A steam system - like all beasts since the Beginning of Time - must breathe, and this system has been holding its breath. There are no vents in the main piping to let air in and out. When a motorized valve opens, the piping cannot breathe out to make room for steam to enter, so steam does not reach the far-flung corners of the building. When the valve closes, the steam in the piping condenses and shrinks to 1/1700 of its original volume, but the piping cannot inhale air to take the place of the steam. Instead, a vacuum forms in the piping and sucks the condensate water back up into the pipes at high speed, where it bangs and clangs inside the piping. Follow my instructions to add vents (and other things), and the only sound you will hear is the gentle and satisfied hiss of the tamed steam-beast."

Gary Elovitz

THE STRANDED ENGINEER

An engineer took his first vacation ever on a cruise ship. It was the experience of his life until a hurricane came up, and the ship sank almost instantly. The man found himself swept up on the shore of an island. He spent the next four months eating bananas, drinking coconut juice, and staring out at the sea, hoping a ship would come to his rescue.

One day, he spotted movement out of the corner of his eye. Could it be true? A ship? Around came a rowboat, rowed by the most gorgeous woman he had ever seen. She was tall and tanned, and her long hair flowed in the breeze. She spotted him waving and yelling and rowed her boat toward him.

In disbelief, he asked, "Where did you come from? How did you get here"? She said, "I rowed from the other side of the island where I landed when my cruise ship sank." "Amazing", he said, "I didn't know anyone else survived. How many of you are there? You were really lucky to have that rowboat wash up with you!"

"It's only me, she said, "and the rowboat didn't wash up. I made it out of materials I found on the island." "But," asked the man, "where did you get tools and hardware?" "Oh," the woman replied. "The south side of the island has an unusual stratum of alluvial rock exposed. I found I could fire it in my kiln and melt it into forgeable iron. I used that for tools and used the tools to make the hardware. But, enough of that", she said. "Where do you live?" The engineer confessed he had been sleeping on the beach.

"Well, then, let's row over to my place," she said. The woman rowed them around to a wharf that led to her place. They walked up to an exquisite bungalow. "It's not much," she said, "but I call it home. Please sit down. Would you a drink?" "No," said the man, "one more coconut juice and I will puke." "Oh, it won't be coconut juice," she replied, "I have a still. How about a Pina Colada?' Trying to hide his amazement, the man accepted, and they sat down on her couch.

After they exchanged stories, the woman asked, "Have you always had a beard?" "No: he replied. "Well if you'd like to shave, there's a man's razor upstairs in the bathroom." The man, no longer questioning anything, went up to the bathroom. In the cabinet was a razor with a bone handle and two shells honed to a hollow ground edge and fastened by a swivel mechanism. The man shaved, showered and went back down stairs. "You look great," she said. "I think I'll go up and slip into something more comfortable."

Soon the woman returned wearing strategically positioned fig leafs and smelling faintly of gardenia. "Tell me," she asked, "we've been out here a very long time. Is there anything you really miss?" The engineer moved close to her and gazed into her eyes. "You mean I can check email here?"

Ken Elovitz

June 2009

INTEGRATED PROJECT DELIVERY

We recently attended a seminar on "Integrated Project Delivery." IPD is the latest buzzword for what is billed as a revolutionary new approach to implementing a construction project without the traditional disputes and finger-pointing. The underlying Idea of IPD is to align the interests of the three groups involved in a project - Owners, Designers, and Contractors - by a negotiated sharing of both risks and rewards as circumstances affect progress and cost. The goal is to encourage the parties to act in the overall interest of the project rather than in their individual self-interests. That alignment of interests can now be possible. proponents claim, because of new and better digital tools to enable collaboration - better ways of communicating project information like CAD, project web sites, and, most recently, BIM - Building Information Modeling.

IPD is not the first "new" project delivery concept aiming to the solve the same problems. We've heard this song before: Before IPD we had CM at risk, joint ventures, partnering, design-build, public-private partnerships, and alliance contracting.

Ken pointed out that even before all these buzzword ideas, we were involved in series of projects for several sophisticated owners that pretty much achieved all those goals. That was certainly true, and it set me to musing about what they did that is different from the construction industry today:

1) Those sophisticated owners didn't expect to be insulated from all risks. They expected the designers to be responsible for the design and the contractors to be responsible for implementing that design at the agreed cost. But they also recognized that since the long term benefits of the project accrued to them, the risks of the unknown and unknowable were theirs.

2) Those owners were professionals: They knew the construction process, they knew the design process, and they knew what they wanted the finished project to do.

3) Those owners led the team: They used pretty much the same designers, contractors, and consultants on every project, but they remained at the helm.

4) They believed in Trust, but they verified.

Once in a while I still have the opportunity to work with clients who operate that way. They know that happy projects with happy results don't come from luck or magic but from competent leadership, sound advice, and commitment to a functional end result.

Dave Elovitz

July 2009

SHEDDING LIGHT

You must have seen compact fluorescent lamps by now. You know - those funny looking spiral shaped tubes that screw into regular light bulb sockets. Maybe you've tried them - then again, maybe not. Compact fluorescents (CFLs) have gotten some bad press recently: some lamps don't seem to give off as much light as promised; sometimes they don't last as long as promised; they start out dim and take a while to get bright; and they contain toxic mercury vapor!

Well, most of the charges against CFLs are basically true. But let's take a closer look:

- CFL packaging tends to exaggerate; for example. offering a 14W CFL as a "60W equivalent." A 14W CFL won't produce quite as much light as a regular 60W bulb. But if a 14W isn't bright enough, use a 19W, or even a 23W! A 23W CFL Will still reduce energy use (and CO-2 emissions) by over 60%

- CFLs are supposed to last years (actually, 8,000 hours of light). I've had mixed results: some last a long time, some don't. Partly it is my own fault for using CFLs in less-than-optimal conditions, and partly it is because quality control has not been as good as it should be.

   But so what if you have to use a larger CFL to get enough light. and it doesn't last as long as it should? A 23W CFL that lasts only 4,000 hours will still save you $25 In electricity costs over the life of the lamp compared with that old 60W bulb, and will reduce CO-2 emissions by 200 lbs. Not bad for an investment of less than $5.

- Today's CFLs do start dim, but they get to full brightness after about a minute. So don't use them where the light will be on only for a few minutes. You get the best "bang for the buck" where the lights will be on for several hours at a time, like In the kitchen, family room, bedrooms, and outside lights. And if you leave the room even for a few minutes, shut off the lights; the old myth that It costs more to shut off the lights than to leave them on is just that - a myth!

- CFLs do contain mercury. But as long as the glass tube is not broken, the mercury is totally contained and safe. So, yes - we have to be more careful with CFLs. Handle them carefully; don't use them in lamps that your young child (or clumsy brother-in-law) might tip over. And make sure to recycle them - some towns and "home centers" already accept CFLs for recycling, and others will certainly follow. But coal-fired power plants also emit mercury into the atmosphere: using CFLs reduces the amount of coal that needs to be burned, reducing mercury emissions.

Someday, I hope that there will be a new type of lamp that will be super-energy-efficient and will not have the drawbacks of CFLs. But until that day, let's use more CFLs - as Voltaire said, let's not let the "perfect" be "the enemy of the good"

Gary Elovitz

August 2009

FROM THE LOVINS

We should have shared this long ago, but a few years ago, cousins Dick and Judy Lovins sent this missive in response to our January 2004 calendar:

'Twas the time of the month,
    when the Energy Guys

Send out their calendar,
    witty and wise,

And before the TV,
    so mundane and boring,

Richard and Judy
    were blissfully snoring

When out at the mailbox,
    we heard such a jiggle,

We knew that the Guys
    had sent us a giggle.

We ran to the box
    With anticipation

And sorted the mail
    with heightened elation.

Sure enough. there it was,
    straight out of Natick,

As merry a mailing
    as if from St. Nick!

Some months it's a discourse,
    and it's really big,

About a malfunctioning
    thingamajig.

The contractor goofed
    when he went to install it,

And believe it or not,
    though we won't recall it. I

We understand some
    and it really intrigues us

And though its not comic,
    it never fatigues us.

So thanks go to Ken,
    Gary, David and Fran

(Who is one of the "guys"
    although not a man)

We beg you, keep sending
    your energy lore

And enjoy all the best
    in two thousand and four!

By Dick and Judy Lovins

Dave Elovitz

September 2009

THE GHOST OF ENERGY USE

Most of the time, our job is to figure out how to make HVAC and electrical systems work. We look for solutions that meet the stated performance criteria but that also are energy efficient.

Sometimes looking for ways to save energy is the core of our assignment. Our answers often involve familiar ideas, like turning off loads when they don't need to run. While that idea seems obvious, identifying loads that run when they aren't needed is not always obvious.

Enter "ghost" energy use. Ghost energy use is energy that equipment uses even when its on/off switch is turned off. Many TVs and other appliances fall into that category. Computers use ghost energy, too. While energy star rated computers and LCD monitors can be set to go to sleep after a selected idle time, these devices still use a small amount of power even when they are asleep. Then there are chargers and power supplies for cell phones and laptop computers. Anything with a transformer uses power whenever it is plugged in, even if the appliance it serves is not connected or being charged.

Each of those loads is small, but they add up. I've seen electric company newsletters that say these ghost or parasitic losses constitute as much as 10% of total electricity use. That number sounded kind of high, so I made some measurements of my own.

A few months ago, I checked the main panel at home with "everything" in the house turned off. I measured 0.89 amps on one leg and 0.49 amps on the other. That current flowed to our electric clocks, chargers, cable box, Tivo unit, etc., etc. The amps I measured convert to 166 VA, which translates to somewhere between 60 and 110 KWH per month. If Patti and I waste around 90 KWH/month on parasitic loads, the corresponding waste at each of our clients' buildings must be thousands of KWH/month. Those utility company reports of up to 10% for ghost energy use are starting to sound believable.

So what can you do about it? For computers and monitors, maybe you can plug them into a power strip, shut down the computer, and turn off the power strip at the end of the day. Unplug chargers and power supplies when they are not in use, or maybe set up a "charging station" with all your chargers plugged into a power strip. Turn the power strip on only when the devices are connected to their chargers. Of course, don't forget to buy efficient products to begin with, and turn the lights off when you leave the room.

Ken Elovitz

hat.

Page 382

October 2009

BURMA SHAVE

In the 1930's, '40s, and '50s, Burma Shave, a popular shaving cream, advertised their product with a series of 5 signs spaced about 100 feet apart alongside the road. The first four lines formed a little poem, the last said, "Burma Shave". Here are a few examples:

Don't lose your head
To gain a minute
You need your head
Your brains are in it
    Burma Shave,

Around the curve
lickety-split
Beautiful car
Wasn't it?
    Burma Shave

Passing school zone
Take it slow
Let our little
Shavers grow
    Burma Shave

We want to be prepared in case that form of advertising comes back, so here are our "Burma Shave" ads:

Thermostats
If set too high
Make your fuel bills
Reach the sky
    Call Energy Guys

If your system
Does not work
We can try
To find the quirk
    Call Energy Guys

Pumps and fans
That make much noise
Are not the kind
The boss enjoys
    Call Energy Guys

Lights that run
All through the night
Tells me something
Is not right
    Call Energy Guys

Fit your fans
With V-F-Ds
You'll save lots of
Bucks with ease
    Call Energy Guys

Gary wrote
'Bout C-F-Ls
Have you tried them
Yet, pray tell?
    Call Energy Guys

If you size fans
Properly
Air will flow
Quite easily
    Call Energy Guys

Energy Guys
Is our name
Engineering
Is our game
    Call Energy Guys

Ken Elovitz

November 2009

PREDICTION

I couldn't help but think of the present worldwide economic situation and the effects of the economic growth of formerly Third World countries when I recently read a quote predicting a then-future globalized economy. The quote starts:

The need of a constantly expanding market for its products chases the bourgeoisie over the whole surface of the globe. It must nestle everywhere, settle everywhere, establish connections everywhere.

The quote goes on to explain that this worldwide commercial activity will result In pulling the rug out from under "old-established national industries" by new industries that no longer depend on using local raw materials or meeting the needs of local markets. These new industries will draw raw materials "from the remotest zones" to meet new wants that require "the products of distant lands and climes" instead of just the production of their homeland industries. The result is communication and commerce in every direction, and "universal interdependence of nations".

Not just in physical products, the prediction continues, but also in intellectual production:

Intellectual creations of individual nations become common property. National one-sidedness and narrow-mindedness become more and more impossible, and from the numerous national and local literatures there arises a world literature.

Rapidly improved means of production and immensely facilitated means of communication draw" even the most barbarian nations into civilization".

Aside from the hint provided by what we consider archaic style and language, it was hard to guess this prediction was written in 1848, not in the immediate face of the commercial resurrection being led by India and China in the 21st Century. Would you have guessed that the author was Karl Marx?

Setting aside the many flaws in the rest of *The Communist Manifesto*, Marx must have had some crystal ball to foresee the impact of the internet and of inexpensive fiber optics linking all the continents!

And yet, isn't the need to make systems work properly and efficiently still the same today, even with all the new high tech tools, as it was at the beginning of the Industrial Revolution?

Dave Elovitz

December 2009

SHOULDA SAID

The stranger to my left turned and introduced himself, He asked me, 'What do you do?' That's a question we get asked often, and it's always hard to answer because we rarely do the things consulting engineers do most. I think I mumbled something about "Studies, Reports, and Opinions". Long after we parted, I realized what I should have said:

We really just have two kinds of clients, clients who have a system that doesn't do what they need, and clients who are afraid they're going to get a system that won't do what they need."

For the clients who have a system that doesn't do what they need, we make our own determination of what it is they need, then analyze the installed system in two ways: What capability the design documents say the system should have, and what performance it actually delivers.
Then we compare the results of those analyses and figure out what needs to be done to modify what exists so it will do what is needed. That group represents about 1/3 of our business. A small portion of that, probably less than 10% of what we do, has to do with being experts for litigation.

Most of the clients who are afraid they're going to get a system that won't do what they need first came to us as clients in the first group, and we had helped them solve a problem. Some were referred by friends or associates who had themselves been in the first group. As one of my first clients said years ago, "Instead of paying you to come in and fix problems after the systems are installed, why don't I hire you to look over the design while it's still on paper and let us know whether we are likely to be happy with it when it's built?" Some of those clients go even further, and involve us as part of the project team right from the beginning, helping them select designers and clarify system goals. That kind of role, design review and being part of the client's team, is about another third of what we do.

A significant number of clients in the second group opt for the design/build method of project delivery. They don't want to write an order for "one pretty good air conditioning system", so they have us develop a design/build "Scope of Work" document that defines the performance the system has to deliver. The Scope of Work uses field measurable terms so we can determine actual system performance at the end of the job.

Well, that's what I should have said, but I dldn't. Now I have.

Dave Elovitz

January 2010

to the tune of
JINGLE BELLS

Weatherstrip the doors
Seal the windows, too
Insulate some more
There's lots that you can do:

Have your burner tuned
To peak efficiency.
Fuel savings soon
Will bolster your esprit!

(Refrain)   Energy Guys are we,
               Saving energy!
               Oh how good it is to be
               Saving energy!

Incandescent lights
Cost too much to run
Make things just as bright
With fluorescent ones.

Your meter will turn less,
And greenhouse gases shrink.
Global warming won't progress;
We'll stay back from the brink.

(Refrain)

Install new controls,
Tighten up your ship_
Meet your budget goals
Assert your leadership!

Measure what you use;
Determine where and why.
Chase the targets that you choose
And make their use less high.

(Refrain)

Sometimes we see jobs
That simply do not work,
And we hear the sobs
Of people gone berserk.

That is when it's time
For us to see what's wrong:
We'll fix it up, then make a rhyme
For a calendar in song!

(Refrain)

Patti, Ken, and Fran,
Dave, Gary, Dena, too
Hope and pray the coming year
Will be so good to you!

To our many friends
All across the land
Goes our wish for health and joy
And a New Year that is grand!

(Refrain)

Dave, Ken, and Gary Elovitz

February 2010

THE TRUTH, THE WHOLE TRUTH

It was long enough ago that it is probably OK to tell the story now. A major shopping center developer called and explained that they had just opened a new mall on labor Day weekend. On opening day, two centrifugal chillers smashed up. They had the chiller manufacturer fly in technicians and parts to rebuild the chillers as quickly as possible. The cost was close to $100,000. Now they wanted to know why the chillers failed. Was it a manufacturing defect? Or something wrong in the installation or startup?

I was very excited at my first assignment for this big, prestigious, national client and was anxious to do a great job so they would call again. I flew down, met with the mall management, listened to the mall's maintenance guy recount the opening day events, examined the installation, and collected failed bearings and compressor components for analysis. After a long discussion in the manager's conference room, I asked if the maintenance man could lead me back to the chiller plant for a final look before I left. He did, and I took advantage of the chance to talk with him alone and ask him to tell me, step by step, everything that had happened that day.

He got to the point in his tale where he showed me how he had wedged the compressor motor starter on the first chiller into the "on" position with a broom handle - with almost immediate disastrous results. Since that didn't work, he did it again on the other chiller with a similar outcome. My heart sank as my chance to be a hero to this new client evaporated: The oil pump on this chiller was a separate device that was not started by the compressor motor starter. He had run these high speed units with no lubrication! I thanked him for his help and left the mall directly.

I found a pay phone nearby and called the client. "Your maintenance guy screwed up, and he caused the smashups," I said. "Do you want me to just get on the plane and make believe I never came here?" "No," came the unexpected response. "Write up just what happened and why, so we can learn from the mistake and avoid having it happen again."

Which I did, and over the next decades, that developer became my largest client as he involved me earlier and earlier in more and more projects. And, I am proud to say, he never again got blind sided by a major HVAC problem.

Dave Elovitz

March 2010

apologies to
THE SCARECROW OF OZ

You can while away your hours,
    A-cleanin' cooling towers
    The facts are very plain:

Soon the sump you'll be scrapin'
    If the algae starts escapln'
    'Fore you flush it down the drain.

You neglected water treatment
    Put in chems, but not as frequent
    As you should, and who's to blame?

When the fill starts to get brittle
    But the budget's very little
    Then you start to feel the pain.

Oh, you could tell me why,
    the pH should be more
That's when you go and measure things
    you've never seen before
Then frown,
    And measure more!

If it looks like you're a nuthin'
    With a head just full of stuffin'
    And its drivin' you insane

Life will get sweet and oh so merry,
    When you find the customary
    Blowdown cycles to maintain.

AND TO THE TINMAN

Yes, it is a fancy kettle,
    But a boiler's made of metal
    And when it's torn apart:

You must look out for scalin'
    Or a leak you'll soon be bailin'
    If you do not play it smart!

Clean every single recess
    Use elbow grease to excess
    Don't forget the underpart

Use a brush, get all the soot out
    That will increase the heat put-out.
    Saving fuel when you restart.

'Cause of you - A spotless flue
    Never had a better draft
Trim the air - adjust with care
    Maximize the C-O-Two!

Keep the flue gas cool when leavin'
    Efficiency achievin'
    And plot the numbers on the chart

Life can't get any better
    In a storm or sunny weather
    When your boiler runs so smart.

Dave Elovitz

April 2010

FAVORITE THINGS

Occupant sensors and LED lighting,
Cost saving measures can be so exciting
VFD systems, the savings they bring.
These are a few of our favorite things.

Condensing boiler with reset controller,
Electric from wind mills and also from solar;
Improper controls we can show how to fix
These are a few of our favorite tricks.

Clean all the coi-ils, clean out the condenser,
Change all the filters 'fore pressure's intenser
Roll back the CFM to just what you need
These are a few of our favorite deeds.

>When the bills show
>You need sa-vings
>Making you look bad,
>We'll help you select from our favorite things
>And your boss will then feel glad!

Right-sizing systems and making adjustments
Perusing designs and then sending our comments
Figuring loads and what the design brings
These are a few of our favorite things.

HRV units, enthalpy recov'ry
How much they save is incredibly lovely
Lets you comply with the outside air rules
They sure are one of our favorite tools

Base CFM on your load calculation;
Simply applying a facile equation,
Capacity matching the room-by-room need
That is the rule you always must heed.

>When the bills show
>You need sa-vings
>Making you look bad,
>We'll help you select from our favorite things
>And your boss will then feel glad!

Tuning up burners for cooler flue gas
Trimming impellers so balance valves bypass
For overhead heat. use low delta-T
These are some things that we like to see!

Don't throttle flow through your cooling towers
To avoid lime from growing like flowers
Control the algae to keep basins clean
You'll get the most from your cooling machine!

VAV boxes that do not have reheat
Must close their dampers all the way complete
Otherwise they can make rooms too cool
Making you look like a bumbling fool!

>When the bills show
>You need sa-vings
>Making you look bad,
>We'll help you select from our favorite things
>And your boss will then feel glad!

Dave Elovitz

GREAT MOLASSES FLOOD

The issue involved the retrofit of a small office building. My search for photos showing shading by adjacent buildings turned up an unrelated, bizarre, and fascinating. story: The ballpark on the opposite side of Commercial Street had been the site of Boston's Great Molasses Flood!

On January 15, 1919, an immense riveted steel tank storing molasses at the Purity Distilling Company burst, releasing a wave of molasses 40 feet high that hurtled outwards at about 35 MPH over the entire area. Nowadays we - who tend to associate Boston with mutual funds, microchips, and microbiology - have forgotten that Colonial Boston was one corner of the Caribbean trade that traded sugar from Jamaica for molasses, which was then turned into rum back in Boston. In 1919, distilleries in Boston, like Purity, were still a major industry, but producing Industrial ethyl alcohol for use in other alcoholic beverages as well as in munitions, instead of rum.

Two plus million gallons of molasses swept nearby buildings off their foundations, crushing them into rubble, buckled the Boston Elevated Railway tracks, and, according to *Dark Tide,* a book by Stephen Puleo, "swept waist deep, covered the street, and bubbled about the wreckage," inundating the entire neighborhood. 21 people and several horses were crushed and drowned in the molasses and 150 people were injured. It took over 87,000 man hours to remove the sticky goo from cobblestone streets, theaters, businesses, automobiles, and homes, and the harbor was still brown with molasses until summer. Even today, some North End residents claim that the area still smells of molasses on a hot summer day!

The cause of the disaster was never really resolved, but the initial point of failure was definitely traced to a manhole cover near the base of the 50 foot high tank. There was speculation that fermentation inside the tank produced carbon dioxide, raising the internal pressure. The sudden rise in the local temperature - from 2F to 41 F - over the previous day was also suspected as a contributing cause. Many pointed to poor construction and inadequate testing. The tank had not been filled with water to test for leaks before it was first filled with molasses and, according to one report, when filled it leaked so badly that it was painted brown to hide the leaks. Local residents even reported having collected leaked molasses for use in their homes!

Dave Elovitz

June 2010

SHUT 'EM OFF!

Years ago, lighting courses taught that it was often less expensive to let fluorescent lights run continuously than to turn them off for brief periods. Back then, fluorescent lamps were expensive and electricity was only a few cents per KWH. Since then, rising electricity costs and falling lamp prices have changed the equation. The question now is how long an off cycle it takes at today's rates for the energy cost savings to balance the extra costs of replacing the lamps more frequently.

Turning fluorescent lamps on and off definitely reduces lamp life. Fluorescent lamps use electrodes to start and run the lamp. A tiny bit of the electrode evaporates at each start. Some of that vapor redeposits on the bulb wall, which is why the ends of "burned out" fluorescent lamps turn black. When enough electrode material has evaporated, the lamp no longer starts. Failed electrodes are the most common cause of fluorescent lamp failure.

Typical fluorescent lamps have a rated life of 20,000 burning hours, based on a cycle of 3 hours on/20 minutes off. 20,000 hours is about 8 years for typical office use. An IEEE study in 1998 showed that switching lights on shorter cycles cuts lamp life to 7400 burning hours. That's six years life because the lamps burn fewer hours per day.

A typical 3-lamp T8 fixture uses 88 watts of electricity. If the lights burn continuously during working hours (2500 hours/year), that fixture uses 220 kwh/year at a cost of about $33. If the lights are off when the office is vacant, the lights run about 5½ hours a day or 1400 hours/year, and the fixture uses 125 kwh/year at a cost of $19. Lamp, labor, and disposal costs are about $8/lamp or $24 for the 3-lamp fixture. That's a lamp cost of $3/year for the fixture that runs 2500 hours and $4/year for the fixture that runs 1400 hours. Annual cost per fixture is $33 for electricity plus $3 for lamps or $36 to run the lights continuously during business hours. If the lights turn on and off as people enter and leave the room, the cost is $19 for electricity plus $4 for lamps or $23.

Everyone agrees that turning the lights off when they are not needed saves energy. Putting today's numbers to the question debunks the old maxim and shows that the energy saved more than outweighs the increase in lamp and labor cost from reduced lamp life. For lights controlled by occupancy sensors, the industry consensus seems to be that a minimum on time of 10 to 15 minutes yields a practical balance between lamp and energy costs.

Ken Elovitz

July 2010

IT'S NOT EASY BEING GREEN

Water dripped from the windows, and mold grew on the walls all winter - not a good sign in upscale urban condominiums. Other experts had looked at the problem and ruled out water leaking through the walls and windows. I arranged to visit several problem units.

As we rode the elevators up into the residential tower, the facility manager told me about the systems in the building. "We have a state-of the-art corridor ventilation system!" he gushed. "It has a total energy recovery wheel- sucks the heat right out of the exhaust and puts it back into the make-up air. Saves us a ton of money on heat!"

·Um-hmm .... " I said, non-committally. (It sounded like a clue to me.) The elevator doors opened and we stepped into the corridor. I noticed two air grilles: one was supply, and I asked about the other: "Is this the exhaust grille for the corridor ventilation system?"

"Yeah," the facility manager replied. "I know it's close to the supply, but this is an old building, and that's the only place we could fit it in."
(Another clue.)

Entering one of the condos, I felt the high humidity wash over me - no wonder they had mold on the walls! Humid air hitting cold walls and windows would condense - just like the "sweat" on cold drinks in the summer.

I checked the bathroom - no bathroom exhaust. (No surprise: exhaust wasn't required when the building was first built if the bathrooms had windows.) Then I saw the stacked washer/dryer. It was a "condensing" dryer with no exhaust duct to outdoors. A few measurements confirmed my suspicions: the dryer may condense some of the moisture from the laundry, but it also releases moisture into the air. I had seen enough.

"The condo units generate a lot of moisture - from breathing, bathing. and the laundry - but there is no ventilation to flush this moisture out of the units. The corridor system moves a lot of air, but with both supply and exhaust in the corridors, none of that air gets into the units. Even worse, the "energy recovery" system recovers not only heat, but moisture as well; about 80% of the moisture exhausted from the corridors comes back in via the supply air.

"I'm afraid we need to shut off your fancy heat recovery system and provide only supply air to the corridors. By not exhausting the corridor, the system will pressurize the corridor slightly, forcing the dry air into the apartments and help flush moisture from them."

Gary Elovitz

August 2010

TOLD YA SO ....

The grand old hotel had a grand old kitchen exhaust system: the system served five different kitchens on three floors, with more than 1000 feet of ductwork in all.

A new "destination" restaurant was slated to open. The tired old seafood place it would replace always had problems with exhaust, and the hotel manager asked me to look at the new restaurant's design. "We really can't have a repeat of the problems we had with the seafood place," he growled.

Fortunately, the restaurant's designers had a balancing technician measure airflow in the exhaust duct after the old hood was demolished. I looked over the report, along with the new restaurant's design drawings; airflow looked good, but the static pressure was quite low. That was a big problem - airflow would certainly go down when the new hood was installed, because any hood produces more resistance to airflow compared with no hood. Worse yet, the new restaurant was planning to use a water wash hood - good for easy maintenance, but the hood requires much more static pressure than a standard hood. They would be lucky to get 60% of their design airflow.

"The hood has already been ordered," the restaurant owner sniffed. "And my engineers checked everything and said it will be fine."

Several months later my phone rang: "There's a problem in the new restaurant!" the hotel manager groused. "The whole place fills with smoke whenever they use their grille! We've tried speeding up our exhaust fan, but all that did was make the building shake." I had to bite my' tongue. Saying "I told you so" might be satisfying, but definitely wouldn't help. After some back-and-forth with the new restaurant, it was clear that replacing their hood was a nonstarter. But after some investigation I came up with a plan.

"Speeding up the fan by itself won't help - that increases the exhaust from all the hoods throughout the hotel equally. Besides, for a fixed system, fan power varies with the cube of airflow. Increasing the exhaust by 25% would require twice the fan horsepower - our fan can't possibly do that. (And your operating costs would go through the roof!) But if we can increase the resistance to airflow In all of the hotel's other hoods to match the new restaurant's hood, that would increase the new restaurant's share of the total exhaust.

"Fortunately, I know of one type of hood filter that can be adjusted to increase its airflow resistance. If we install these adjustable filters in the hotel's hoods, we'll increase the exhaust Just from the new restaurant."

Gary Elovitz

September 2010

TOO MUCH IS NOT ENOUGH

Every winter we get at least one of these calls: "It's stifling upstairs, but we can't get the downstairs above 60!"

Sure enough, with a couple of questions we learn that the townhouse condos have gas furnaces in the attic, supply ductwork to grilles in the ceilings of all the rooms, and one central return grille in the ceiling of the top floor.

I made a number of measurements at the site: with outdoor temperature at 28°F, it was 55°F at the floor and 64°F at the ceiling downstairs. Upstairs it was about 68°F at the floor and 85°F at the ceiling. The supply air was a blistering 155°F! Clearly, the very hot supply air was simply too buoyant to mix in the room, so the heat could not penetrate to floor level.

Based on my measurements, I figured that the airflow was less than 2/3 of the design. The fans were running at top speed, and pressure measurements suggested that the ductwork was significantly undersized. Increasing airflow was out of the question.

However, low airflow was only partly responsible for the high supply air temperatures. My calculations indicated that the furnaces were more than twice as big as they needed to be. The drawings showed that the designer had significantly oversized the furnaces, but the furnaces were even larger than that - the installing contractor had added a little extra, "just to be on the safe side."

I made two suggestions:

(1) De-rate the furnaces: Because the furnaces were so grossly oversized, reducing the gas input would reduce the supply air temperature. But we could not go too low on the gas input without risking condensation in the furnace, which would shorten the life of the furnace. Supply air would still be over 100°F.

(2) Move the main return air grille from the upstairs ceiling to downstairs, near floor level. That would draw the coolest air in the apartment as return air (rather than the hottest air at the ceiling), which would reduce supply air temperature and the overall temperature difference from floor to floor. Getting a big return air duct down from the upstairs ceiling to the downstairs floor was not easy, but with a little guidance, the General Contractor came up with a pretty clever route that had minimal impact on the occupied space.

The changes worked like a charm, and cut the temperature rise from the downstairs floor to the upstairs ceiling down to 8°F.

Gary Elovitz

October 2010

CORRIDORS

Funny how things seem to come in clusters: The last couple of weeks seem to have been Apartment Corridor Week. Four different clients with four different projects but all with the same complaint - uncomfortable corridors. Probably time for a review of the basics for heating-and ventilating corridors in apartment buildings.

The thing about heating and cooling loads that too many designers don't seem to think about when they design the HVAC system for a corridor is that corridor space heating and cooling loads are very small. The most common corridor layouts are surrounded by conditioned space on all sides, except maybe an insulated roof above the top floor corridor. Most corridors have no exterior exposure and no windows. In fact, many corridors need so little cooling - the only heat gain is from the lights - that it is not uncommon to find only heating - with no mechanical cooling.

A couple of things can complicate matters:

> The corridor needs so little heating or sensible cooling that the supply air temperature needs to be only a couple of degrees above or below the desired corridor temperature: pretty much the same supply air temperature all year around. But many designers use the corridor system to provide makeup air to replace exhaust from the apartments. A large amount of outside air can need a large amount of heating to make near-room-temperature supply air.

Many designers use the corridor system to serve not only the corridors, but also the first floor entry, sometimes including a lobby where occupants might spend enough time that temperature control becomes important. While corridors need about the same supply air temperature all year around, entries and lobbies usually have a lot of outdoor exposure and a significant amount of glass. They need real space heat in winter. If the supply air temperature that is right for the entry lobby is used for the corridors, the corridors will really suffer. And vice-versa.

Fortunately, corridor temperature control is not as critical as controlling the temperature inside the apartments. No one spends much time in the corridors, so a wider range of temperature will be acceptable. As long as entry/lobby areas have separate temperature control from corridor areas, a corridor supply at a constant 68°F - 70°F seems to keep everybody happy.

Dave Elovitz

November 2010

PUNS FOR THE EDUCATED MIND

My cousin Julian called these to my attention:

1. The roundest knight at King Arthur's round table was Sir Cumference. He acquired his size from too much pi.

2. I thought I saw an eye doctor on an Alaskan island, but it turned out to be an optical Aleutian.

3. She was only a whiskey maker, but he loved her still.

4. A rubber band pistol was confiscated from algebra class, because it was a weapon of math disruption.

5. No matter how much you push the envelope, it will still be stationery.

6. A dog gave birth to puppies near the road and was cited for littering.

7. A grenade thrown into a kitchen in France would result in Linoleum Blownapart.

8. Two silk worms had a race. They ended up in a tie.

9. When cannibals ate a missionary, they got a taste of religion.

10. Atheism is a non-prophet organization.

11. Two hats were hanging on a hat rack in the hallway. One hat said to the other, "You stay here. I'll go on a head."

12. I wondered why the baseball kept getting bigger. Then it hit me.

13. A sign on the lawn at a drug rehab center said, "Keep off the grass."

14. The short fortune-teller who escaped from prison was a small medium at large.

15. The man who survived mustard gas and pepper spray is now a seasoned veteran.

16. A backward poet writes inverse.

17. In a democracy, it's your vote that counts. In feudalism, it's your count that votes.

Dave Elovitz

December 2010

to the tune of
OL' MAN RIVER

I can't fix it.
I'm sick of trying:
The help's all sweating;
Just sit there sighing.
That cooling unit, old cooling unit
Is just not doing its job.

Here's the boss now.
He don't look happy.
"We must do something;
We can't do nothing
Just get some cooling.
Just get some cooling turned on."

    Brand new unit, I made it plain,
    No more trouble like this again.
    Make less noise,
    Use less juice,
    Extra ventilation
    Makes the place smell new.

A roof-top unit,
New roof-top unit,
Don't need no chiller
Nor cooling tower.
New rooftop unit
Will just keep humming along.

Just change the filters,
V-belts need checking,
They're nice and simple.
Won't be much trouble
With rooftop unit
To keep it humming along.

    Controls guy.
    No sweat, no strain.
    Hooks It to the time clock
    And Auf Wiedersehen!
    Thermostat:
    Pick cool or heat
    Dial the degrees,
    It just can't be bea-eat!

Plunk and dunk it.
We got the unit.
Took two hours
To start and tune it.
That rooftop unit.
New rooftop unit, come on.

I'm a hero:
The fan keeps turning.
And come the winter,
It'll be gas burning.
The place is comfy,
The place is comfy and calm_

              Dave Elovitz

January 2011

to the tune of
WE THREE KINGS

We three guys, of energy are
Solving problems, near and far.
    Feeling warmish?
      That's not normish.
N'r'gy Guys, solve problems bizarre.

When hot water, doesn't come quick
Your hot water, system is sick.
    Don't just suffer;
      We've cured tougher.
Solving puzzles, fits with our shtick.

If your heating. won't measure up
We're the guys, to help you warm up.
    Give a holler;
      Help will foller.
We'll find out, what needs tuning up.

Cooling towers, need lots of care:
Water treatment, dirt from the air.
    Clean the strainers;
      Check blowdown drainer.
Bio-check, Disease Legionnaire.

Need to shave, your energy bills?
Let us locate energy spills.
    'Til you're knowing
      Where it's going.
You can't cure the system's ills.

What about the hu-mid-i-ty?
Is It what you would like to see?
    Is it higher
      Than you desire?
Can we fix it? Yes, yessiree.

There's no need, to sit there and sweat;
High rise tow'r, or small maisonnette.
    Check supply air;
      What needs repair?
Make it right, so cooling they get.

Friends and clients, both fa-ar and near,
Thank you for, another good year.
    Not the greatest
      But we made it;
Not much new, to engineer.

Twenty-eleven, waits in the wings
For you we wish, all wonderful things.
    Gary, Dena,
      Patti, Ken,
Fran and Dave, all join in to sing

Joyeaux Noel, and Bonne Anee
May New Year, the greatest be.
    We wish you a healthy year
      For you and all that you hold dear;
A year of peace, and joy, and glee.

Dave, Ken, and Gary Elovitz

February 2011

THE COST TO BE GREEN

The Associated Industries of Massachusetts (AIM) provides information to Massachusetts employers about doing business in Massachusetts. One of their projects investigated the cost to consumers of state mandated energy programs. Here are their findings for 1000 kwh of electricity:

Cape wind premium:
$3.95 - $4.63
Energy efficiency programs:
$4.36 - $5.56
Greenhouse gas surcharge:
$1.24 - $1.24
Solar energy:
$1.41 - $2.83
Total $10.96 - $14.26

For the average Massachusetts customer who pays $0.15/kwh for electricity. state mandated energy programs add 7% to 9% to the customer's final cost. A typical residential customer uses 500 kWh/month, so each household pays about $12 50/month for green mandates.

Natural gas customers also pay extra to be green. AIM estimated a cost of $25-$35 tacked on the cost of 1000 therms of natural gas for energy efficiency programs. For an average Massachusetts household that pays $1.30/therm of gas, state mandated energy programs add about 2% to the customer's cost. A typical residential customer uses about 1200 therms/year, so we each pay about $3/month in green mandates on our gas bills.

You can run your own calculation by visiting www.aimnet.org. Click on the link on the right hand side of the page.

Once you have that Information, what can you do with it?

First, take advantage of the programs and subsidies available to you. If you have not had a home energy audit recently, sign up for one. The audit is free, and you might be eligible for some help paying for insulation and weather stripping. The electric company might give you some energy saving compact fluorescent lamps.

Next. while you can't do much about the cost of electricity and gas, you can do something about the amount of it you use. Turn off lights and appliances when they are not in use. Use a programmable thermostat to cut back heating and cooling while you sleep or when you are not home.

Finally, make sure the energy you use gets used wisely. Consider energy efficiency and performance when designing projects or replacing equipment.

Ken Elovitz

March 2011

THE $124,000 RULE

A colleague recently sent me an excerpt from an article about construction defects. It's called "The $124,000 Rule". The article postulates that the cost to correct a defect grows by a factor of 10 with each step in the design and construction process:

- $12.40 if you catch the problem in the design stage before any construction work begins.

- $124 if you catch tile problem after it has been installed, but before you move onto the next step.

- $1,240 if you find the problem after the work is completed, but before the customer moves in.

- $12,400 if the problem is found after the customer moves in and the team moves to correct it right away.

- $124,000 when the problem becomes a construction defect and ends up in court.

Clever, and cute, but the real question is what people can do to nip a potential problem in the bud and solve it earlier In the process.

On almost every job there's pressure to "get the contractor going". But going where? Don't get confused between activity and results. There's little point in starting down the road until you know where you want the road to take you.

Then there's the frequent push to "get the contractor involved in the design". It's certainly true that contractors often have good ideas about how to do the work. After all, doing the work is their expertise. But once again, there's little point in talking about how to do the job before you have tied down what the job is supposed to do. Designers need to figure out what the end result is supposed to be before suggestions about how to get there can improve the process. Inevitably, focusing on the best way to do something before you know what that something is results in a compromised end product

On almost any project, a little extra time spent up front checking and getting the design and details right generally pays dividends later on.

Ken Elovitz

April 2011

PHILATELICA

We have been mailing these calendars every month since January 1980. The project started as a handy way to keep track of my appointments, but it quickly grew into a good way to stay in touch with friends and clients.

This month's calendar is issue #376; so it does not surprise me that people often mention our calendars when they talk with me. Mostly people say something about the stories, and a few of you even use the calendars the way I do to keep track of appointments. But what really surprises me is how many people over the years have commented on the postage stamps! Quite a few have even thanked me (on behalf of a stamp collecting nephew or grandson, I suppose) for sending such a variety of attractive issues.

I have not had a stamp collection for almost 70 years, but when I was a kid, almost every kid had a stamp album and looked forward to adding colorful new issues from exotic places. I don't know if kids still do that any more, but even if most of them don't, I have always thought postage meter imprints are ugly and that it was worth the little extra effort to select from the great variety of really attractive stamps now available from the Postal Service.

So four times a year, when the mailman delivers a new USA Philatelic catalog, we huddle over the bright color pages and pick out the stamps we are going to order for the next three months' calendars plus whatever other stamps we think we'll need for general use. Just because I want the envelopes that bring you our monthly greeting to look a little nicer. Mostly, we try not to repeat, but every once in a while I confess that we have been so struck by a stamp that we forget we already used it and order it again. That's why Winslow Homer's "Boys in a Pasture" on this month's envelope might look familiar!

What prompted me to talk to you about stamps is that I just read a little booklet, The Best of Beyond the Perf, where I learned that USA Philatelic, the department of the Postal Service responsible for all the new stamps, has a website, *www.BeyondthePerf.com,* with the stories behind how topics for new stamps are selected and how specific stamps got designed. So, if you are among those who notice and enjoy the stamp designs we select each month, or even if you are just curious, why not spend a few minutes at that website and find out how your favorite stamps got designed and created.

And if you have a grandchild, niece, or nephew who collects stamps, tip 'em off about the website. They'll be impressed.

Dave Elovitz

May 2011

REGULATIONS

In a January 18, 2011 opinion piece in *The Wall Street Journal,* President Obama wrote about his initiative to reform federal regulations. He described his executive order 13563, which is designed to "root out regulations that conflict, that are not worth the cost, or that are just plain dumb". We could use some of that thinking here in Massachusetts:

(1) High efficiency condensing boilers and furnaces are some of the tools we can use to save energy. Some of you have these products in your homes. They squeeze additional energy out of the fuel by capturing heat from water vapor in the flue gas. The result is a small amount of condensate that usually goes down the drain. The condensate is acidic, but it is no more aggressive than *Coca-Cola.* Nevertheless, regulations require that it be neutralized. That's not so bad unless you live in one of the 650,000 housing units in Massachusetts served by an on site sewage disposal (septic) system. Our plumbing and environmental codes consider this relatively benign condensate (it's nearly pure water) an industrial waste to be taken by a hazardous waste hauler or disposed in a special underground injection well (with associated permits and costs). Do our regulators think this extra requirement will encourage us to use the latest technology?

(2) In the 1980's, many studies showed that having tenants pay their own utility costs would save a lot of energy. Those studies gave rise to a sub-metering industry that let landlords allocate energy use (electricity and fuel for heat) according to measured use. Sounds like a good idea. Unless you heat with oil in Massachusetts. There the state Sanitary Code requires the landlord to provide the oil for heating and hot water unless there is a separate oil tank for each dwelling unit. As the landlord found out in a recent lawsuit, that means no sub-metering residential hot water heat if the boiler burns oil. The rule applies even if sub-metering would save the tenant money. The rule does not apply to buildings heated with gas or electricity. Where's the sense in that regulatory scheme?

(3) Massachusetts nursing home regulations call for lighting "in each patient or resident room to provide an adequate, uniform distribution of light". That's reasonable. What's not reasonable is the next section that says, "No electric bulb under 60 watts shall be used for illumination for patients' or residents' use." I like plenty of light when I'm reading, but I get it from compact fluorescents that give me more light than a 60-watt incandescent for many fewer watts. Apparently Massachusetts regulators don't want nursing homes to save electricity. No wonder nursing home costs are so high.

Ken Elovitz

June 2011

GETTING OUT OF HOT WATER

We get quite a few calls about "not enough hot water." Here's the approach we take when those calls come in. First, determine exactly what the problem is. Is there low flow from the hot water outlet? Is the water not hot enough? Or does it have to run too long (minutes, not seconds) before it finally runs hot? Does it happen only during the busiest time? Or does it happen when no one else is likely to be using hot water?

Next, determine whether the problem is production or distribution. Is the water hot when it leaves the water heater? If not, that's a production problem. Depending on the type of heater, the question is whether the water heater is firing or whether it is getting enough heating water from the boiler. If the water heater isn't really trying to make hot water, the water can't be very hot at the outlets.

If the water heater makes hot water with heating hot water from the boiler, a pump has to circulate boiler water through a heat exchanger, and the boiler has to fire to make the water hot. Generally, the domestic hot water temperature will be about 20°F lower than the temperature coming from the boiler. If an outdoor reset control reduces the heating water temperature in mild weather, you may need a control to remind the boiler to raise the boiler temperature when you need DHW.

If the water is hot leaving the heater, but not at the outlet, that's a distribution problem. The question is, where does it get not hot enough? If the system has a tempering valve, that's a good place to start: If the water is hot at the inlet but lukewarm at the outlet, that valve is the problem. The ideal system makes and stores hot water at a high enough temperature (probably 140°F or higher) to kill bacteria and has a tempering valve to deliver hot water at a low enough temperature to avoid the risk of scalding, maybe 120°F or a little less.

Is the water sizzling at the heater, but runs cool at the tap for a long time? That's a symptom that you do not have a DHW recirculating system, or it does not operate effectively. It takes a lavatory faucet about 2½ minutes to empty all the cool water out of a 100 feet of pipe, so that's 2½ minutes of wasted time (and wasted water) before hot water from the heater gets to the tap. A DHW recirculation system keeps the pipes full of hot water to solve that problem. The recirculation system should flow enough water back to the heater from every branch so the return water is no more than 20F cooler than when it left the heater.

Now you know what we do when we get "no hot water" calls. So file this away for the next time you have a DHW complaint.

Dave Elovitz

WASTE OF MONEY

The contractor was about to start installing new coils to replace coils that froze. That's when the Director of Engineering called for a conference call to go over my design. "Why did you add those strainers?" he asked. "That's just extra cost for nothing."

"Well," I considered, "One possible reason the old coils might have frozen is if debris in the piping partially blocked water flow through the coils. There is a pretty long history of dirty water in this system. These are 100% outside air coils; if debris blocks one of the small tubes the new coils could freeze, too."

"Waste of money," growled the engineer. "Maybe there were problems before I came here, but I know what I'm doing. I watch the water treatment program and check the chemicals myself. I've got the system running clean now." But the General Manager was nervous about another freeze-up, and the strainers got installed.

The day after the coils went on-line I met the contractor in the field. "Something funny is going on," the foreman greeted us as we got there. "When I put the coils into operation yesterday everything looked fine, but now the pressures are all out of whack, and now we're getting hardly any heat from the coils."

After a quick look at the pressure gauges I had a thought. "Just out of curiosity," I asked the foreman, "Could you just pull out a couple of strainer baskets so I can take a look?" (The strainer baskets are removable perforated cylinders that sit inside the strainer bodies: Water flows into the center of the cylinder and out through the perforations into the strainer body. If there is any sludge or debris, it builds up on the inside surface of the baskets.) This is what I saw:

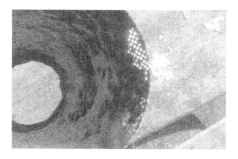

That's about half an inch of sludge that built up in about 24 hours of operation. If the strainers weren't there, that sludge would be lining the insides of the tubes in the coil, where it would impede heat transfer and possibly block the flow through some of the tubes completely, leading to another freeze this winter.

Gary Elovitz

August 2011

SPRING SHOWERS

You would think that picking out a showerhead should be a pretty simple matter. But you have probably reckoned without the U.S. Department of Energy and Plumbing Manufacturers International. The DOE, charged with enforcing the 1992 Energy Policy and Conservation Act (which includes a 2.5 GPM maximum allowable flow from a showerhead) decided, almost 20 years after the passage of said act, that it needed to define showerhead for the purposes of enforcing the law. DOE issued a "draft interpretive rule" unilaterally redefining "showerheads" as "shower valves." I kid you not!

The new definition was aimed at limiting each shower compartment to a single shower head using no more than 2.5 GPM. if you have been to a plumbing showroom - or a home improvement store - in recent years you would be aware that the plumbing industry is anxious to convince you that you cannot possibly get clean from just one spray and that not being simultaneously wetted on all body surfaces with sprays from every possible direction - sort of like in a car wash - was less than civilized. On learning of the proposed definition, plumbing manufacturers erupted with a maelstrom of protests. Their industry association stormed DOE with claims that the proposed new definition would not only deprive homeowners of their basic civil rights, it would deprive the plumbing industry of some $400 million per year. DOE finally withdrew the proposed definition in order to "avoid unnecessary economic disruption."

In fact. DOE apparently decided that the world did not need a definition of "showerhead" after all. Instead, we would have to make due with what they called "a brief enforcement guidance".

For purposes of the maximum water use standards, a single showerhead is multiple spraying components sold together as a single unit designed to spray water on a single bather. (I wonder how that will relate to Levi's new water conservation campaign urging each of us to shower with a friend?) Incidentally. tub spouts, locker room showers, and nozzles where water can be diverted to a hand sprayer but the sprayer and the main nozzle cannot both run at the same time, are definitely not covered by the enforcement guidance.

Aren't you glad the DOE is keeping its focus and not getting distracted by things like the implications of the Japan nuclear plants?

Dave Elovitz

September 2011

BATTERY LIFE

Back in 2001, a lot of our friends teased Franny and me a bit because Franny was one of the first people in the area to buy a Prius hybrid. I drove hers just once and ordered one for me. Ten years later, we still love those same cars, nice and roomy inside but very small and easy to park outside. I confess that filling up once every other week, and for less than $25 when gas was close to $4 a gallon, didn't make us too unhappy either. After the 8 year warranty on the traction battery expired, we were always aware that financial catastrophe might be lurking out there. Yet neither of us could work up much enthusiasm for trading for a new model. Despite occasional inquiries from friends, neither of us had decided what we would do when facing thousands of dollars for a new battery.

Do any of you remember what it was like around here last January 23 and 24? 0°F on Sunday and -5°F at 7 AM Monday. I went out to start the car in the morning and it caught, then almost immediately died. Same thing, two or three times. Everyone knows that cold weather and batteries don't get along: I hoped against hope that was all it was and that my trusty steed would struggle back to life when the temperature eased in the afternoon. But deep in my heart, I didn't really believe it. Three or four o'clock, and same thing. I explained the situation to the Toyota dealer and then had my car towed to the shop. I didn't sleep well that night: What should I do when they told me how much it was going to cost for a new traction battery? I couldn't expect much of a trade in without one, yet did it make sense to pour that kind of money into a 10 year old car?

The service writer was very apologetic when he called Tuesday noon: They couldn't have the car ready until Thursday, and the repair was going to cost $1500. He must have thought I was some kind of nut when I sounded so delighted. After expecting to replace the traction battery, $1500 was chump change. Here's what the repair invoice said when I picked the car up, and I quote:

> *Found mouse ate through the air filter and got into throttle body, contaminating throttle body motor and MAF sensor. ..... Replaced throttle body and MAF meter. Replaced air filter and cleared code. All systems functioning well at this time.*

They found lots of fur, and a hole the size of a baseball chewed through the air filter, but no bones. Br'er Mouse was long gone by the time the technicians got under the hood. That was months ago (knock, knock), and my Prius is still purring along. I still don't know what I'll do when the battery finally goes.

Dave Elovitz

October 2011

HEATING TIPS

Winter is coming! We got so many comments on our June calendar about domestic hot water systems, Ken asked me to write about how you can make your heating plant more cost effective.

Let's start with the burner on your boiler or furnace: The burner mixes fuel and air and ignites the mixture into a flame. The fuel - which is largely carbon and hydrogen - and air - which is largely oxygen and nitrogen - become hot water vapor and carbon dioxide and, unavoidably, hot nitrogen. The flame passes through a heat exchanger that transfers heat into water (a boiler) or air (a furnace) and then leaves up the chimney as flue gas.

The ideal burner takes in just enough air so all the carbon becomes carbon dioxide and all the hydrogen becomes water. Nitrogen just goes along for a free ride. Too little air? All the fuel is not burned, and some of the carbon only gets to be carbon monoxide. Too much air? A lot of the energy in the fuel goes into making hot nitrogen that just goes up the chimney. Tuning the burner for the optimum fuel/air mixture uses a simple test that measures the chemistry and temperature of the flue gas, not just eyeballing the color of the flame.

The other key component is the heat exchanger. The ideal heat exchanger transfers all the heat it can out of the flame without cooling it so much that the water vapor starts to condense. Impurities in the fuel can turn that condensed water into acid that can damage the heat exchanger and the chimney system. Some new super-efficient boilers - condensing boilers - are constructed of special materials and use special materials in the chimney system so acid doesn't damage them. Those boilers can take extra heat out of the flame by cooling the flue gases until they condense. Condensing pulls extra useful heat from the same amount of fuel, but only while the temperature of the water leaving the boiler is low enough to cool the flue gas down to where it condenses. Keeping the heat exchanger surface clean of soot, dirt, and scale lets the water in a boiler or air in a furnace cool the flue gases as close to the leaving water or air temperature as possible.

Aside from replacing your old faithful boiler or furnace with a fancy new condensing model, you can get the most out of what you have by having it serviced and tuned up annually. A free utility company home energy audit often includes measuring combustion efficiency. If efficiency is not optimum, it's time to adjust the mixture or clean the heat exchanger, but "don't try this at home". Call a service contractor to make the adjustments properly.

Dave Elovitz

November 2011

BOIL AWAY

"Our boilers had cracked sections twice last year. The consultant says the boilers are too small, they cracked from thermal shock, and we need new boilers. What should we do?"

Analyzing the gas bills let me develop a formula that predicts heating load at any outdoor temperature. The formula showed that the boilers had 50% excess capacity: two boilers could each carry 3/4 of the load, providing reasonable redundancy. They didn't need new boilers for capacity.

That was the easy part. The harder question was why the boiler sections cracked. At the site I saw that only one of the two boilers had cracked, and the cracks were in sections 5, 6, and 7 of an 18 section boiler. Thermal shock occurs when cold water enters a hot boiler. Water enters the boiler at the back, so thermal shock usually affects the last few sections. With failures occurring in the middle of the boiler, thermal shock was unlikely.

On top of that, the way the boilers were piped, flow did not divide evenly between the two boilers. Less "cold" system return water would flow through Boiler 2 than Boiler 1, so Boiler 2 would be less susceptible to thermal shock than Boiler 1. Yet Boiler 2 is the one that cracked. Again, not thermal shock.

Then I saw that the boilers had on/off control. Each boiler fired at 5 million BTU/hour or was off. 5 million is a big capacity step. Except on the coldest days, the boilers would bang on and off again and again. So I had three clues: failures in the middle of the boiler, failure in the boiler with less water flow so less ability to absorb heat, and frequent cycling at a high firing rate.

Boilers experience cyclic stresses every time the burner starts and stops. Those stresses occur because the boiler structure resists the expansion and contraction that would otherwise occur from temperature change. The higher the firing rate, the higher the stresses. Cold return water increases the stress because of the greater temperature difference between the fire side of the casting and the inside surface exposed to the water.

Taken together, the clues said thermal fatigue. Fatigue affects materials subjected to cyclic loads. Classic fatigue failures occur from physical stress, but fatigue failures can result from any type of cyclic load, including repeated temperature change.

The solution was easy: Add a low fire controller to reduce cycling by letting the boiler run at low fire on startup and at light load, jumping up to high fire only at high load.

Ken Elovitz

December 2011

TOO MUCH PRESSURE

"I'd like you to take a leek at a noisy HVAC system," the general contractor said. "I think we need to add a variable speed drive."

The HVAC system was a constant volume rooftop unit with dampers that shut off airflow to each room when its thermostat was satisfied. With a constant speed fan, closing .off .one zone forces more air to the other zones. That excess airflow can generate noise. To avoid that problem, the system has a bypass duct and damper. The bypass provides a path for air not needed in the zones to bypass from the supply back to the return. The bypass controls let a constant volume fan provide variable air volume to the zones.

The room certainly was noisy, and the HVAC system was a definite contributor. The first test set all the thermostats to full cooling. The unit delivered the design 4000 CFM airflow. The pressure was 0.20' wg at the sensor in the duct. Noise was acceptable at this full load condition, so fan speed and capacity was not the problem.

Since the first test showed that the noise occurred only at part lead, the next test was to evaluate airflow, pressure, and noise with one room calling for cooling and the other zones satisfied. The bypass damper opened as expected, but the .outlets were way too loud. The pressure in the duct rose to 0.33' wg instead .of the 0.20' wg we saw before. That extra pressure pushed more than 1000 CFM into a zone that was designed for only 800 CFM. No wonder the outlets were noisy!

The bypass damper was wide open, so why did the pressure, airflow, and noise in the room increase, and what could I do about it?

The key to this puzzle was recognizing that the bypass duct and the duct to the rooms are parallel paths. With parallel paths, the pressure drop from point A to point B has to be the same, regardless .of which path the air takes. The airflow will divide to make the pressure drops equal. That's why we got extra air and too much noise in the room. Since the fan was constant volume, not enough airflow through the bypass caused too much air to flow to the room when only one zone called for air.

The solution? Install a bigger bypass. Pressure drop calculations for the two parallel paths showed that changing the 16" bypass to 20" would make the pressure drop of 3200 CFM flowing through the bypass equal to the pressure drop .of 800 CFM flowing through the supply and return ducts for the zone. The same total CFM as before, but now divided the way we wanted. Sending the extra air through the bypass get rid of the extra noise.

Ken Elovitz

January 2012

to the tune of
SANTA CLAUSE IS COMIN' TO TOWN

You're riddled with doubt,
What else can you try?
Boss man may pout,
If you can't tell why,
Power costs are not comin' down.

You checked all the bills;
The cause mystifies.
You gotta add skills:
Call Energy Guys.
Power costs have gotta come down.

> We'll find where you are wastin'.
> Show you the steps to take.
> We'll figure out what to control
> And what savings you can make!

Use our know-how
And you can predict.
What you must do
So problems get licked.
Power costs are gonna come down.

You better not wait,
Take action that's wise.
Beat the high rates,
Call Energy Guys:
Fu-el costs have gotta come down.

> We'll check for insulation
> And sequence your controls.
> Get burners tuned and filters changed;
> We will help you meet your goals.

We'll track what you use
And then analyze.
Point out abuse;
Show where you can slice.
Fu-el costs are gonna come down.

Both Gary and Ken,
And David all cry:
We wish you the best,
I'm telling you why,
Two Thousand Twelve is comin' to town.

Dave, Ken, and Gary Elovitz

February 2012

READ THIS BOOK

Over the last 32 years, we have written about a huge variety of topics that we thought you would find of interest, but - as best I can recall - this is the first time I am sending you a book review. Well, maybe it is not really a book review: I can't distill 717 pages into a 425 word calendar. Whatever your interest in energy, whether it is:

- Why your home heating oil costs over $4.00 per gallon;

- Why, after decades of commanding a premium price over oil, natural gas is now cheaper per BTU than oil;

- Where are today's sources of petroleum?

- Who controls them?

- How long will they last?

- Is Global Warming real? Man-made?

- Must the US and China come into conflict over oil?

I urge you to read *The Quest - Energy, Security, and the Remaking of the Modern World* by Daniel Yergin. A master story teller, Yergin takes us on a fascinating journey from the very different 1991 world of oil that he explained in *The Prize* to the new and even more complex energy world we live in today. Not just oil, but nuclear and coal electricity, the new unconventional sources of petroleum, and natural gas, their impacts on the global economy and geopolitics, and the impacts of the global economy and geopolitics on energy supplies.

As I read, the last 20 years - and the extraordinary changes in the world of energy during those 20 years - came alive and began to make sense as *The Quest* explained, in turn: The transformation of oil from an energy source to a market commodity, which Yergin dubs the financialization of oil; The collapse of the Soviet Union and the rebirth of Russia as a major oil producer; The rise of China as an energy user and energy resource; The impact of "aboveground risks" on new resources; The gripping "inside" stories of what happened with Deepwater Horizon and Fukushima; And so much more.

If you are like me, you will keep saying to yourself, "I lived through all this. I read about it in the Globe every day. How could I have missed what it all meant?"

Dave Elovitz

March 2012

BIG QUESTION

Friend and colleague Tim Foulkes asked an interesting question the other day. "If someone really wants to watch their energy use, which would use less energy - running the bathroom exhaust fan for 15 minutes after showering, or opening the bathroom window for 5 minutes?"

Before you read any further - which do you think?

We're engineers, so let's analyze: An 80 CFM bathroom fan running for 15 minutes exhausts 1,200 cubic feet of air. That's about 3 times the volume of a typical 50 sq. ft. bathroom. When that 1,200 cubic feet of air leaves the house, the same volume of cold outdoor air has to find its way into the house. It takes energy to heat that cold air to room temperature.

If you open the bathroom window (but close the bathroom door), the maximum amount of cold outdoor air that can come into the bathroom is one "bathroom-volume". That's because once all the air in the bathroom has been exchanged for outdoor air, each new cubic foot of cold outdoor air that comes into the bathroom has to push out a cubic foot of cold outdoor air that already came in through the window. On that basis, running the exhaust fan for 15 minutes seems to use 3 times as much energy as opening the window regardless of how long the window is open!

Is that realistic? A house always has some air leakage unless it is really tight construction. Running the bathroom fan mostly changes where the exfiltration occurs (i.e., through the fan rather than up into the attic). Changing the path of exfiltration or exhaust probably changes the rate of infiltration only slightly. (The infiltration rate depends on pressure difference between indoors and out, and running the fan doesn't change indoor pressure much, if at all.) So running the bath fan might not increase infiltration or change energy use much at all.

If the bathroom door is open or has a big gap at the bottom, the open window could let in more than one bathroom-volume of air because some of that cold outdoor air escapes to the rest of the house rather than being trapped in the bathroom. So in reality, opening the bathroom window probably uses more energy than running the fan.

Now that we've presented this exhaustive analysis, how much energy is at stake? If you run your 80 CFM bath fan for 15 minutes - and even assuming that additional outside air is brought in to replace all of that exhaust - how much would that energy cost? Only a penny or two. (Based on 30F outdoors and natural gas heat.) Aren't you glad Tim asked?

Gary Elovitz

April 2012

REFLECTIONS

Tenants on the second floor complained that noise from the boiler kept them up at night, and I was on the case. As I approached the boiler room door at the back of the building, I could hear the roar of the boiler through the vent openings. But it was unlikely that noise from those vents was the problem - the building was "U" shaped, and the complaints came from an apartment at the bottom of the "U". The boiler room vents were at the top of the "U", about 30 ft. away, and the vents faced away from the apartment windows.

Inside the boiler room the noise was pretty loud - especially from the flue pipe connecting the boiler to the chimney. Maybe the noise was traveling up through the roof of the boiler room. I climbed up on the roof of the boiler room, which stuck out into the courtyard formed by the "U", putting it closer to the apartment windows. I could really hear the noise here. But it didn't seem to be coming up from the boiler room below. Remembering an old mechanic's trick, I used my screwdriver as a stethoscope and couldn't hear anything through the boiler room roof or through the chimney that runs up the side of the courtyard.

Looking up, though, I could just barely see a rain cap at the top of the chimney. Could the noise be reflecting back into the courtyard from the underside of that rain cap? I had to take a look.

I climbed up six flights of stairs to the roof. As the door to the roof came into view a few steps from the top, I stopped short. The door was locked, of course. (I had wondered why there were two keys on the boiler room key ring. Now I knew the answer, but since I had left the keys in the boiler room that was cold comfort.) I guess I would have to trudge all the way down and all the way back up. But first - I remembered that the boiler room keys used the same "blank" as my house key. It was a long-shot (about 100,000 to 1, in fact), but I was tired enough to give it a try. I slid my key into the lock... and it turned! What a lucky break!

The boiler fired up a minute or so later, and I had my answer: the metal rain-cap - which was basically a big inverted bowl - reflected the stack noise down into the courtyard, working like a satellite dish in reverse. A satellite dish gathers radio waves and concentrates them at its focal point. My rain cap took a noise source at its focal point and broadcast it to the apartments below. Simply changing the stack discharge would quiet those noise complaints.

Gary Elovitz

May 2012

BLOWIN' IN THE WIND

Friend and young engineer Adam Miller shared this story with us.

A toothpaste factory had a problem: they sometimes shipped boxes with no tube inside. People with experience designing production lines know how difficult it is to have everything happen with exact timing so every unit is perfect 100% of the time. Most manufacturers solve that problem with strategically placed quality assurance checks to kick out defective products before they get to the customer.

Understanding the importance of not shipping empty boxes, the CEO of the toothpaste factory gathered top management together to start a new project to avoid shipping any empty boxes. The in-house engineering department was too busy to take on a project of this magnitude, so management hired an external company to solve their empty box problem.

Six months and $8 million later they had a fantastic solution - on time, on budget, and high quality. Plus, everyone on the project had a great time. The solution was a high-tech precision scale that would stop the line, sound a bell, and flash a light whenever a toothpaste box coming down the line weighed less than it should. A worker simply had to walk over, pull the defective box, and re-start the line.

Not long afterward, the CEO reviewed project: amazing results! Not a single empty box after the scales were installed. Very few customer complaints, and they were gaining market share. "That's money well spent!" the CEO said. Then he looked closely at the other statistics in the report.

After 3 weeks of operation, the new scale had not picked up any defects. Past experience said the scale should pick up at least a dozen empty boxes a day. Maybe there was something wrong with the report. The CEO asked the engineers to investigate, and they came back saying the report was correct. The scales weren't picking up any defects, because all boxes that got to that point in the production line were good.

That just didn't seem right, so the CEO walked over to the production line where the precision scale was installed. A few feet upstream of the scale, he saw a $20 desk fan, blowing empty boxes off the belt and into a bin.

"Oh, that," one of the workers said. "One of the guys put it there 'cause he was tired of walking over every time the bell rang".

Dave Elovitz

June 2012

EVER WONDER?

Parade magazine that comes with my *Sunday Globe* recently provided some fascinating explanations:

Why do your fingers and toes wrinkle in the water? The most popular theory is that the top layer of your skin absorbs more water than the layers beneath, increasing its surface area. But the now baggy skin is still attached to the non-bloated layers beneath, so the top layer bunches up forming wrinkles. Since fingers and toes have thick shells of dead keratin (an absorbent protein) their skin takes on more water so it wrinkles a lot.

What makes you cry when you chop onions? When you slice into an onion, a chemical called a lacrimator is expelled into the air. Once the lacrimator reaches your eyes and nose, it breaks down into irritating compounds that attack nerve endings there. Your eyes produce extra tears, and your nose might run, to flush away these compounds. Some claim that pre-chilling the onions in ice water for 30 to 60 minutes before slicing will slow the release of the lacrimators, reducing the number of offending molecules in the air, so fewer tears!

Why do you huff and puff climbing stairs when you can easily run a mile on a treadmill? You know that pulling a heavy rolling suitcase along a level path is a lot easier than picking one up. Running on a treadmill, you barely lift any of your body weight. But walking up a typical staircase requires that you move 70% of your body weight against gravity.

Why is there crusty dirt in your eyes when you wake up in the morning? That gunk is just dried tears. Your tear glands continue to "wash" irritants out your eyes while you sleep. While awake, blinking wipes the tears away before they can build up. But your eyes stay closed at night, so tears dry up and accumulate in the corners of your eyes.

Why do mosquitoes find some people more attractive than others? Mosquitoes are drawn by a variety of signals, including heat, carbon dioxide, movement, and the smell of skin secretions like lactic acid. One study even found that mosquitoes tend to prefer people who have recently drunk a beer. No one knows precisely what combination mosquitoes find most irresistible, but if you seem to be the meal of choice, slather on repellant, and pass up the brew.

Dave Elovitz

July 2012

ANOTHER FREEZE

It was a problem we've seen before. A pipe froze during unusually cold weather, and the homeowner's insurance company paid a lot of money for the clean up. Now they want to sue everyone who ever had anything to do with the house and get their money back.

The freeze occurred at a washing machine outlet box that was exposed to freezing conditions because of a small gap in the building insulation. An ice plug formed in a copper tube and pushed a soldered joint apart. The insurance company's expert said the joint was defective because he conducted a peel test that showed solder did not fill the entire capillary space in the socket of the fitting. There were three things wrong with using that theory to tag the plumber for the loss:

(1) The house was completed in mid-2006, and the failure didn't occur until January 2009, so the joint held without leaking for 2« years in normal service. That "defective" solder joint was strong enough for the job it was supposed to do.

(2) There was no freeze in January 2007 or January 2008 when outdoor temperatures were as cold or colder than the day of the freeze in 2009.

(3) If the solder joint had been strong enough to hold against the pressure created by the expanding ice plug, the pipe would have split instead of the joint coming apart, and the result would have been the same.

The insurance company also criticized the design of the house, saying the eave space behind the washing machine outlet box should have been insulated and not a ventilated attic space. That's a designer's choice; a building can work either way.

Then they said the small gap in the building insulation violated the plumber's duty under the plumbing code to insulate pipes that might be exposed to freezing temperatures. But the plumber had insulated the pipes, and the building insulation was installed between the pipes and the ventilated attic. So the pipes were insulated, and they were on the warm side of the building insulation.

What really happened? In 2007 and 2008, the homeowners were away for the winter and had a management company watch the house. The thermostat was set at 68F during very cold weather, as would be smart to do in an empty house where pipes might freeze. In 2009, the owners' son was in the house, and he lowered the thermostat to 55F when he went out. That 13F difference in room temperature was likely why the small gap in the building insulation allowed the pipe to freeze in 2009 but not in 2007 and 2008.

Ken Elovitz

August 2012

PUMP IT UP

"Our cooling tower isn't doing the job," the property manager explained. "We think the last 5-ton unit we added pushed us over the edge. We have a proposal to retrofit the tower to 170 tons, but before we drop $50,000, we want you to check it out."

First I ruled out the obvious: The pumps and the tower fan rotated in the right direction. Next I traced out the piping and measured pressures at strategic points. Something didn't look right. In fact very little looked right. And every time I rechecked, the pressures were different. The building engineer agreed to get a balancing technician to read the circuit setters to measure flow.

Meanwhile, I found the design drawings at the site and took some data from them. The connected load was 96 tons. Tenants had added about 10 tons of new load. Even without accounting for load diversity, the tower should have walked away with the job. That's when it got strange. The tower was selected for a flow of 375 gpm, but the pump nameplate read only 150 gpm. The pipe sizes on the drawings were more in line with 150 gpm than 375 gpm.

The balancing contractor measured 220 gpm of flow. He had the same problem reading pressures as I did. To make matters worse, the data from the site didn't match up with my calculations. The building engineer kept pushing me to stop fooling around and tell him what size pumps to install to fix his problem. I insisted I couldn't figure out what to do until my calculations and the field data lined up.

As I wrote up my findings in a progress report, I realized what I wished had come to mind that day I was at the site: Pressures bouncing all over the place. A weeping relief valve. Classic signs of a failed expansion tank. I added checking the expansion tank to the recommendations in my report.

A few days later, I got the news: The expansion tank was indeed water logged. The owner replaced it and brought the balancing contractor back to recheck flows. Everything now made sense on the pumping side.

With the new expansion tank installed, the pump pushed 275 gpm through the system. I never did figure out why the building had a pump sized for 150 gpm and a cooling tower selected for 375 gpm. But I did figure out that the tower capacity with 275 gpm flow was 110 tons - enough to meet the load. Running the numbers to figure out what's happening sure beat spending tens of thousands of dollars on cooling towers or piping.

Ken Elovitz

September 2012

COMPANY NEWS

On May 14th, Dave Elovitz passed away after a long career and a brief illness. After a close call in April, we were blessed with 4 weeks of progress in rehab when family and friends from all over came to visit, share stories, and cheer him on. He had so many visitors that the receptionist once asked who this guy was and how she could meet him.

Dad graduated from WPI in 1953. We don't know his GPA or class rank, but we do know that generations of fraternity brothers used his lab reports as a reference. I also know that back then students who had achieved a certain grade level during the course could accept their grade based on class work and exams and opt out of the final. I think it was an electrical engineering class when Dad knew he had a 95 going into the final so couldn't understand why his name wasn't on the opt out list. He found out he had to take the final so the professor could use his test as an answer key.

Dad began his career as a test engineer at Pratt & Whitney Aircraft working on a nuclear aircraft engine. It was soon clear that he was destined to be the consummate engineer. So much so that he managed to convince my mother that my birth announcement should be in the form of an engineering test report.

He worked for a handful of companies as a facilities engineer and as chief engineer for a mechanical contractor. He started Energy Economics around 1975 and mailed the first of these monthly calendars in January 1980, so this is number 393 in the series. Sometimes the message was a case history, sometimes it was a just chance to share a story or evoke a smile. The stated purpose was to keep in touch with clients and friends, but we all knew he hoped the calendars would remind people that he was growing an engineering business. I remember someone once asked for the name of the service that provides these clever calendars. The guy was amazed to learn that we do it ourselves.

Dad was a colleague and a mentor, known and respected throughout the industry. He was a true friend to many. His advice was often sought, always generously given, and more often than not, followed with good results. He helped many people get their start in business, and he gave many others the chance to prove themselves, sometimes when no one else would give them that chance. I am one of those people.

Gary and I will continue to run Energy Economics and are committed to the same standards our father set for himself and instilled in us.

Ken Elovitz

October 2012

SHAKE, RATTLE, AND ROLL

"We need to add springs to the air-handler on the left side of the 8th floor," the property manager moaned. "The people in the offices on both sides of the mechanical closet say they get nauseous and can't work."

That was puzzling: the air-handler was only 5 tons - not much bigger than one you might have at home. Also, there were at least a dozen similar units in the building and no noise or vibration complaints about any of them. If this unit was vibrating, something had to be wrong that springs would not fix.

The property manager, a service technician, and I convened on-site the next morning. The unit was rattling, the walls of the mechanical closet were shaking, the ceilings in the adjacent offices were thumping, and the diffusers were whistling. "Everything was way out of alignment when I got here," the tech offered. "You should've heard it then! But I re-aligned the fan motor, and now everything is just right." May be, but clearly something was still wrong.

We took the cover off the air-handler: the motor was flexing about 1/16" up and down on its base - not a huge amount, but it doesn't take much when something is spinning 30 times every second. I remembered some advice my friend (and crack mechanic) Bobby Beauchemin once gave me: don't ignore the motor sheave.

"Something here is slightly out of round," I began. "It could be the fan motor, the fan shaft, one of the sheaves, or even a bump on the fan belt. You already changed the belt, so that's unlikely. There's a good chance that adjustable sheave is our culprit. Adjustable sheaves simplify fine-tuning fan speed, but they're not always perfectly balanced. Let's try using a fixed-pitch sheave.

"But that's not all," I continued. "I heard a lot of air noise in the offices, and sure enough, every damper I checked was choked way down - I bet the rest are, too. It looks like the air balance guy took the lazy way out - he was supposed to adjust the fan speed so that at least one damper is wide open, but he just cranked down all the dampers instead. The fan is backed way up on its curve, maybe even operating on an unstable part of its curve. That means excess noise and vibration, even when everything is perfectly round. So you should have a proper air balance to determine the right fan speed before you replace the sheaves."

Two days later, with a slower fan and a new sheave, the office workers were able to sleep in peace. (It was a state office, after all.)

Gary Elovitz

November 2012

WINTER FUN

As the cold weather moves in and we get ready for winter, I thought I would tell you about some low cost/no cost indoor activities that Patti and I enjoyed last winter.

The Culinary Arts Museum at Johnson & Wales in Providence was a great find. Besides showing the history of cooking appliances that many of us remember, there's an informative exhibit about the role of African Americans in the development of the food service industry. If you go, you'll learn that the traditional chefs coat is double breasted so the chef can button it the other way and have a clean side in front when he goes out to the dining room.

Providence has some other little known treasures to help you fill the day. The Haffenreffer museum at Brown University is a small anthropology exhibit. You'll be interested to learn about the things Roger Williams and his companions found when they settled in Rhode Island. It's on the first floor of Manning Hall on the Brown campus, and it's free. Around the corner on College Street you can visit the List museum, also free and open from 1-4 PM on weekends. The exhibits rotate, and some are student curated, so it's a nice add to your visit.

Closer to Boston, we recommend the MIT museum. It's free to Massachusetts residents on Sundays if you enter between 10 AM and noon. The main museum is on Mass. Ave. near State and has all kinds of technology related exhibits, including some you are encouraged to touch. Don't limit your trip to the main museum. The List Visual Arts Center at 20 Ames St. (near Main St.) is free. Let's say the experience was "interesting". And if you walk from Mass. Ave. through MIT's "infinite" corridor on your way to the List, you can stop at the Edgerton corridor gallery and learn about Doc Edgerton's experiments with strobes. Remember that picture of a bullet passing through an apple?

We also recommend the Davis Museum at Wellesley College (free). The cafeteria in the Wang Student Center (near the Route 135 entrance) is a great place for Saturday brunch. An all you can eat buffet was $7.50, I think. I considered staying the whole afternoon.

A few places we didn't get to visit last winter are on our list for this year: The Rose Art museum at Brandeis (free, I think), the Worcester Art Museum (free on the first Saturday morning of the month, and, according to Jared Bowen of WGBH, every bit as good as the Boston Museum of Fine Arts), and the RISD Museum in Providence (free the last Saturday of the month, but check to be sure). Perhaps we'll compare notes next fall.

Ken Elovitz

December 2012

CAN YOU HEAR ME NOW?

"We got a new air conditioning unit for our sanctuary," the synagogue president reported, "but it's too noisy - some people say they can't hear the rabbi's sermon. Can you give me some quick advice?"

"I see," I said. "So you want to know how to make it noisier, so everybody will be happy? But seriously, there's a pretty simple solution. The rooftop unit has two steps of cooling but a single speed fan, so it runs with full airflow even when it needs only half its cooling capacity, and that's most of the time. You only need the second stage of cooling when the sanctuary is packed with people on a hot, sunny day.

"The noise you described is almost certainly from the fan. That means it would be an easy fix to add variable speed control so the fan runs at low speed (and low noise) on low cooling and jumps up to high speed only when the second stage of cooling comes on. Admittedly, on those rare occasions when you need the second stage of cooling, you'll still have some noise, but a really full room is less noise-sensitive, in part because the crowd absorbs some of the noise."

The president liked my idea and said he would ask the HVAC contractor for a proposal to implement it. A few days later he forwarded an e-mail from the HVAC contractor: "The manufacturer does not make nor offer an aftermarket vfd for the unit, and they don't have instructions for adding a vfd to the unit." He asked if I could intervene.

"I don't know that I can intervene," I responded, "but I think I can solve your problem. Years ago, manufacturers and contractors were open to engineered solutions, but these days no one seems to want to touch anything that requires thinking. But this is HVAC, not rocket science, and engineering solutions is what we do. I can draw up a design for installing a variable speed drive with the necessary controls. I'll include enough detail for a qualified electrician to do the work. Once the work is done, I'll be there when the electrician puts the unit through its paces so we can see that the unit works correctly with the new speed control."

In pretty short order (and for pretty short money), I delivered the design and the electrician installed the speed control. The sanctuary was cool and quiet when the rooftop unit ran on low cool. We even found we could trim the fan speed at full load, knocking off a few decibels without giving up cooling capacity for those hot days. Now the rabbi has to work a little harder on his sermon.

Gary Elovitz

January 2013

ROCK OF AGES
(MA'OZ TZUR)

Every year about this time
We compose a little rhyme.
We want you to think of us
When your systems put up a fuss.
   EnergyGuys are we;
   Engineers for H, V, and A/C.

(refrain:)
   Let us know what went wrong,
   EnergyGuys do the rest.
   We're the ones you should call.
   We'll be there at your behest.

Boilers, chillers, pumps and fans
B-T-Us and octave bands.
Those are things that we adore.
We are techies to the core.
   Lots of calculations
   Lead to our elation.

(refrain)

Summer was so very hot.
A/C ran an awful lot.
'Lectric bills achieved new highs.
Paid 'em off with woeful sighs.
   Winter soon will be here.
   That means gas bills ever more so dear.

(refrain)

Romney and Obama sparred.
Each one campaigned oh so hard.
Now that the election's done,
We must act in unison.
   Brown fought 'Lizbeth Warren
   Campaigns got so borin'.

(refrain)

When you're hot or cold or wet,
We'll be there to help, you bet!
We solve problems one-two-three,
And we do it creatively.
   We explore your systems.
   Tell you how to fix them.

(refrain)

Twenty twelve is almost done.
Hope thirteen brings lots of fun.
Ken and Gary send this song;
May you prosper all year long.
   May we all be healthy
   And be blessed with family and friends.

   Here's from us to all of you,
   Have a happy new year, too.
   Here's from us to all of you,
   All our best the whole year through.

Ken and Gary Elovitz

February 2013

RANDOM KINDS OF FACTNESS*

In the 1880's, about 15% of city dwellers in America has access to an indoor bathroom. As late as the 1970's, some apartment buildings in Boston's North End still did not have bathrooms.

The US Mint pays about half a cent to mint a penny, 2-1/2 cents to mint a nickel or a dime, and one cent to mint a quarter.

A dime has 118 ridges around its edge.

The Capitol Building in Washington, DC has 365 steps - one for each day of the year.

The crack in the Liberty Bell occurred on July 8, 1835 when the bell was being rung for the funeral of Chief Justice John Marshall.

The first 8 presidents of the United States were not born in the United States. They were born in England's American colonies before the United States was established.

Five American presidents had beards. Four had moustaches but no beard. All nine were Republicans.

Four state capitals are named after presidents: Lincoln, Nebraska; Jackson, Mississippi; Madison, Wisconsin; Jefferson City, Missouri.

Maine is the only state with a one-syllable name.

William Shakespeare has the most published plays (37) in the English language. He had three brothers: Gilbert, Richard, and Edmund.

L. Frank Baum (his first name was Lyman) claimed to have gotten the name for his imaginary land of Oz from a file cabinet labeled O-Z. He wrote 14 novels about the land of Oz.

Ayn Rand, author of *The Fountainhead* and *Atlas Shrugged*, is a pen name for Alissa Rosenbaum.

What do Elvis and Liberace have in common? Both had a twin who died during birth.

The Incas all had the same blood type (O positive).

*mostly from a book of the same title by Erin Barrett & Jack Mingo

Ken Elovitz

March 2013

YOU CAN'T FIGHT CITY HALL

It should have been a simple project. The basement space of the downtown high-rise had been a health/fitness club for years. The new owner just wanted to change the name, subdivide one room, splash on some fresh paint, bring in new exercise equipment, and throw open the doors. The project should have sailed through with a "short-form" building permit, and the space could be open in a couple of weeks.

Thirty days later, the City sent back a form letter - among the requirements: a plan showing fire alarm and emergency lighting, and a plan showing that the space meets the egress (exit) requirements. The owner needed professional help.

I got working on the fire alarm and emergency lighting plan, and the owner got an architect to mark travel distances on the plan. Still pretty straightforward, until the architect sent an e-mail: "The space does not comply with the required travel distances in the Code. The rear section of the locker room is about 25 ft. beyond the maximum allowable travel distance. The only way we can comply is to wall off the back section of the locker room."

That didn't seem right. Could the space have been so far out of compliance all those years? While the architect started to draw up plans to wall off part of the locker room (and the owner nearly in tears), I did a little research. It turns out the issue was not travel distance, but the "common path of travel." A few minutes looking through old code books and I had my answer:

"When the space was built in the '90s there was no 'common path of travel' requirement," I explained. "That was added to the code fairly recently. Total travel distances haven't changed since forever. The important point is that the rules for remodeling existing space are different from the rules for new construction. Existing spaces do not have to comply with all the rules for new construction - that would be an unjustifiable burden and would discourage people from getting permits for renovations. The 'Existing Buildings Code' specifically states which features of an existing space need to be changed during a renovation, and 'common path of travel' is not one of them. The architect should be able to show that the space complies with the 'Existing Buildings Code', which is the applicable code for a renovation project."

With that cleared up, the owner brought the fire alarm/emergency lighting and egress plans over to City Hall ... where they were promptly rejected. The City didn't like that the two plans were on different sized paper. You know what they say about fighting City Hall.

Gary Elovitz

April 2013

UP IN THE ATTIC

A long time general contractor client called in a panic. "We need your help on some heat pump shutdowns in an assisted living facility. The heat pumps are in the attic, and they go off on high temperature. We think there's a design flaw. And we're especially concerned because we have the exact same design in another building."

It didn't seem right to me that the attic had anything to do with the problem. The compressors are refrigerant cooled, so ambient temperature should not affect them. The fan motors are in the conditioned air stream, so they also should not care about being in an attic. From the description of the problem, I thought water flow was the most likely culprit, so I included pipe calculations in my proposal along with a systematic check of the refrigeration circuit.

The project manager responded to my proposal by email. "Do you really expect to spend that much time diagnosing the problem? We are 95% sure the problem is incorrectly specified units. I think we need help assessing the options for addressing the problem we know we have."

"I'm not convinced that high temperature in the attic is the root of the problem," I replied. "I need to find out what is, and that requires a systematic analysis. Until I know exactly what causes the problem, I can't tell you how to solve it."

My calculations showed the loop pump had more than enough capacity, and the balancing report readings were consistent with my calculations. A few measurements at the site confirmed the water side was OK. Then the mechanic and I went up to the attic and got to work on the refrigeration circuit. It took a while to set up our gages and thermocouples so we could monitor the system. It ran for about a half hour before it shut down. The suction pressure was low enough to freeze water. That explained the low pressure lock out. But the suction line temperature was high. So was the discharge pressure, but the compressor discharge temperature looked OK. That all said faulty expansion valve, not ambient temperature. The proposals to ventilate the attic were unlikely to solve the problem.

The contractor ordered new expansion valves. Within a few days, the units were up and running with the new valves. A message arrived in my email inbox on Monday morning. "The building was very comfortable over the weekend. If this trend stands up, the faulty expansion valves seem to have been our true problem over these many months."

Ken Elovitz

May 2013

PAPER OR AIR?

The other day I was in a restroom and saw this sign on the electric hand dryer: "This dryer can dry 25 pairs of hands for the cost of one paper towel." I thought, "Can that be true?" So I did a little math:

A hand dryer uses about 1200 Watts (or 1.2 KW), and I found that it takes about 25 seconds to dry my hands under the dryer. (I timed it, of course.) 1.2 KW x 25 seconds, divided by 3,600 seconds per hour = 0.00833 KWH. At the current average rate of $0.15/KWH, that is $0.00125. (That's 1/8 of one cent.)

OK, how about paper towels? A quick Google search indicated that a box of 2,400 standard bathroom towels costs about $30. That's $0.0125 apiece. Almost exactly 10 times what I figured for the air dryer.

But if you're like me, one paper towel is never enough. (I usually use two.) So, two paper towels cost 20 times as much as using the air dryer. That's pretty close to the 25 claimed by the sign.

But I was just getting started.

The electric dryer itself costs a lot more than the paper towel dispenser and has to be installed and wired by an electrician. If a standard electric hand dryer costs $500 (installed), lasts about 10 years, and gets used 250 times over the normal business day, that works out to about $0.0008/use. The paper towel dispenser is relatively cheap, but it needs to be refilled frequently, and the used paper towels have to be collected and disposed of. If a janitor making close to minimum wage spends 5 minutes refilling the dispenser with 250 towels and emptying the trash, that works out to about $0.007 per hand drying - that's 8 times more than the cost for the electric hand dryer.

But hold on a second - maybe a whole bunch of seconds. I found that I can dry my hands with paper towels in about 10 seconds - 15 seconds faster than the air dryer. (I timed that, too.) What is the extra 15 seconds worth? If you value your time at minimum wage, 15 seconds is worth about $0.04 (about 3 paper towels worth). And if you value your time at professional rates, those 15 seconds are worth 40 paper towels or more!

So electric hand dryers are certainly more economical for the building owner (and maybe better for the environment), but at the cost of a little extra time. Tough call? Next time you see an electric hand dryer next to a paper towel dispenser, you can always just wipe your hands on your pants.

Gary Elovitz

June 2013

THE NEXT GENERATION

As you read this, I just finished my first year at WPI. For those who might wonder how I could have been in practice for 30 some years without a college degree, let me explain. This was my first year *teaching* in WPI's new Architectural Engineering program. Because my duties at WPI are part time, I am keeping my "day job" at Energy Economics.

Architectural engineers study the engineering disciplines that support architecture and building construction. My students are juniors and seniors. Most of them will soon enter the work force, perhaps as engineers in a design firm, project managers or mechanical coordinators for contractors, or facilities engineers.

One challenge for me is to teach students how to use the fundamentals they learned in courses like fluids and thermodynamics to solve real world problems. When the students come to my class, they know how to use CAD (computer aided drafting). They seem surprised that engineers have to see in their mind's eye how the lines on the computer screen relate to ducts and pipes in buildings.

The students also have to adjust their thinking from getting the right answer on the test to understanding how pieces fit together into a functioning HVAC system. Before they take my class, the students seem to expect homework assignments and tests to be the same as the examples in the text book, just with different numbers. I teach them that engineers have to understand the formulas they learn in school and why those formulas work. Then I show the students how to apply the formulas to real building problems.

In the fall, I teach the basics of heating and cooling load calculations and how to convert loads to air and water flows that deliver the required capacity. The next topic is duct and pipe pressure drop, where the students learn the role of friction in real world applications. Then it's fan and pump curves - selecting to meet the calculated pressure drop and diagnosing field data. In my spring class, I add the concepts of diversity and solar variation that affect cooling loads. Then I have a section on psychrometrics or the physics of air and one on building energy use. While my two courses are not enough to let students design the next Hancock tower, they provide a good foundation for entering the engineering work place.

When I stumbled upon the ad for this new position, the job description called my name. It's an opportunity for me to share what I know and to influence the future of the industry. With my first batch of students now "out of the oven", it's good to know that the type of engineering Gary and I do will not become a lost art.

Ken Elovitz

July 2013

A QUICK TRIP

The newly opened, upscale restaurant had three 36 KW electric water heaters. They thought they would never run out of hot water. What a surprise when one of the brand new heaters "shorted out". The loss of hot water would have been bad enough, but the water heater didn't just trip its own 60 amp breaker. It also blew the 400 amp main fuse, plunging the whole place into darkness. The first time it happened, the contractor replaced the fuses, chalking the problem up to a start up anomaly.

When it happened again 10 days later, the general contractor called the water heater service tech. The first thing that guy did was turn the heater back on, blowing the main fuse a third time. That gave him the bright idea to check the wiring in the water heater. He found that the heater had been miswired at the factory. He fixed the wiring, but since the main fuse had blown, he announced that "the building's electrical system, or at least the portion supporting the heaters, may not have been sized properly to handle the load".

The water heater salesman was in the spotlight and asked for help. He was confident the water heater was fixed and wouldn't trip again, but everyone was irate that the main fuse blew. Shouldn't the breaker at the water heater be the only one to trip? And if by some chance the water heater does fail again, how can we be sure the main fuse won't blow?

The time to blow a fuse or trip a breaker depends on current flow. The higher the current, the faster the trip. Each device has its own characteristic. A 400 amp fuse should hold 400 amps almost indefinitely, but a 60 amp breaker exposed to 400 amps should trip in 5 to 25 seconds. Either device should trip in hundredths of a second if exposed to thousands of amps, as might occur in case of a short circuit.

How do engineers design systems so short circuits trip the fuse or breaker nearest the load and not the main? We perform coordination studies, which is a fancy way of saying we plot the performance curves for all devices in a circuit on the same graph. In this case, the curve for the 400 amp main fuse crossed the curve for the 60 amp breaker at 4800 amps. That means any current over 4800 amps will blow the 400 amp main fuse before the 60 amp local breaker can trip. A short circuit could easily exceed 4800 amps, so another short at the water heater was bound to blow the main again. But if they replace the 60 amp breaker with a faster device like a 60 amp fuse of the same type as the 400 amp main fuse, another short at the water heater won't leave the whole place in the dark.

Ken Elovitz

August 2013

UP OR DOWN?

The historic renovation of a small office building was supposed to be a model of comfort and energy efficiency. In the first few years, it achieved neither. Losing out on energy efficiency wasn't so bad, but the consistent over heating in the areas with in-floor radiant heat drove the facility manager nuts.

A previous study had recommended adjusting the water temperature control so the radiant heat would always offset 3/4 of the heating load. With that approach, the radiant heat would run continuously during the heating season, keeping the floor warm without overheating those rooms that have internal heat gain from lights, people and equipment. Fancoils would trim the heating capacity, adding heat if needed during occupied hours and shutting off at night. No need for night setback on the radiant, because it will heat the space to 52F or 55F when the fancoils are off. A simple and elegant solution, but the building was still too hot.

The service contractor was convinced the piping was wrong. The heating system had a transfer loop in the boiler room. Two boilers, each with its own pump, add heat to the loop as needed. Seven loads (fancoils, baseboard, and in-floor radiant manifolds), each with their own pumps, withdraw heat from the loop as needed. The wrinkle was that the design called for a pump to circulate water through the transfer loop, but that pump was never installed. As a result, flow in the transfer loop and the water temperature available to each of the seven loads would vary depending on which pumps were on. I could see how omitting the transfer pump might result in under heating, but the complaint was over heating. The missing pump did not explain the problem.

The next step was a site visit to check the thermostats and control valves. Raising the thermostat setpoint created a call for heat, applying power to the valve actuator. The valve stem moved up, and the pipes got hot, so we knew the valve was open. Lowering the thermostat took power off the actuator. The stem moved down, so the valve closed. That all looked OK - we didn't find a mechanical problem. But, wait. The energy management computer had it backwards! The computer showed the valve closed when it was actually open. Every time the room temperature reached setpoint, the program thought it was commanding the valve closed but actually drove it open, adding more heat to the room. A simple programming change was all it took to stop the over heating.

Ken Elovitz

September 2013

QUIET AND PEACE?

"The neighbors are complaining about noise from our new rooftop unit," the synagogue president sighed. "We need to know whether we are violating any noise ordinance."

That was pretty simple: the City's ordinance said noise more than 5 decibels (dB) above the background noise level is considered noise pollution. I chose a quiet day, measured background noise level at the neighbor's property line, then turned on the rooftop unit and measured the noise level again. The increase was about 4 dB, so the synagogue was in compliance. I made my report and figured I would never hear from them again.

A couple of months later I got another call. "The neighbors are still complaining about the noise, and they went to the Aldermen. The Aldermen are pressuring us to do something. Our contractor got a proposal to install sound-absorbing panels around the rooftop unit on the roof. Before we spend 5 grand, we need to be sure it will work. Can you look at it?"

The first thing I noticed was that the proposal was from a manufacturer's rep who sold acoustical panels, not from an acoustical engineer. But I was even more skeptical about the details: the proposal was to install 3 foot high acoustical panels around the rooftop unit. This "fence" would block the unit from view from next door, but noise doesn't always follow line of sight: sound waves can diffract around obstacles, so the fence would not solve the neighbors' problem.

I took a closer look. Most of the noise came from the big propeller-type condenser fans. Like most commercial rooftop units, these fans had stamped aluminum blades. Manufacturers use these fans because they're cheap, and every penny counts when you compete to sell equipment. However, these fans are not very efficient, and that makes them noisy. For units on the roofs of strip-malls and office buildings, outdoor noise is not much of an issue, so the cheap fans are not a problem. But this synagogue was in the middle of a residential neighborhood.

If we could replace the noisy fan blades with more efficient blades, we could reduce the noise at the source. I contacted a fan blade manufacturer, gave them my parameters (airflow, fan diameter) and asked them to select a quieter fan. They came up with a fan that would move the same amount of air at 30% slower speed and about 40% less power. We could easily slow the fans down with a low cost speed controller. The new blades would be a lot quieter, and the total cost - installed - was under $700.

With the new fan blades installed, my noise meter did not even register an increase in noise when the rooftop unit started. The synagogue even got a thank-you note!

Gary Elovitz

October 2013

THE ENERGY IN FOOD

As Energy Guys, we're always thinking about BTUs and KWH, mostly as they relate to HVAC and electrical systems in buildings. That's far from the only measure of energy use in our lives. Anyone shopping for a new car probably pays at least some attention to miles per gallon. And some of our colleagues who emphasize sustainability try to compare the energy input of different construction materials.

Does anyone think about the amount of energy it takes to create the food we eat? The editors of IEEE's *Spectrum* magazine did. Their June 2013 issue reports on a study from the KTH Royal Institute of Technology in Stockholm on the energy input (including pesticides, fertilizer, harvesting, and transportation) for various foods.

Shelled shrimp was at the top of the list, with an energy input of 220 megajoules/kilogram (MJ/kg). One megajoule is the amount of energy in 30 ml or 1-1/2 teaspoons of gasoline, so getting a pound of shrimp from sea to table requires the energy content of about a tablespoon of gasoline. The high number for shrimp is mostly due to fuel intensive use of running a shrimp boat.

Beef is further down the list but is still pretty energy intensive, coming in at 70 MJ/kg. On the other hand, milk is one of the most energy efficient animal products. Ice cream wasn't rated.

Poultry eaters don't do too badly. Chicken has an energy content of 40 MJ/kg. Vegetarians don't necessarily get off easy. Farm raised salmon is over 80 MJ/kg - even more than beef.

Substituting fruits and vegetables for meat and fish, as the dietitians tell us to do, is a mixed bag. Tropical fruits like mangoes are the second highest item on the list at 110 MJ/kg. Their high number comes from transportation costs because they are shipped by air. The energy input for tomatoes depends on the time of year. Those greenhouse tomatoes we get in winter cost 60 MJ/kg, but the fresh, locally grown tomatoes we eat in summer are way down at 5. A similar story for apples. Imported apples come in at 9, but locally grown apples are the lowest item on the list at 3.5 MJ/kg.

So what do I make of all of this? I think it's saying I need another slice of apple pie.

Ken Elovitz

November 2013

WHEN ALL ELSE FAILS

The new wing of the assisted living facility opened in late spring. The mechanical contractor was a regular visitor during the summer, responding to cooling outages. He attributed them to normal start up "glitches" and assured the owner that everything was OK. Then when fall arrived, the building was sweltering. The facility manager was convinced the design and installation were flawed and refused to release final payment.

I got to the site on a sunny, 40 degree day in December. The south facing rooms were warm. Solar gain explained that problem but not the sauna-like conditions in the common areas. Even though the fancoil unit fans were off, hot air wafted out of the supply air grilles.

Popping out a few ceiling tiles gave me access to the fancoils above the ceiling. Hot water rushed through the coils. Wriggling between the ceiling grid and the pipes, I could see that each fancoil had a control valve. If the space was so hot, why didn't the control valve close? It probably would have, if the actuator had been wired. "Oh, we didn't forget about those valves," the contractor explained. "Last summer we kept getting low flow alarms at the chiller. So we disconnected those valves to keep the chiller running. We just forgot to hook them back up when we did the summer/winter changeover."

The idea of disconnecting and reconnecting valve actuators didn't impress me as a smart practice. There had to be an engineering explanation for the problem and an engineering solution for it.

The building had 27 fancoil units. 20 had 2-way control valves, so their water flow varied with the load. Anywhere from 29 gpm if all called for cooling, all the way down to zero if none called. The other 7 fancoils had 3-way valves. The flow through those fancoils was a constant 20 gpm: Their valves would direct water through the coils on a call for cooling and around the coil if the thermostat was satisfied. Engineers sometimes put a few 3-way valves at the ends of the branches on 2-way valve systems so water does not stagnate and warm up during light loads.

The 3-way valves provided 20 gpm of flow, but the 2-way valves could go to zero, so the system flow could be as low as 20 gpm. But the manufacturer's data showed that the chiller needed at least 23 gpm of flow for adequate tube velocity and load to prevent freezing. No wonder the chiller tripped out!

The fix was easy: change the 2-way valves on 4 more fancoils to 3-way valves to up the minimum flow by 6 gpm. Another case of "when all else fails, read the directions".

Ken Elovitz

December 2013

A SWITCH IN TIME

"We had a meeting about the problems with the new dryer venting system. The dryer vendor says his dryers are fine, the HVAC guy says he installed everything as designed, and the engineer says his design is OK. But my people can't get the dryer to run. Can you give us a hand?"

Three commercial gas dryers in the basement vent to a common exhaust stack with a fan on the roof, 4 floors up. The fan is controlled to maintain a slight suction where the dryers connect to the stack. The laundry workers said the dryers would not start unless the exhaust fan was tuned off. So every time they wanted to start a dryer, they had to shut off the exhaust fan, start the dryer, and then restart the exhaust fan.

I asked the worker to try starting a dryer without shutting off the fan. The dryer fan ran for a minute or so, then shut down and the dryer displayed: "Ignition Failure". That at least pointed me in the direction of the gas burner. I started looking around the burner, and had the laundry worker start and stop the dryer with the roof fan both off and on, so I could look for differences. My attention was pretty quickly drawn to a 3" metal disk hanging in front of a hole next to the dryer's air intake. When the dryer's internal fan started, the dryer fan's suction pulled the disk in and covered the hole, tripping a microswitch in the process. This was clearly a safety feature: if the disk didn't trip the microswitch, that would mean the dryer fan was not running (so the dryer shouldn't light the gas). The manufacturer didn't want anyone to be able to bypass the safety control by securing the microswitch in the tripped position permanently, so the dryer's controls do not let the dryer start unless the microswitch is in the "off" position before the dryer fan starts.

I noticed that with the roof fan on and the dryer off, there was a weak suction through the dryer that pulled the metal disk about half-way towards the hole, just enough to trip the microswitch! The microswitch just needed a little adjustment so that it would not trip until the disk was almost touching the hole.

I probably should have written up a report of what I found, which probably would have led to another big meeting on-site, and the dryer vendor might eventually have sent a technician to adjust the microswitches. Or I could make slight adjustment to the switches myself, check the dryers to make sure they all ran OK, and call it a day. Which would you do?

Gary Elovitz

January 2014

to the tune of
WHITE CHRISTMAS

I'm dreaming of a mild winter
To keep my heating bill at bay.
   With no major blizzards
   or power failures
And snow that quickly melts away.

Long drought gave way to massive flooding.
And wildfires raged throughout the west.
   But our New England
   was spared much trouble
As hurricanes gave us a rest.

Last April was the Boston bombing.
In Oct. the government shut down.
   In between was summer,
   Oh what a bummer
When giant A/C bills came round.

If your building is a real nightmare;
All of your systems misbehave.
   You've got steam traps hissin',
   And fan belts missin'
We can help you save the day.

I'm dreaming of new solar panels
With every payment check I write.
   Can I save the day
   With PV arrays?
Will S-RECs give me R-O-I?

Don't forget about A/C maintenance:
Keep filters clean and ductwork tight.
   To your bearings listen;
   Don't be remiss in
Adjusting everything just right.

Pumps running 24-7,
Could it be time for VFDs?
   And as water heaters
   Spin those fuel meters,
Wishing won't make them go away.

Plug chargers into strips with switches
And then turn off those extra lights.
   Drop incandescents
   And use fluorescents
Saves watts and still be bright.

I had to take a tepid shower.
My water heater runs a lot.
   With some piping movements
   And pump improvements
I got that water hot.

If you encounter heating troubles
If A/C simply will not cool.
   Then by all means call us;
   and we'll bring solace
For we'll apply the proper tools

With hopes that you will all stay healthy
And that your homes be warm at night
Comes this wish from Gary and Ken
That your year be prosperous and bright.

Ken and Gary Elovitz

February 2014

DIAL IT DOWN

"We have a job with radiant floors," the general contractor's email read. "They are having a lot of trouble and want us to pay to add bypasses at each manifold. Can you help us figure out what really needs to be done?"

The project was a product manufacturer's training center. The design drawings called for hot water supply and return stubs to connect to radiant heat "by others". The manufacturer designed the radiant heating system to showcase its products.

Soon after the building opened, excess heat poured out of the radiant floor. To disable the sauna, the training manager eventually shut the radiant heat off. To make matters worse, the interface between the main hot water loop and the mixing stations that control the radiant heat did not have the manufacturer's recommended crossover piping. Apparently the base building designer thought the crossover was part of the manufacturer-designed radiant heat, and the manufacturer considered the crossover a part of the base building system. Neither one talked to the heating contractor, who installed the system his own way.

The training manager insisted the only way to make the system work was to install the crossovers. He also said the system was useless as a product demonstration unless it was installed the way the manufacturer recommends. The contractor insisted he installed the parts and pieces the manufacturer supplied in accordance with the plans he bid. He didn't see why he should add the crossovers unless he got paid extra. Impasse!

The hot water system had a variable speed pump that maintained a constant pressure difference between supply and return to push design water flow to each load. The pump was set for 15 psi difference at the very end of the system. That means the pressure difference was at least 15 psi everywhere else, including at the mixing stations. The problem was that the 3-way valves in the mixing stations were rated for 7 psi pressure difference. At any more than 7 psi difference, the force of the water trying to open the valve would overcome the closing force of the actuator, pushing hot water into the sub-loop whether needed or not. Ergo, over heating.

Analysis showed that the reheat coils on the job need less than 5 psi pressure difference to achieve their design flow (and design heating capacity). Restoring control to the radiant sub-loops was simple: just dial the pressure difference down to 6 psi - enough for the reheat coils with a little cushion, but not enough to blow the 3-way valves off their seats.

Ken Elovitz

March 2014

HUMIDOR

Walking through the corridor of the newly completed apartment building felt like trudging through a rain forest. My handy meter read 75% relative humidity; I almost expected to see water dripping from the wall sconces and hear the chatter of monkeys in distant trees.

It wasn't supposed to be like this. I know that humidity control in apartment corridors is a challenge, especially when the corridor system handles a lot of outdoor air for ventilation. There isn't a lot of cooling load in the corridor, but outdoor air can be quite humid, so the design called for a rooftop unit with two stages of cooling and "hot gas reheat". When the space needs dehumidification but not much cooling, one cooling compressor would run, making half of the cooling coil cold enough to condense moisture from the air. Instead of discharging the waste heat outdoors (like we do for a standard cooling system), this unit had another refrigerant coil right after the cooling coil to reheat the supply air enough to keep the corridor comfortable. So - up to the roof for a closer look.

At first glance everything looked OK. One compressor was running and one was idle. The suction pipe on the active compressor was certainly cold enough, so I knew that refrigerant charge was not the problem. But there was barely a trickle of water dripping from the condensate drain.

When I opened the access door to the coil section, I could see condensate glistening on the whole cooling coil; there was clearly plenty of condensation going on. But wait a minute - why was the whole coil wet? Only one compressor was running, so only half the coil should have been wet.

I checked the suction pipes from the coil; the top half of the coil was cold, and the bottom half was not, so the active compressor was the one that fed the top half of the coil. Aha! The top half of the coil was active, and it was cold enough to condense a lot of moisture from the air. But that moisture had to drip down through the bottom half of the coil to get to the drain. Warm air flowing over the wet (but idle) bottom half of the coil reevaporated most of the moisture into the supply air stream! That moisture made the air leaving the rooftop unit higher relative humidity than the outdoor air (because it was cooler). All we needed to do was fix the controls so the compressor for the lower half of the coil would run first. Then the corridor would be cool and dry.

Gary Elovitz

April 2014

BEING GREEN

Checking out at the store, the young cashier suggested to the much older woman that the woman should bring her own grocery bags because plastic bags aren't good for the environment.

The woman apologized and explained, "We didn't have this 'green thing' back in my earlier days."

The young clerk responded, "That's our problem today. Your generation didn't care enough to save our environment for future generations."

Maybe the clerk was right. Back in the '50s and '60s we didn't have recycling. Instead, we returned milk, soda, and beer bottles to the store. The store sent them back to the plant to be washed and sterilized and refilled, so we reused the same bottles over and over.

No one took their own bags to the grocery store. Instead, grocery stores hired teenagers to bag our groceries in brown paper bags that we reused in many ways. Most memorable besides household trash bags was using brown paper bags as covers for our schoolbooks. Then we personalized our books on the brown paper bags. But that wasn't officially "green".

Our houses had one TV, not one TV in every room. And the TV had a screen the size of a handkerchief, not a screen the size of the state of Montana. We dried clothes on a line, not in an energy-gobbling machine -- wind and solar power really did dry our clothes back in our early days.

When we packaged a fragile item to send in the mail, we cushioned it with wadded up newspapers, not Styrofoam or plastic bubble wrap. We didn't call it recycling, but that's what it was.

Back then, kids rode their bikes or walked to school instead of turning their moms into a 24-hour taxi service in the family's $45,000 SUV or van.

Isn't it sad the current generation laments how wasteful we old folks were just because we didn't have the "green thing" back then?

We don't like being old, so it doesn't take much to tip us over ... especially when it comes from tattooed, multiple pierced, wisenheimers who can't make change without the cash register telling them how much.

(from an old time engineer who often shares knowledge and wisdom with me)

Ken Elovitz

May 2014

COLD IS RELATIVE

As you read this message in April or May, remember how cold it was last winter? Cold, however, is a relative thing:

60 above zero:
  Floridians turn on the heat.
  South Dakotans plant gardens.

50 above zero:
  Californians shiver uncontrollably.
  People in Sioux Falls sunbathe.

40 above zero:
  Italian & English cars won't start.
  South Dakotans drive with the windows down.

32 above zero:
  Distilled water freezes.
  The water in Pierre gets thicker.

20 above zero:
  Floridians don coats, thermal underwear, gloves, wool hats.

  South Dakotans throw on a flannel shirt.

15 above zero:
  New York landlords finally turn up the heat.

  South Dakotans have the last cookout before it gets cold.

Zero :
  People in Miami all die.
  South Dakotans close the windows.

10 below zero:
  Californians fly to Mexico.
  South Dakotans get out winter coats.

25 below zero:
  Hollywood disintegrates.

  The Girl Scouts in South Dakota are selling cookies door to door.

40 below zero:
  Washington DC runs out of hot air.

  South Dakotans let the dogs sleep indoors.

100 below zero:
  Santa Claus abandons the North Pole.

  South Dakotans get upset because the
    mini-van won't start.

459.67 below zero (absolute zero):
  All atomic motion stops.

  South Dakotans start asking, "Cold 'nuff fer ya?"

And last not but not least, at 500 below zero:
  Hell freezes over.

  South Dakota public schools open 2 hours late.

Ken Elovitz

June 2014

DIVINE INTERVENTION

The brand new synagogue building bustled with activity. There was only one problem - energy use was through the roof. The chairman of the "environment" committee was bewildered. "We compared energy use in our new building with our old building, and the increase is huge, even after adjusting for size. The engineers told us the new VAV systems would be super energy efficient, but we're using more energy! Our gas use has really exploded. We're even using a lot of gas in the summer when the heat should be off! Can you help us?"

"I think I can," I replied. (In fact I was pretty sure I could, if my initial hunch proved right.) "Send me the plans and I'll take a look."

It took less than an hour to confirm my hunch: the minimum airflows for the VAV boxes were all set at 50% of the maximum airflows, and all of the VAV boxes had hot water reheat coils. VAV systems save energy by reducing the amount of cooling delivered to each space when there is less load. But with minimum airflows at 50%, any time the cooling load in a zone is less than half of the peak, the system still delivers 50% of peak cooling capacity. The hot water heat has to come on to keep the space from overcooling. Some minimum level of airflow is usually needed for ventilation, and sometimes higher airflows are needed for heating, but 50% is pretty high, and my analysis confirmed that the minimum airflows for ventilation could be much lower.

To be sure I was on the right track, I needed to see how the system was really running. Fortunately, the building has a computerized control system that gave me (with the right passwords) access over the Internet and a high degree of control.

I logged in from my office and confirmed not only that the boxes were operating with high minimum airflows (per the original design), but that the standard programming for the boxes allowed for separate settings for "minimum" airflow and "heating" airflow. That meant the boxes could close to a low minimum airflow to meet ventilation needs, and then open up to a higher "heating airflow" when the space needed heat. Since the engineers didn't specify different minimum and heating airflows, those two values were set the same in each box.

It was pretty simple to change the minimum airflow settings to what was needed for ventilation and change the heating airflow settings to what was needed for heating. The result? Gas use dropped by 40%, and electricity use dropped by about 15%, saving over $20,000 in one year.

Gary Elovitz

July 2014

STEP BY STEP

The new academic building failed its smoke control test. Investigation revealed that the engineer forgot to call for two smoke exhaust fans to have generator power. That oversight triggered a review of the entire generator power system. The university's engineer said they had to replace the existing 600 kW generator with a 1000 kW machine. The designer was adamant that 600 kW would do the job. Who was right?

The answer requires knowing a little about generators and motors. When the lights go out, the electrical code says egress lights and the fire alarm system have to be back on line within 10 seconds. The code allows 60 seconds to restore power to other required loads like smoke exhaust fans and fire pumps.

When a load switches on or off, the change in current causes a corresponding change in voltage. Utility power sources are usually "stiff" enough that the voltage change is imperceptible. But sometimes even a household load like a washing machine or a garbage disposer makes the lights flicker. The lights dim when the voltage dips and return to normal when the voltage bounces back. The amount of flicker depends on the size of the load compared to the size of the power system. For the most part, flicker is little more than an annoyance.

On-site generators are much smaller than utility sources so are more susceptible to voltage dip. One way to reduce voltage dip is to start motors in steps, letting the generator recover from one "hit" before starting the next group. That's what the design engineer did for the new academic building: Step 1 was the lights and fire alarm, step 2 the fire pump, and step 3 the smoke exhaust.

The fire pump is a big load, and engineers know that starting the biggest load first usually results in a smaller generator. But fire pumps are a special case. The electrical code limits voltage dip for fire pumps to 15%. Once the fire pump is on line, the voltage dip is limited to 15% for subsequent steps. With the egress lights and fire pump running, the 600 kW generator could not start the smoke exhaust fans with less than 15% voltage dip.

Fire pumps and smoke exhaust fans have the same priority. There is no rule about which to start first. What if we start the smoke exhaust fans in step 2, allowing the customary 25% voltage dip, and make the fire pump step 3, with its 15% voltage dip limit?

Sure enough, the special 15% voltage dip limit for fire pumps made the 600 kW generator look too small. Simply making the fire pump step 3 instead of step 2 showed that the 600 kW machine could do the job.

Ken Elovitz

August 2014

NO FREE LUNCH

The internet is an interesting place, packed with of all kinds of surprising and useful information. For example, I have seen a number of links to articles and videos that show how you can heat your home for "pennies a day" with simple "tea light" candles, a few flower-pots, and some inexpensive hardware. (Try Googling "flower pot heater".) This cried out for more investigation, and I put on my thinking cap.

The basic principle is to put something with a bit of thermal mass over a candle to keep the heat from the candle down in the occupied zone of the room, rather than allowing the heat from the flame to rise straight up to the ceiling where it doesn't provide much benefit. That makes sense, as far as it goes.

A tea light candle weighs about 1/2 oz. and burns for about 3 hours. Paraffin wax has a heating value of about 19,000 Btu/lb, so each tea light can release 622 Btus of energy (if it burns completely, which it probably won't). That's about the same amount of heat that a 50 watt electric light bulb generates. Could you really shut off your heating system and heat your home with a 50 watt light bulb (or two, or three)?

Then there's cost.

A therm of natural gas contains 100,000 Btus, so a tea light is equivalent to 1/161 therm of gas in terms of energy content. At current gas prices - say $1.50/therm - a tea light contains the same amount of energy as about $0.0093 (a little less than 1 cent) worth of gas. Unless you can buy tea lights for less than a penny apiece, it will be cheaper to heat your house by with gas. (Needless to say, it is a lot safer to run your boiler or furnace than to keep several candles burning.) If you have oil heat, the economics of candle heaters are slightly better - since oil is about twice the cost per Btu as gas, you could spend up to 2 cents per tea light for them to be competitive with oil.

What about electric heat? The energy content of our tea light is roughly equal to 0.16 KWH of electricity. At current rates (roughly $0.15/KWH) that is worth about 2.4 cents.

So, how much do tea lights cost? The cheapest I could find online was $10 for 125 candles - 8 cents per candle. So heating with tea lights is 8 times more expensive than gas and 3 - 4 times more than oil or electricity. Tea lights may be more romantic than your furnace, but I wouldn't suggest stocking up on tea lights for the winter.

Next month on the internet - how to get your car to run on plain water.

Gary Elovitz

September 2014

WORDS OF WISDOM

Bad decisions make good stories. Otherwise stated: Good judgment comes from experience; experience comes from bad judgment.

How many times is it appropriate to ask a speaker to repeat before you just nod and smile because you still didn't hear or understand what they said?

Time is more precious than money. Once you've spent it, it's gone, and you can never get it back.

It's unwise to pay too much, but it is worse to pay too little. If you pay too much, you lose a little money. But if you pay too little, you sometimes lose everything because the thing you bought does not do what it was bought to do.

WHAT'S UP

We've all heard that "if" is the little word with the big meaning. How many of us have stopped to think about another 2-letter word with multiple meanings?

We call UP our friends. And we brighten UP a room or polish UP the silver. We warm UP the leftovers and clean UP the kitchen. We lock UP the house and some guys fix UP the old car.

People stir UP trouble, line UP for tickets, work UP an appetite, and think UP excuses. To be dressed is one thing, but to be dressed UP is special.

A drain must be opened UP because it is stopped UP. We open UP a store in the morning but we close it UP at night.

We seem to be pretty mixed UP about UP! To learn about the proper uses of UP, look it UP in the dictionary. In a desk-sized dictionary, it takes UP almost a quarter of the page and can add UP to about thirty definitions. If you are UP to it, you might try building UP a list of the many ways UP is used. It will take UP a lot of time, but if you don't give UP, you might wind UP with a hundred or more.

When rain is on the way, we say the sky is clouding UP. When the sun comes out we say it is clearing UP. When it doesn't rain for a while, things dry UP.

I checked with Gary, and he didn't come UP with anything to add, and I've used UP my space, so I'll wrap it UP for now.

Ken Elovitz

October 2014

WD-40

In November 2005, our calendar message told you about duct tape, sometimes known as the universal sticking agent. Despite its name, duct tape is good for joining almost anything except HVAC ductwork, though I have a distinct memory as an 8- or 9-year old helping my father assemble ductwork in a neighbor's attic using 4" wide grey duct tape. Maybe the formulation changed.

Removing today's duct tape leaves behind a sticky residue. Getting rid of that residue calls for application of the universal unsticking agent, better known as WD-40.

Besides unsticking duct tape, here are some other uses for WD-40 that have been reported:

- Unstick a stuck zipper
- Lubricate locks
- Smooth the operation of sliding doors and windows
- Ease umbrella operation
- Loosen rusted nuts and bolts
- Speed operation of prosthetic limbs
- Polish shoes (don't know if it waterproofs them as well)
- Prevent silver from tarnishing
- Prevent bathroom mirrors from fogging
- Clean grout
- Clean guitar strings
- Clean stainless steel sinks
- Remove crayon marks from walls
- Remove graffiti and other paint
- Remove lipstick stains
- Remove scuffs from floors
- Remove chewing gum
- Remove construction adhesive from hands
- Quiet squeaky rocking chairs
- Quiet squeaky door hinges
- Repel pigeons
- Keep flies off cows
- Dry out wet cell phones
- Drive moisture out of wet wiring
- Cleans and protects windshield wiper blades
- Degrease barbecue grilles
- Heal insect bites

WD-40 was invented in the 1950s in a search for rust prevention solvents and degreasers for the aerospace industry. WD stands for "Water Displacement", and 40 supposedly means the product was the 40th attempt.

WD-40, along with duct tape, is one of the most versatile products of our time. For more about WD-40, including the rest of the list of more than 2000 uses for the product, visit their web site at wd40.com.

Ken Elovitz

November 2014

PRESSURE!

In May 2010, we wrote about the Great Molasses Flood in Boston in January 1919. The flood released a 40 foot high wave of molasses that moved as fast as 35 MPH through the streets of Boston's North End. 21 people and several horses died, either crushed or drowned in molasses.

Another food-related disaster occurred 200 years ago last month. The Horseshoe Brewery in London had a wooden vat held together by iron hoops. The vat contained 135,000 Imperial gallons of beer until about 6 PM on October 17, 1814. That was when one of the hoops flew apart. The explosion sent beer, metal, and wood across the area. A chain reaction damaged other vats at the brewery and released a total of 323,000 gallons of beer that demolished 2 houses. 9 deaths were attributed to drowning or injuries from the flood. One person died of alcohol poisoning.

Those two incidents were both pressure vessel failures. We tend to think of pressure vessels in terms of industrial processes and heating equipment like boilers and water heaters. But we are surrounded by many types of pressure vessels. Think of a kernel of popcorn. The shell is a tiny pressure vessel that contains a small amount of water. When heated to 400 degrees, the water converts to steam and expands. Then the "pressure vessel" bursts. The kernel turns inside out and becomes nearly 40 times its original size.

Did you ever think of a watermelon as a pressure vessel? In the spring of 2011, farmers in Danyang in eastern China sprayed their crops with a growth regulator. The chemical made the fruit expand so rapidly that internal pressure built up beyond what the rind could contain, and the watermelons exploded. The farmers wound up with 115 acres of exploding watermelon "land mines". There were no reported injuries, just a loss of valuable crops.

Nowadays, engineers have safety codes and can specify devices like pressure relief valves to help protect against damaging boiler explosions. When properly sized and installed, pressure relief valves sense the pressure inside a vessel and release expanding fluid before the pressure builds beyond what the vessel can contain. These automatic devices are a huge improvement over the days when boiler safety relied primarily on the skill and attention of the operator.

To learn more about exciting and interesting pressure vessel failures, check out the book *Blowback* by Paul Brennan, "an anecdotal look at pressure equipment and other harmless devices that can kill you!"

Ken Elovitz

December 2014

FREE LUNCH

Ever take a close look at your electric or gas bill and notice a small "Energy Conservation" charge - about a dollar or two a month? That money goes into a fund that the utilities use to pay for energy conservation projects for customers, including rebates for new high efficiency equipment. In Massachusetts the program is administered by MassSave; many other states have similar programs.

One program available to residential customers in Massachusetts is a free energy audit. I signed up for an energy audit last winter (through NextStepLiving.com). The technician came equipped with an infrared camera and checked all my exterior walls for insulation and air leaks. Since it was cold outside, the infrared camera showed warm and cold spots on the walls quite vividly. My walls were pretty well insulated, though the technician found a few small areas that the insulators had missed. (I don't know what they do if it isn't cold outside - it is definitely easier to check for insulation when there is a big temperature difference between indoors and outdoors.) The technician also measured the efficiency of my boiler and looked for old inefficient incandescent lights to replace with efficient lighting. (He didn't find any, of course.)

The technician also came equipped with a laptop and printer, and at the end of the assessment he printed out specific proposals for adding insulation and doing air-sealing, complete with detailed pricing. The utility pays a portion of the cost of the insulation and air-sealing, so the energy savings typically recover the cost of the insulation and air-sealing in 3 years or less. For larger projects, the utility can arrange no-interest loans to spread the cost out over a number of years.

Another program is installation of ductless mini-split systems that provide both heat and air-conditioning. If your home has oil heat or electric heat - and especially if you don't already have central air-conditioning or your systems are very old - ductless mini-splits could be a good way to reduce heating costs and add air-conditioning at the same time.

The technician left me with proposals for some insulation and air-sealing upgrades, plus a "goody bag" with a bunch of compact fluorescent and LED lamps and a "smart" power strip that turns off unneeded loads - like a DVD player and cable box - when the main load (TV or PC) is shut off.

So if your home is poorly insulated, or your heating system is old and inefficient, or you haven't switched to efficient lighting, investing a few hours of your time in a free energy audit could yield warm returns.

Gary Elovitz

January 2015

'TWAS THE NIGHT BEFORE

"Twas nigh onto New Year's
and all through the town,
The two Energy Guys
had to write something down.

The calendar message
was due in a week,
So Ken said to Gary,
"of what shall we speak?"

As Twenty-fourteen
brings itself to a close,
We look back at the year
with its highs and its lows.

Olympics in Sochi!
It looked like they might
Not get it together - but
they came out alright.

Skiers raced downhill
at speeds 80-plus.
Not bad when you know they
were skiing on slush.

But then came Crimea,
unrest in Ukraine.
Followed by Gaza;
Mid-East up in flame.

And after a summer
as hot as could be,
The electric suppliers
near doubled their fees.

The virus ebola
took over the news.
If you had a fever
your flight was refused.

The stock market climbed
'til it reached a new high.

Then in October
it fell from the sky.

November's election
brought numerous ads.
Each candidate claimed,
"My opponent's a cad!"

As winter rolls in,
what will be in store?
Snow piled so high
I can't open my door?

But (no thanks to OPEC)
our fuel bills are down,
With luck we'll pull through
until spring comes around.

What a year we lived through!
Some good; a lot frightful.
Is it too much to ask
that next year be delightful?

So the Energy Guys
want to say in your ear,
"Best wishes to all
for a wonderful year!"

Ken and Gary Elovitz

February 2015

LONG DISTANCE CALLING

Our friends from Phoenix called one evening hoping I could help with a noise problem in their new air conditioning unit. Their service contractor had enough foresight to recommend they replace an aging 30-year old unit before it failed. But beginning on day one, the new unit was so noisy that it often woke our friends up in the middle of the night.

The service contractor responded to the noise complaint by installing a sound blanket. The sound blanket helped but did not solve the problem. Our friends didn't know what to do, so they called an EnergyGuy for advice.

Trying to diagnose a system I had never seen with a noise problem I had never heard wasn't easy. I had to figure out whether the noise might be the compressor, the condenser fan outdoors, vibration broadcasting into the room below, or the supply fan in the unit. After several questions and answers, I narrowed the problem down to something in the ductwork - maybe even something as simple as a damper blade rattling. My advice was to have the mechanic take the diffuser core out of the guest room outlet and stick his head into the duct to see what's going on.

A few weeks later, a card showed up in my mailbox.

"We passed along your suggestions to the tech who came the next week. As soon as he actually heard the air coming through the guest room vent, he said the noise was due to its being the closest outlet and getting the full thrust of the air. He inserted a plate to distribute the air along the ductwork, and it was like magic! There was immediate quiet in the guest room without loss of cooling there or anywhere along the line. Your willingness to talk to us and your calm analysis gave us much optimism. We're now happy with our unit."

The bottom line is a principle that Gary and I learned long ago and try to remember to apply: symptoms (in this case, the noise) provide clues to help identify and find the problem, but treating symptoms (in this case, the sound blanket) probably won't do the job until you find the underlying problem.

Ken Elovitz

March 2015

MORE RANDOM KINDS OF FACTNESS*

Ian Fleming was an avid bird watcher. He named his famous spy, James Bond, after the author of his favorite bird identification book, *Birds of the West Indies*. Fleming also wrote the children's classic, *Chitty Chitty Bang Bang*.

Blood travels at about 0.7 mph. It takes about 20 seconds for a drop of blood to travel from the heart, through the body, and back to the heart.

Particles ejected from a sneeze can move at more than 100 mph.

Pencils were originally cut in a hexagonal shape so they would not roll off tables.

In 1858 Joseph Rechendorfer received a patent for attaching an eraser to the end of a pencil.

The official name for the part of a jigsaw puzzle piece that sticks out is a nub. The part that accepts the nub is called a void.

Levi's jeans first came on the market in the 1800's. Lee Jeans were the first to have a zip fly in 1926.

The pressure inside a champagne bottle is about 90 pounds per square inch.

The Bible mentions two nuts by name: almonds and pistachios.

In 1582, Pope Gregory XIII eliminated the 10 days from October 5-14 to get the calendar back in sync with the solar year.

A phone number that spells a word is called a numerym.

Keystones, quoins, dentils, lozenges, and garlands are all types of architectural details.

A round window is called an oculus (Latin for "eye").

There is no English word that rhymes with orange. Nor with purple.

Dr. Seuss created the word "nerd" in *If I Ran the Zoo*. His nerds were animals, not geeks.

mostly from a book of the same title by Erin Barrett & Jack Mingo

Ken Elovitz

April 2015

INSPIRATION

Never interrupt someone doing what you said couldn't be done.
Amelia Earhart

If you do what you've always done, You'll get what you've always gotten.
Anonymous

It is not because things are difficult that we do not dare; it is because we do not dare that they are difficult.
Seneca

A ship in harbour is safe, but that is not what ships are built for.
William Shedd

A good goal is like a strenuous exercise - it makes you stretch.
Mary Kay Ash

It's hard to beat a person who never gives up.
Babe Ruth

There is only one thing more painful than learning from experience and that is not learning from experience.
Archibald MacLeish

Failure is the opportunity to begin again, more intelligently.
Henry Ford

A mind that is stretched by a new experience can never go back to its old dimensions.
Oliver Wendell Holmes

I never let schooling interfere with my education.
Mark Twain

We do not see things the way they are; we see them the way we are.
Anais Nin

One day Alice came to a fork in the road and saw a Cheshire cat in a tree. "Which road do I take?" she asked. "Where do you want to go?" was his response. "I don't know," Alice answered. "Then," said the cat, "it doesn't matter."
Lewis Carroll

No one can make you feel inferior without your consent.
Eleanor Roosevelt

Some trees grow very tall and straight and large in the forest close to each other, but some must stand by themselves or they won't grow at all.
Oliver Wendell Holmes

Every expert was once a beginner.
Donna Erickson

Ken Elovitz

May 2015

DID YOU KNOW

More than 150 million people were born in 2014. In the US, 10 babies were born every minute.

The 25% of the population in China with the highest IQs is more people than the entire population of North America. They have more honors kids than we have kids.

China will soon be the number 1 English speaking country in the world.

1 out of 2 workers has been in their current job less than 5 years. 1 in 4 has been there less than a year.

The top 10 jobs in demand in 2013 did not exist in 2004.

We are currently preparing students for jobs that do not yet exist using technologies that do not yet exist to solve problems we do not know are problems.

We have 4 generations working side by side, and they grew up communicating very differently: Traditionalists (write me), Boomers (call me), Gen Xers (email me), and Millennials (text me).

2.4 billion people use the Internet. Wikipedia has articles in 270 languages.

1 out of 6 couples who married in the US last year met on line.

1 in 5 divorces are blamed on Facebook. Facebook has become the divorce lawyer's best friend.

More than 4000 books are published each day. A week's worth of New York Times contains more information than our 18th century forebears were likely to encounter in a lifetime.

3.5 zetabytes (3.5×10^{21}) of unique new information will be created in the world this year. That is more information than was created in the previous 5000 years.

The amount of new technical information doubles every 2 years. For students starting a 4-year college degree today, half of what they learn in their first year will be outdated by their third year.

(from *https://www.youtube.com/watch?v=XrJjfDUzD7M&feature=youtu.be*)

Ken Elovitz

STRANGEST DREAM
(inspiration from Ed McCurdy)

Last night I had the strangest dream
I'd ever dreamed before.
The architects had all agreed
To praise engineers ever more.

Our buildings all would work so well,
Temperatures would be fine.
The owners would express their thanks
To engineer firms like mine.

Each building would have ample space
To run our ducts and pipes.
Equipment rooms would be so nice
With units of every type.

Our chillers and their water pumps
Would hum and purr with ease.
Clean air would flow into the space;
The users all would be pleased.

Last night I had the strangest dream
I'd ever dreamed before.
The engineers that I would meet
Were marked by the smiles that they wore.

The GCs and the other subs
Would help the H-V-A-C.
And when the job was fin'ly done
Everything worked perfectly.

Inspectors would all read the plans,
And when they visit the site
They'd see the job was neat and clean
And all the work was done right.

As tenants occupied the space
They'd turn to each other and say,
We thank the engineers who let us
Have such a wonderful day.

Last night I had the strangest dream
I'd ever dreamed before
The architects had all agreed
To praise engineers ever more.

Ken Elovitz

July 2015

COOL HISTORY

As the summer's hottest days approach, it's nice to imagine enjoying a refreshing, ice cold drink and then think back to the early days of the household refrigerators that make those drinks cold for us.

I've known people who call their refrigerator a "frigidaire". General Motors established Frigidaire Corp. in 1919 so GM dealers could supplement their car sales after World War I. Due to technical problems with the product, Frigidaire was not successful until 1921.

Kelvinator marketed their first successful electric refrigerator in 1919. The product was a two-part system with the "ice box" in the kitchen and the refrigeration unit in the basement. In 1925 Kelvinator sold a combination unit with a refrigeration unit and a food storage compartment in one cabinet.

Today's HVAC engineers associate the name Servel with natural gas powered residential air conditioners, but the company name is an acronym for "Serve Electrically". Servel sold electric refrigerators in 1923. Servel's gas refrigerators were made by Electrolux, a Swedish company, and came to market in 1926. Gas refrigerators were popular in rural areas that did not have electricity.

Copeland Corporation, now a leading compressor manufacturer and a division of Emerson Electric, marketed their first electric refrigerator in 1923. Copeland stopped making household refrigerators in 1953.

General Electric was known for its "Monitor Top" refrigerator made between 1925 and 1927. The "Monitor Top" was a round refrigeration unit that sat on top of the refrigerator. It was the first completely sealed refrigeration unit, eliminating the need for oiling and changing belts.

Westinghouse was the second company to use a sealed refrigeration unit, introducing their product in 1931. Westinghouse advertised that the even, low temperature in their refrigerator could help with baking by keeping yeast and pastry doughs fresh for several days.

Crosley Radio Corporation entered the refrigerator market in 1933. The inventor was the first to put shelves in the refrigerator door. The Crosley "Shelvador", introduced in 1934, became the model for modern refrigerators and was sold through the 1970s.

Today's refrigerators are not just sealed, efficient, nearly maintenance free appliances that keep our food cold. Many also automatically make ice and dispense both ice and those cold drinks we enjoy so much.

Ken Elovitz

August 2015

SAVAGE WISDOM

Recently - between cat videos - I have been listening to "YouTube" recordings of conversations between interesting people. I listened to a couple of conversations with Adam Savage of "Mythbusters" - the one without the beret. He had so many things to say that resonated with me that I wanted to share (in slightly paraphrased form):

I said to one of my kids: Think about yourself 5 years ago, what an idiot you were back then. Now get used to it - that's going to keep happening for the rest of your life.

At some point in every project - and believe me, this is true on everything I've ever built - I get to a point where I feel like: "When is someone going to tap me on the shoulder and tell me to go home. I clearly have no idea what I am doing." No one is tapping me on the shoulder? OK, I'll keep going.

[On being a generalist]: Sewing and welding are the same thing - they're just planes meeting each other.

An expert craftsman is not somebody who doesn't make mistakes. Experts make the same mistakes as anybody does - they can just smell them coming a lot farther away.

Me again - One of the conversations was between Adam Savage and John Cleese (of Monty Python), and John Cleese has this to say:

I came across this wonderful bit of research by a guy at Cornell named David Dunning. He's interested in self assessment - how good people are at knowing how good they are at doing things. Dunning discovered that to be able to know how good you are at something requires almost exactly the same aptitudes as it does to be good at that thing in the first place. And it follows as a corollary that if you're absolutely no good at something, you lack exactly the skills you need to know that you're no good at it. So there are thousands of people out there who have no idea what they're doing, and have absolutely no idea that they have no idea!

Gary Elovitz

Page 453

September 2015

THE LOWLY TOILET

The toilet may be the most ridiculed yet most important appliance in our homes. The contribution of indoor plumbing to public health is probably unparalleled, except perhaps, by its contribution to convenience.

Until the 1970's, household toilets used between 5 and 7 gallons of water for each flush. The oil embargoes of the 1970's led to lines at gas stations, explosion of energy prices, and an interest in energy conservation that made headlines. Around the same time, a quieter interest in water conservation began to develop. One result was an improvement in toilet design that cut water use almost in half to 3-1/2 gallons per flush.

Water conservation for toilets really took off when the federal Energy Policy Act of 1992, [42 USC □6295(j)], mandated a reduction in toilet water use to 1.6 gallons per flush. Manufacturers rushed to market with products that met the new mandate but sometimes did not work very well. Double flushing became common. Bowl cleanliness suffered. Drain line clogging became more frequent.

In response to user complaints about poor flushing performance, an industry group formed an organization that developed the Maximum Performance (MaP) test protocol for toilets. MaP testing uses soybean paste and toilet paper to simulate real world demands. Results for more than 3000 toilets are freely available at www.map-testing.com

MaP tests the ability to remove solid waste, but it does not evaluate how well the flushing action cleans the inside of the bowl. Toilets sold in the US must meet minimum standards for bowl washing. In my opinion, those with the larger water spots seem to perform better than those with small water spots.

Standard toilet height is 15" to the top of the bowl (not including the seat). Manufacturers now also offer toilets with names like "right height" or "chair height" that are 16"-18" tall. Like curb cuts, ramps, and railings that were originally developed for those who need assistance, the taller toilet bowl can be a benefit for all of us.

Today's 1.6 gallon per flush toilets are much improved, and the industry now offers "high efficiency toilets" that cut water use by 20% to 1.28 gallons per flush. If you are replacing an old 3-1/2 or 5-7 gallon flush toilet with a new 1.6 or 1.28 toilet, consider the options - performance, bowl shape, height. Also know that many water department conservation programs offer rebates to help with the cost. Be sure to check their requirements before doing the work.

Ken Elovitz

October 2015

FOLLOW THE BOUNCING BALL

One of the rules for steam heat is to keep steam where steam belongs and condensate where condensate belongs. Back in the steam heating hey day, manufacturers and installers has some ingenious ways to accomplish that goal. One was a radiator return elbow that used a stainless steel ball a little smaller than a dime as a check valve. If someone closed the steam supply valve on a radiator, the steam in the radiator would condense, forming a vacuum. The vacuum would pull the stainless steel ball against a seat so the vacuum could not suck condensate from the return pipes into the radiator.

Those clever solutions can be lurking problems for 21st century people who reuse old steam radiators and piping when they convert to hot water heat. The problem this time was a loud clatter every time the pump cycled on. It was repeatable, but it lasted only a few seconds each time. That was long enough to track it down to a general area but never long enough to find out which pipe or valve or radiator was the source.

The owner was a "hands on" type who tried everything he could think of to find the noise. He regularly bled radiators to make sure there was no air. He used an infrared camera to look for a cold pipe inside the walls, thinking a cold pipe might show where the noise maker was restricting flow and rattling in the pipe. The installer was sure the problem was oversized pipes that the homeowner insisted on keeping, though he couldn't explain how oversized pipes would cause the noise.

Then one day the owner and a contractor drained down the system to make a repair. As they filled the system back up, the clattering noise greeted them. This time, the noise continued the whole time the system took to fill. That was finally enough time to track down the source of the noise. Using a simple stethoscope to listen inside pipes and radiators, the homeowner located the culprit! He disconnected some pipes and found an old radiator return elbow that the installer forgot to remove during the conversion to hot water heat. The stainless steel ball was trapped inside the radiator. Every time the pump started, the ball banged against the radiator wall for a few seconds until pressures stablized. With the little ball out of the radiator and tucked safely in the owner's pocket, the system was finally quiet.

It's always rewarding to solve a problem. It's even more fun when the search uncovers a little surprise.

Ken Elovitz

November 2015

BE SURE BRAIN IS ENGAGED

Lehigh classmate and now British architect Doug Bennett shared these excerpts from the book *Disorder in the American Courts*. They remind us to be sure brain is engaged before putting mouth into gear.

Attorney: What gear were you in at the moment of the impact?
Witness: Gucci sweats and Reeboks.

Attorney: Is your appearance here this morning pursuant to a deposition notice which I sent to your attorney?
Witness: No, this is how I dress when I go to work.

Attorney: This myasthenia gravis, does it affect your memory at all?
Witness: Yes.
Attorney: And in what ways does it affect your memory?
Witness: I forget.
Attorney: You forget? Can you give us an example of something you forgot?

Attorney: Are you sexually active?
Witness: No, I just lie there.

Attorney: What was the first thing your husband said to you that morning?
Witness: He said, "Where am I, Cathy?"
Attorney: And why did that upset you?
Witness: My name is Susan!

A few questions that don't need answers:

Attorney: Now doctor, isn't it true that when a person dies in his sleep, he doesn't know about it until the next morning?

Attorney: The youngest son, the twenty-one year-old, how old is he?

Attorney: Were you present when your picture was taken?

Attorney: How was your first marriage terminated?
Witness: By death.
Attorney: And by whose death was it terminated?

Attorney: Can you describe the individual?
Witness: He was about medium height and had a beard.
Attorney: Was this a male or a female?

Attorney: Doctor, how many of your autopsies have you performed on dead people?

Attorney: Are you qualified to give a urine sample?

Ken Elovitz

December 2015

NO FLOW NO GO

The new luxury high rise condominium building had been plagued by problems. First a 6" cold water line broke over night during construction, flooding several floors. Then an elbow in a plastic heating line let loose over Christmas weekend, creating another flood. The contractor repaired the damage and finished the job, but only 3 of the 50 condominium units sold. During that first hot summer, the chiller routinely shut down, leaving the residents without air conditioning.

The chilled water system was a variable speed primary pumping design - typical of modern, energy conscious practice. As individual fancoil units shut off, their valves would close, raising system pressure. Automatic controls responded to reduce the pump speed to maintain a pressure differential of 20 psi. If the pump slowed to 35%, the pressure increase from any more fancoil valves closing would open a bypass valve to keep the differential pressure from going above 20 psi. All of that was as it should be. So why did the chiller keep cutting out?

Analyzing the pump curve showed that it had to run at about 60% speed to develop 20 psi differential pressure. But if the pump stayed above 60% speed to maintain differential pressure, the bypass valve would never open. With the bypass closed, total system flow was only as much as needed to satisfy the fancoils that call for cooling. In a building with only 3 of 50 apartments occupied, that flow could be pretty small.

The O&M manual for the chiller lists a minimum allowable flow of 493 GPM, but the control scheme only sensed pressure. It had no way to maintain that minimum flow through the chiller. As a result, when only a few fancoils called and system flow was only a fraction of the required 493 GPM, the chiller's safety controls would shut it down.

The solution was simple: add a flow meter and incorporate a second control loop to use the bypass valve to keep flow through the chiller above 493 GPM. The two control loops have to operate independently, so the bypass valve can open to achieve 493 GPM flow even if the pump speed has not dropped to the 35% minimum. The flow measurement does not need to be precise, so an ultrasonic flow meter mounted to the outside of the pipe would work just fine and make the installation easier and less expensive.

A few weeks later, an email arrived in my inbox. "We got a change order for the chiller and, knock on wood, it works like a champ! Thanks."

Ken Elovitz

January 2016

to the tune of
JINGLE BELLS

Energy Guys are we
We're the guys to call
Room too cold? Bill too high?
We can solve it all.

   Ohhh! (repeat)

Last year's record snow
Made the piles so high
Often asked myself
Will they melt by July?

Gasoline stayed low.
'Lectric rates did slack.
But with muggy summer nights
Huge A/C bills were back.

   Ohhh! Energy guys are we ...

Check your pumps and fans.
Try out VFDs.
Look at all your lights.
Get some LEDs.

Calculate the load,
Size your boiler right
You'll save fuel and still stay warm
Despite the coldest nights!

   Ohhh! Energy guys are we ...

We heard Donald Trump
On the campaign stump
Lashing out at Bush.
Who will win that push?

China's market crashed.
Our stocks took a bath.
Who knows when the Fed will raise
Our interest rates at last?

   Ohhh! Energy guys are we ...

Seal up leaking ducts.
Check the filters, too.
Punch those boiler tubes.
Don't waste heat up the flue.

Change the oil when due.
Keep air filters clean.
Fill those tires up with air
And you'll save gasoline!

   Ohhh! Energy guys are we ...

This time every year
We prepare a verse;
Look back at the year,
Some good; some could be worse.

Ken and Gary hope
Your year is fat, not lean.
We wish you the very best
For two thousand sixteen!

   Ohhh! Energy guys are we ...

Ken and Gary Elovitz

February 2016

GET STEAMED

The 4-story condo built in the 1880s had most of its original steam radiators. New owners in Unit 15 were cold unless the central thermostat was set so high that everyone else had to open their windows. Unit 15 complained to the city health department and threatened to sue the trustees.

"We never had this problem until the new owners moved in," a trustee reported. "It's because Unit 15 removed a radiator in the kitchen and vaulted their ceiling," another was sure. "You have to Squick and skim the boiler," a contractor pronounced. The problem could be any of those, or it could be none of them. It needed a systematic approach.

First I compared the boiler steam output to the capacity of the connected radiators. A steam boiler has to produce steam as fast as the connected radiators (and pipes) can condense it. Otherwise, the system might "run out of steam" before every radiator gets hot. This boiler had plenty of capacity, so the boiler was not the problem.

Next I checked steam delivery. Once the boiler built steam pressure, I followed the mains through the basement to the end of the line. The steam pipes got hot right away, and it wasn't long before the condensate return lines got warm, proving that steam was going out to the loads and giving up its heat to the rooms. The distribution looked OK - steam got in, air got out, and condensate returned.

Steam made it to the top of the riser much faster than I could climb 4 flights of stairs to Unit 15. Unit 15 had 2 radiators connected to the same riser. The one on the left got hot right away. Air streamed out of the vent. The owners were happy with the heat in that room. That means the boiler, the mains, and the riser were not the problem.

The room on the right was always cold. No wonder, because its radiator was slow to heat. Only 2 sections were hot after the boiler was on for more than 10 minutes. The air vent sounded like a dog panting. Eventually the radiator heated all the way across, and the room got warm.

A "panting" radiator means steam cannot get in. The branch line might be plugged with sludge or the radiator valve might be faulty. A "breath" of steam enters the radiator, pushing some air out. New steam cannot overcome the resistance in the line until steam in the radiator condenses, reducing the pressure enough for another "breath" of steam to enter. A few hours and a few dollars for a pipe fitter to replace a short branch line and a valve restored heat to Unit 15 and resolved the feud.

Ken Elovitz

March 2016

PAPER OR PLASTIC

We've all heard it at the supermarket, but what about disposable drinking cups? The environmentally conscious might instinctively choose paper over foam. A 1991 study by Martin Hocking of the University of Victoria, British Columbia, tries to compare the overall environmental impact of paper vs. polystyrene foam drinking cups and calls that instinct into question.

Raw Materials: A polystyrene cup uses 3.4g of petroleum products, but a paper cup is not petroleum free. Manufacturing a paper cup uses an average of 2g of petroleum for energy; even more if the cup has a plastic coating to make is suitable for hot drinks. And the heavier paper cup (10.1g compared to 1.5g for the foam) requires more petroleum for shipping.

Energy Input: Manufacturing a foam cup uses about 13,000 pounds of steam and 280 kWh of electricity. That's a lot of BTUs. But the paper cup needs even more: 23,000 pounds of steam and 980 kWh of electricity. Hocking shows paper cups are almost twice as energy intensive as foam cups.

Air Emissions: Manufacturing a paper cup releases small amounts of chlorine, chlorine dioxide and sulfides, mostly from the paper manufacturing process. Manufacturing a foam cups releases none of those. However, manufacturing a foam cup releases up to 100 pounds of pentane used as a blowing agent.

Wastewater: Manufacturing a paper cup generates much more waste water than manufacturing a foam cup. The paper cup releases about 30,000 gallons of water laden with 20 pounds of suspended solids compared to only 660 gallons of water with about a pound of suspended solids for the foam cup.

Recycling: Polystyrene foam is not commonly accepted for curb side recycling in our area, but the material is inherently recyclable. Paper is easily recycled as well, but hot melt adhesives used to assemble the cup and coatings that make the paper cup suitable for hot drinks add complexity to the recycling process of recycling a paper cup.

Waste to Energy: If trash goes to a waste-to-energy plant instead of to a landfill, burning the foam cup yields twice as much energy per pound as the paper cup. Be careful about Hocking's numbers, though. A paper cup weighs 10g compared to 1.5g for the foam cup. Therefore, on a per cup basis, the paper cup offers 3 times the recoverable energy potential of a foam cup.

In the end, what's the answer? Like so many questions, it depends!!

Ken Elovitz

April 2016

THE SAMURAI

There's a tale from Japan about a Samurai who was very hot headed and full of pride. If anyone dared make fun of him or challenge him in any way, his eyes would blaze. The Samurai would draw his sword, and in a flash the challenger would be dead. So the Samurai was much feared by the locals.

One day, a great teacher, old and wise, came to town. He knew much about the universe and a great deal about the mind and the spirit. The townsfolk were happy to bring the teacher food and offer him shelter so he would share his great learning with them.

One day, the Samurai approached the teacher. Towering over the old man, the Samurai gruffly demanded, "Tell me something, Teacher. You know the secrets of the universe, do you? I want to know about heaven and hell. How can I avoid hell and go to heaven?"

The teacher looked up at the Samurai with bright eyes and spat at him. "You great oaf!" the teacher scoffed. "How could a pig headed fool like you ever expect to go to heaven?"

The Samurai's blood began to boil. "How dare he!" the Samurai thought to himself. With a roar of pure rage the Samurai drew his sword and raised it high.

At that moment, with the sword about to fall, the teacher calmly said, "That ... is hell."

The Samurai stopped in mid-strike and lowered his sword as a profound realization came over him. The Teacher smiled and said, "And that .... is heaven."

ELECTRICITY EXPLAINED

Recent research into the workings of electric and electronic devices revealed what makes them all work -- smoke. The theory is proved by noting that every time you let the smoke out of an integrated circuit, it stops working. The same is true for any electrically operated equipment. In the case of a major electrical failure, the smoke escaping from the device can ignite other combustibles and start the house on fire.

Wires carry the smoke from the source to the device. If the wire springs a leak, it lets the smoke out, and the device stops working.

Large equipment requires more smoke than small items, and that is why larger devices require larger wires.

Ken Elovitz

May 2016

WASTE NOT, WANT NOT

"Our utility bills are too high!" the caller insisted. "We have a super efficient ground source heat pump system, yet we spend $1000/month on utilities!" $1000/month translates to 5000 kWh, so the problem was more than rate escalation. If I could find as little as ~~~10% or 15% in savings, I'd be a hero.

Graphing electricity use against temperature produced a true scatter plot - data all over the place. Buildings with electric heating and cooling usually show a "V" pattern - high for winter heat, lower in mid seasons when load is light, and high again for summer cooling. These bills did not have even a hint of a pattern. Something was running when it shouldn't.

The system was sophisticated. A well water loop served 6 water to air heat pumps for the house and one water to water heat pump for domestic hot water, a pool heater, and a unit heater in the garage.

The well pump was a good candidate for extra energy use. Even a modest load that operates 24/7 could eat up a lot of extra kWh. But this pump was on a variable speed drive and came on only when one of the heat pumps ran, so that wasn't the problem.

I was looking for 2000 kWh/month. That much electricity would heat more than 13,000 gallons of hot water. Even 2 clean freaks would not use that much hot water in a month. Maybe the system circulates water out to the garage all the time for freeze protection. The pump would use a lot of electricity, and continuous water flow through the unit heater could "leak" a lot of heat that the air conditioner has to cancel all summer long. But the pipes were insulated, the unit heater had a control valve that closed when the thermostat satisfied, and the pump ran only when one of its loads called for heat.

As I traced out the piping, I noticed the water to water heat pump ran in very short cycles - a minute on, 5 minutes off. Again and again. I had to know why. The rat's nest of wiring held the key. The contractor wired a sensor on the heat pump outlet to maintain temperature even when no loads call. With no water circulating, the pipe cooled off in just a few minutes, so the compressor restarted. Only a little heat dumped into the basement each time, but there were a lot of compressor starts, and it went on 24 hours a day, every day.

A simple wiring change to enable the heat pump only when one of its loads calls for heat was all it took to knock a thousand kWh off the bill each month.

Ken Elovitz

June 2016

BROKEN WINDOWS?

The project had reached the finger-pointing stage: "You provided the wrong glass!" "You installed the glass backwards!" (Someone said that the direction the low-e coating faces affects performance.) "The glass manufacturer must have sent us defective glass!"

Regardless of the glass, the house got much too warm in the afternoons, when the sun shined through the huge west-facing windows overlooking Buzzards Bay. My job was to find out why.

The glass that was supplied had a slightly higher Solar Heat Gain Coefficient than the specified glass, but my cooling load calculations showed that the difference was nowhere near enough to account for how warm the house got. I also analyzed the difference in glass performance based on the direction of the "low-e" coating. It turns out there is a difference, but it is very small - not much more than a rounding error.

Could the manufacturer really have sent the wrong glass? It seemed unlikely. Removing the 8 ft. tall windows and sending them for laboratory testing didn't seem realistic. I needed a way to test them in the field.

It's easy to tell low-e glass from uncoated glass: hold a flashlight at an angle to the glass - the reflections in low-e glass will have a slight greenish tinge. That's a good test, but it would not be enough.

There is no way to measure the Solar Heat Gain Coefficient (SHGC) of the glass in the field, but maybe I could measure Visual Light Transmittance (VLT) with reasonable accuracy. VLT is not the same as SHGC, but there is a pretty good correlation between the two values. I experimented with a light meter - comparing light measurements through a closed window and from the same place with the window open. I tried that in a bunch of places, with several kinds of glass, and the results were promising - my measurements were within 10% of the published VLT data. I was ready to visit the house.

The flashlight test showed greenish reflections - the glass definitely had the low-e coating. Light meter measurements through the window glass and with the windows open also revealed VLT values that were pretty close to the published data. With those results I could state that the glass was OK.

But exonerating the glass didn't make the house any more comfortable. Now what? Analyzing the whole system showed the real culprit was the chilled water piping. I'll tell you about that next time.

Gary Elovitz

July 2016

THE DEVIL IN THE DETAILS

Last month's story showed that faulty windows were not to blame for the a/c problems in a Cape Cod beach house. But if it wasn't the windows, what was it?

The HVAC system was a little unusual: a chiller made chilled water that was pumped to four air-handling units. On paper it all seemed to work - chiller, air-handlers, pumps and piping all seemed to be sized correctly. I started looking at details:

The chilled water flow to the air-handlers changes as different zones call for cooling and then are satisfied. The system had a "buffer tank" - basically a 10 gallon bucket full of cold water - to allow the chiller (which needs fairly constant water flow) to operate properly. It works like this: The "system" pump draws chilled water (at 45F) from the bottom of the tank, pumps it through the air-handlers, and returns warmer (55F) water to the top of the tank. On the other side of the buffer tank, the chiller pump draws 55F water from the top of the tank, sends it to through the chiller and returns cold (45F) water into the bottom of the buffer tank. The water in the tank remains stratified - cold at the bottom, warmer at the top.

A thermometer on the buffer tank showed that the system was delivering 45F water to the air-handler coils - so far so good. But on close inspection I saw that the chiller was pumping cold water into the top of the tank, not the bottom! Instead of keeping the tank stratified, the water in the tank was well mixed. In order to send 45F water to the coils, the 55F return water from the coils had to mix with much colder water from the chiller. The chiller had to make 35F chilled water! But it's much harder for the chiller to make 35F water, and that caused the chiller's capacity to drop by about 1/3. Any time the load in the house exceeded 2/3 of the design load, the chiller couldn't keep up, and the house would get warm. (It was cloudy when I was there, so the load was low and the system was able to make 45F water.)

The chiller piping was not the end of the story. Several of the air-handler coils were piped backwards. The direction of the water flow through the coils makes a difference - if the coldest (entering) water is not exposed to the coldest (leaving) air, the coil makes about 10% less cooling. Not a big deal most of the time, but at times of maximum load that last 10% can make a big difference!

A little repiping at the buffer tank and the coils got the engineer and the contractor out of hot water.

Gary Elovitz

August 2016

THE HOLY MAN AND THE SNAKE

Long ago, a certain snake was terrorizing a village. Although people were not its natural prey, the snake would take pleasure in biting and threatening children - and even adults - who tried without success to bait it, catch it, or kill it. In spite of being so powerful and feared by everyone, the snake felt rather lonely and depressed.

One day a holy man passed through the village. The snake made straight for him and was about to bite him, but the wise one remained calm and unworried. The snake was so surprised that it hesitated. "Why do you not cower before me like the other people?" the snake inquired.

"Because I don't cower," the wise man replied. "But I can see you are ill at ease. What ails you?"

The snake, being more used to biting with its fangs than talking with its tongue, took some time to feel ready to enter the conversation. After all, it had its reputation to consider.

"Oh, I don't know. I guess I'm lonely. I'm the only snake in town, and everyone is afraid of me. What should I do?"

"It doesn't take a genius to know what to do," the holy man replied. "Stop biting everyone, and little by little people will begin to trust you. You will make friends and become part of the community."

"Thank you, Wise Man! I will do as you say. I'll stop lashing out and not bite anything that isn't in proper relation to me in the food chain," the snake said in gratitude. They parted company, and the holy man went on his way.

Months later, the wise man happened to approach the same village where he was confronted by a sorry sight. The local children were tossing the once-vicious snake in the air like an old stick, laughing at it, and kicking it like a football. When they got bored, they dumped the snake in a heap by the side of the road.

After the children left, the snake saw the holy man. "Hey! How dare you show your face to me? I followed your stupid advice, and I've become a ridiculed plaything. Ever since I stopped biting, no one gives me a shred of respect!"

"Ah!" said the holy man. "The fault was not in the advice but in the following of it. I told you not to bite. I never said not to hiss once in a while!"

Ken Elovitz

September 2016

WHAT A STINK!

"The stench from cooking fish in Unit 5 stinks up the whole place! We spent a pile of dough upgrading our kitchen, and the problem is worse than ever!"

When the old brownstone was converted to condos in the 1980s, each unit had a typical household kitchen hood. A fan on the roof ran continuously and kept the building free of cooking odors. Over time, unit owners upgraded their kitchens, adding new range hoods with powerful exhaust fans. A riser duct and roof fan that once matched the capacity of the 3 kitchen fans was now barely enough for even 1 of the souped up hoods.

Each hood had a pressure switch that turned on the roof fan whenever a unit hood was on. The roof fan pulled air into the riser duct and pushed it outdoors. A new back draft damper provided an extra measure of protection against exhaust air from Unit 5 feeding into Unit 1. But Unit 1 continued to complain.

Testing confirmed that the pressure switches worked. A little air leakage at the connections to the backdraft damper might have let some smelly air leak into the common plenum, but Unit 1 still complained after the contractor resealed the duct. There had to be a big air leak somewhere; the task was to find it.
A smoke test would have set off alarms, but a bottle of lavender air freshener came in handy. A generous spritz into the Unit 5 hood, and lavender was distinct on the roof. That proved the roof fan did its job. We were barely off the roof before Unit 1 demanded to know who was fumigating their space.

A closer look at the backdraft damper showed a screw on the back side stuck too far into the connector and kept the damper from closing tight. That explained how smelly air got past the damper. But how did the smelly air get out of the duct and into the first floor kitchen?

Removing some cabinet panels in Unit 1's kitchen showed that poor workmanship caused Unit 1's problem. Their contractor did not install the recommended duct collar, so the backdraft damper in the hood was stuck half way open. The stuck damper not only let smelly air from the main exhaust duct leak back into Unit 1, it created a restriction that kept Unit 1's hood from working properly. And the 2 inch gap between the flex duct and the hood made for even more leakage. A proper duct collar freed the backdraft damper and provided a place to connect the flex duct with a tight seal.

Once Unit 1 fixed his own hood, cooking fish in Unit 5 was no longer a problem.

Ken Elovitz

October 2016

FIBONACCI

A recent trade magazine article featured the Early Childhood Center in Earl Shapiro Hall at the University of Chicago. That new building has a series of solar panels arranged according to the Fibonacci sequence to form a spiral on the side of the building. What's the Fibonacci sequence and why would anyone care? The Fibonacci sequence is a series of numbers discovered by Italian mathematician Leonardo Fibonacci in the 13th century. The sequence is 0, 1, 1, 2, 3, 5, 8, 13, 21, 34, 55, 89, and on to infinity. Each new number is the sum of the two previous numbers, so the next number after 89 will be 89 + 55 = 144.

Scientists say the outline of a Nautilus shell, the cloud pattern of hurricanes we see on TV weather reports, and the seed pod pattern of sunflowers all have a Fibonacci spiral shape. Some investment analysts use the Fibonacci sequence to try to predict stock market movement and decide when to buy and sell.

The Fibonacci sequence has some properties that intrigue number geeks. Once you get well into the sequence, the ratio of one number to its successor is 0.618; the ratio of one number to its predecessor is 1.618.

Mathematicians call 1.618 the Golden ratio. The Golden ratio comes from dividing a line into two pieces where the ratio of the long piece to the short piece and the ratio of the whole to the long piece is the same - 1.618.

Some people say the Golden ratio is particularly eye pleasing and that's why the ratio of the width to height of the ancient Greek Parthenon is 1.618. Those same commentators say architects design building facades to that ratio even today. Other people dispute the theory and say it's just coincidence like the fact that the ratio between miles and kilometers, 1.625 is very close to the Golden ratio. Whether it's science or sorcery, these numbers are fun to play with and think about.

GREAT BIKE

Two engineering students were walking across campus when one said, "Where did you get such a great bike?" The second engineer replied, "Well, I was walking along yesterday minding my own business when a beautiful woman rode up on this bike. She threw the bike to the ground, took off all her clothes and said, "Take what you want."

The second engineer nodded approvingly, "Good choice; the clothes probably wouldn't have fit."

Ken Elovitz

November 2016

WHY SO HUMID?

"This place is a rain forest," the caller grouched. "I have humidity meters everywhere, and they're all over 70%! We didn't have this problem before the new HVAC went in. My service guy says it's because the systems are over sized." Calculating loads and comparing them to the capacity confirmed the unit was too big. But analysis showed that the indoor moisture level was much higher than it should have been, even with an over sized unit that short cycles.

The building is next to a lake. Maybe the ground was wet, letting water vapor infiltrate up through the slab. "Can't be," the client said. "We poured a new slab; put in a layer of plastic and a drainage trench all around." Curing concrete and drywall mud release lots of moisture into a new space and can cause high humidity the first year. But this was well into year two. The construction materials surely had dried out by then, so that wasn't it.

An outdoor air damper that stays open over night can let large amounts of moist, night air into a building and overload dehumidification capacity, especially if the toilet exhaust runs all night. A site visit showed proper exhaust fan and damper operation and control, killing that idea. There were only a few people in the building and no unusual moisture sources like cooking or fountains.

The HVAC unit cooled OK, it just didn't get moisture out the way it should. "I'll need your mechanic to help me take detailed measurements so we can figure out exactly what the refrigeration side is doing." We found a hot, humid day and let the unit run for a good half hour while we set up gages and thermo-couples for a systematic check of the four refrigeration system components: evaporator (cooling coil), compressor, condenser (heat rejection coil), and metering device.

The evaporator had high supply air temperature. That explained cooling without dehumidifying but didn't explain why. The compressor showed the unit did not work as hard at it should. Confirmation that the unit did not dehumidify, but no reason why.

The condenser analysis looked like refrigerant backed up in the condenser coil instead of feeding into the evaporator. That can happen if a mechanic tries to fix a unit by adding refrigerant when that's not the problem. But there was no record of refrigerant additions, and the pressures at the condenser did not suggest excess refrigerant.

That left the metering device. A metering device that does not feed properly will not let the evaporator coil get cold enough to pull moisture out of the air. Replacing the metering device got the unit running like a champ, wringing rivers of water out of the air. Another reminder of the need to follow all the steps to find and fix the real problem.

Ken Elovitz

December 2016

BRAIN TEASERS

111,111,111 x 111,111,111 = 12,345,678,987,654,321

Take 1000 and add 40 to it. Now add another 1000. Then add 30 and another 1000. Now add 20. Add another 1000. Finally, add 10. Did you get 5000? The correct answer is 4100. Check it with a calculator!

(1) What are 10 parts of the body that have only 3 letters?

(2) What is it that occurs once in a second, once in a month, and once in a century but never in a year or a week?

(3) How quickly can you find out what's unusual about this paragraph? It looks so ordinary you'd think nothing was wrong with it at all - and, in fact, nothing is. But it is unusual. Why? Study it, think about it, and you might find out. No doubt, it you work at it for a bit, it will dawn on you. Who knows until you try? So hop to it, and try your skill.

(4) If a man and a half can eat a pie and a half in a minute and a half, how many men would it take to each 60 pies in 30 minutes?

Answers:
(1) Arm, leg, toe, hip, eye, ear, lip, gum, rib, jaw
(2) The letter "n"
(3) The paragraph does not have a single letter "e", the most common letter in the English language.
(4) 3 men. For one man to eat one pie still takes a minute and a half. In 30 minutes, one man can eat 20 pies, so it takes 3 men to eat 60 pies in 30 minutes.

THE SQUIRREL

The squirrel had a problem. He had been burying his acorns in the same place for years, but recently they started disappearing.

The squirrel watched the spot all day, but saw no animal dig up his treasure. Yet when he checked, the acorns would be gone.

He piled up stones, as heavy as he could carry, on top of the site to protect it. But still, the acorns disappeared.

He was beginning to despair, so he asked the wise old owl for help. She thought about it for a moment and suggested he try digging a wider hole.

Bemused, the squirrel went back and began digging, only to uncover a small tunnel. And out of that tunnel peered a rather well-fed mouse. The squirrel's mistake was to assume everything is as it appears on the surface.

Ken Elovitz

January 2017

CALENDAR GUYS
(inspiration from Neil Sedaka)

I wish more engineers were like the calendar guys;
Yeah, those calendar guys.
I love to get my mailing from the calendar guys;
Each and every month of the year!

(January)
Had a winter without snow;
(February)
Made it easy to come and go.
(March)
Antonin Scalia died; Can the Iphone lock be pried?
(April)
Tubman named for the 20 buck bill.

   Yeah, yeah! A word to the wise:
   I love my monthly mailing from the calendar guys.
   Every month (every month)
   Every month (every month)
   Of the year!

(May)
Riots over shootings and cops;
(June)
Brits decided they will leave the clan; wonder how they'll carry out the plan.
(July)
Conventions filled with politics
Trump or Hil'ry: who'll be our pick?
(August)
Had a drought the year throughout, while the South was flooded out.

   Yeah, yeah! ....

(September)
Had the first candidate debate; which way will our nation turn?
(October)
Dylan won the Nobel prize; Samsung phones all began to burn.
(November)
We give thanks and share some cheer;
(December)
Wishing you a great new year!

   Yeah, yeah! You open my eyes!
   I love my monthly mailing from the calendar guys
   Every month (every month)
   Every month (every month)
   Of the year!

Ken and Gary Elovitz

February 2017

FIRST IMPRESSIONS

Occam's Razor is a principle attributed to 14th century philosopher William of Ockham. The rule says that when two different theories explain a phenomenon, the simpler is likely the better. Albert Einstein is credited with a similar observation: Everything should be as simple as possible, but no simpler. Those principles are good to keep in mind when solving HVAC problems.

"The tenant in 804 is complaining about low airflow," the building owner reported. "We went up there, and the volume coming out of the grille seems a lot less than other the other apartments."

"I'd start with the simple things - check for dirt on the coil and in the fan," I told him. "Dirt on the coil blocks airflow, dirt on the fan blades changes the aerodynamics of the fan. You could also check the fan motor current. Low amps would confirm low airflow."

They must not have cleaned the coils, because a week later they invited me to the site to take some data. Supply air temperature of 57F was fine. Measured supply air flow was 147 CFM, and measured return air flow was 157 CFM. The two measurements were within instrument tolerance but well below the design 500 CFM. By comparison, measurements in the apartment directly below with the same size unit were 231 CFM supply and 262 CFM return. The absolute numbers in Apt. 704 looked low as well, but Apt. 804 clearly had low airflow.

Supply fan amps in Apt. 804 measured 0.84 compared to 0.94 rated and 1.04 in Apt. 704, providing another indication of low airflow.

A fan running backwards will have low airflow, but we reconfirmed that the fan turned the right way. We noticed that 804's fan had the motor on the left (coil side), while 704 fan had the motor on the right. Maybe a "wrong hand" fan meant the motor in 804 obstructed airflow into the fan. But putting the 704 fan into 804 did not change the readings.

Another week went by, and an email showed up. "Today I was able to conduct a camera inspection on the fan coil unit in 804 and found the coils on the back side of the unit about 75% blocked. I should have done what you said in the first place and started by cleaning the coil."

As Ockham and Einstein would say ......

Ken Elovitz

Page 471

March 2017

END OF AN ERA

In October 1980, my father got the idea to use a monthly calendar with a message on the back to keep in touch with friends and clients. Sometimes the message was a case history; sometimes it was a bit of wisdom or a bit of humor.

I knew I was a full fledged "Energy Guy" when I wrote my first calendar message in June 1981. Gary came on board a few years later and published his first story in October 1985.

We hoped our calendars would bring a smile or provoke a thought, and, of course, remind people about the engineering services we provide. And a few people like me even use the calendars to keep track of appointments.

Over the years we received many compliments on our calendars. Someone once even asked where he could sign up for the service that sends them out. He was amazed that we do it all ourselves (with the help of a local printer and the US Postal Service).

In 447 monthly issues, we have not repeated a calendar message, and, since we started using commemorative stamps, I don't think we've repeated a stamp.

My father often claimed the phone would routinely ring with potential clients looking to engage our services a day or two after each calendar mailing. These days, both new and existing clients seem more inclined to send an email than pick up the phone. And people seem to look to an Internet search for news and information rather to a monthly newsletter.

Times change. And much has changed for us. One of those changes is that this will be our last calendar.

We thank our readers for their kind words, encouragement, and support. Although we will no longer send our monthly calendars, Gary and I will maintain the engineering part of our "calendars and engineering" business, always looking for the kind of challenges that gave rise to an interesting story on the back of a monthly calendar mailing.

Ken Elovitz

Made in the USA
Columbia, SC
17 June 2019